GUIDELINES FOR

Safe Storage and Handling of Reactive Materials

Publications Available from the
CENTER FOR CHEMICAL PROCESS SAFETY
of the
AMERICAN INSTITUTE OF CHEMICAL ENGINEERS

Guidelines for Safe Storage and Handling of Reactive Materials
Guidelines for Technical Planning for On-Site Emergencies
Guidelines for Process Safety Documentation
Guidelines for Safe Process Operations and Maintenance
Guidelines for Process Safety Fundamantals in General Plant Operations
Guidelines for Chemical Reactivity Evaluation and Application to Process Design
Tools for Màking Acute Risk Decisions with Chemical Process Safety Applications
Guidelines for Preventing Human Error in Process Safety
Guidelines for Evaluating the Characteristics of Vapor Cloud Explosions, Flash Fires, and BLEVEs
Guidelines for Implementing Process Safety Management Systems
Guidelines for Safe Automation of Chemical Processes
Guidelines for Engineering Design for Process Safety
Guidelines for Auditing Process Safety Management Systems
Guidelines for Investigating Chemical Process Incidents
Guidelines for Hazard Evaluation Procedures, Second Edition with Worked Examples
Plant Guidelines for Technical Management of Chemical Process Safety, Rev. Ed.
Guidelines for Technical Management of Chemical Process Safety
Guidelines for Chemical Process Quantitative Risk Analysis
Guidelines for Process Equipment Reliability Data, with Data Tables
Guidelines for Vapor Release Mitigation
Guidelines for Safe Storage and Handling of High Toxic Hazard Materials
Guidelines for Use of Vapor Cloud Dispersion Models
Understanding Atmospheric Dispersion of Accidental Releases
Expert Systems in Process Safety
Concentration Fluctuations and Averaging Time in Vapor Clouds
Safety, Health, and Loss Prevention in Chemical Processes: Problems for Undergraduate Engineering Curricula
Safety, Health, and Loss Prevention in Chemical Processes: Problems for Undergraduate Engineering Curricula—Instructor's Guide
Workbook of Test Cases for Vapor Cloud Source Dispersion Models
Proceedings of the International Symposium and Workshop on Safe Chemical Process Automation, 1994
Proceedings of the International Process Safety Management Conference and Workshop, 1993
Proceedings of the International Conference on Hazard Identification and Risk Analysis, Human Factors, and Human Reliability in Process Safety, 1992
Proceedings of the International Conference/Workshop on Modeling and Mitigating the Consequences of Accidental Releases of Hazardous Materials, 1991.
Proceedings of the International Symposium on Runaway Reactions, 1989
CCPS/AIChE Directory of Chemical Process Safety Services

GUIDELINES FOR

Safe Storage and Handling of Reactive Materials

CENTER FOR CHEMICAL PROCESS SAFETY
of the
AMERICAN INSTITUTE OF CHEMICAL ENGINEERS
345 East 47th Street, New York, New York 10017

Library of Congress Cataloging-in-Publication Data

Guidelines for safe storage and handling of reactive materials / Center for Chemical Process Safety of the American Institute of Chemical Engineers.
p. cm.
Includes bibliographical references and index.
ISBN 0–8169–0629–7 (hc)
1. Chemicals—Storage—Safety measures. I. American Institute of Chemical Engineers. Center for Chemical Process Safety.
TP201.G853 1995
660'.2804—dc20 94–2481
CIP

Contents

Preface

The Center for Chemical Process Safety (CCPS) was established in 1985 by the American Institute of Chemical Engineers (AIChE) for the express purpose of assisting industry in avoiding or mitigating catastrophic chemical accidents. To achieve this goal, CCPS has focused its work on four areas:

- Establishing and publishing the latest scientific, engineering, and management practices for prevention and mitigation of incidents involving toxic, flammable, and/or reactive material.
- Encouraging the use of such information by dissemination through publications, seminars, symposia, and continuing education programs for engineers.
- Advancing the state of the art in engineering practices and technical management through research in prevention and mitigation of catastrophic events.
- Developing and encouraging the use of undergraduate engineering curricula that will improve the safety, knowledge, and consciousness of engineers.

In 1988, *Guidelines for Safe Storage and Handling of High Toxic Hazard Materials* was published. A more recent work, *Guidelines for Chemical Reactivity Evaluation and Applications to Process Design*, gives details of current methods for evaluating chemical reactivity and the use of evaluation results in the engineering design of reactive chemical processes. This document, *Guidelines for Safe Storage and Handling of Reactive Materials*, builds on the preceding CCPS guidelines, but nevertheless is intended as a stand-alone resource for persons responsible for reactive chemical handling. Many books and articles have been written on chemical reactivity, and the intent of this book is not to give an exhaustive discussion of reactivity. Rather, the purpose of this book is to summarize current process industry practices for designing and operating facilities to safely store and handle reactive materials.

The current book is the result of a project begun in 1992 in which a committee of process safety professionals representing CCPS sponsor companies worked with Battelle's Process Safety and Risk Management group to develop this document. The project included an extensive survey of CCPS sponsor companies and their current practices related to the safe storage and handling of reactive materials. The survey results are included as part of this text.

The safe storage and handling of reactive materials requires a sound and responsible management philosophy, together with a combination of superior siting, design, fabrication, erection, inspection, monitoring, maintenance, operation, and management of such facilities. These elements are necessary parts of a reliable system to prevent equipment or human failures that might lead to a reactive chemical incident such as a vessel rupture explosion. These Guidelines deal with each of the above elements, with emphasis on design considerations.

These Guidelines are technical in nature. They are intended for use by engineers and other persons familiar with the manufacture and use of chemicals. They include discussion of some of the current industry practices for controlling reactivity hazards, both for existing facilities and for plants presently being designed. They are *not* a "standard," and make no attempt to cover all the legal requirements that may relate to the construction and operation of facilities for the storage and handling of reactive chemicals. Meeting such legal requirements is a minimum basis for design and operation of all facilities. These Guidelines highlight and supplement those basic requirements that are particularly important to the safe storage and handling of reactive chemicals. Thus, they should be applied with engineering judgment as well as a knowledge of the hazards and properties of each particular reactive chemical.

Existing facilities may have been designed and constructed to earlier versions of codes and standards, and thus may not fully reflect current practices. When major modifications or additions are made to older facilities, the new portions should meet current design practices for new facilities. However, it is the responsibility of management to decide whether additional safety-related design changes in older facilities are necessary and warranted. Nevertheless, the management of existing facilities for the storage and handling of reactive chemicals should apply current standards and safety practices to their operating, maintenance, management, and emergency procedures and should also reassess safety monitoring and control systems to see whether enhancement of such systems is needed to meet current levels of good practice.

Acknowledgments

The American Institute of Chemical Engineers wishes to thank the Center for Chemical Process Safety (CCPS) and those involved in its operation, including its many sponsors whose funding made this project possible. Thanks are due to the members of its Technical Steering Committee who conceived the idea and supported this project, and to the members of the Subcommittee on Reactive Materials Storage for their dedicated efforts, technical contributions, and enthusiasm.

The members of the Subcommittee on Reactive Materials Storage are:

Robert W. Nelson,	Industrial Risk Insurers (Chairman)
Laurence G. Britton,	Union Carbide Corporation
David L. Halsted,	Monsanto Chemical Company
F. Owen Kubias,	CCPS Staff Consultant
Albert Ness,	Rohm and Haas Company
Matt R. Reyne,	E. I. du Pont de Nemours & Co.
Norman E. Scheffler,	The Dow Chemical Company
Jan C. Windhorst,	Novacor Chemicals

John V. Birtwistle (Monsanto Chemical Company), Stanley J. Schechter (Rohm and Haas Company), and Stanley M. Englund (The Dow Chemical Company) also served on the subcommittee during its early work. The members of this subcommittee especially wish to thank their employers for providing the time to participate in this project.

The Battelle project manager and principal author of this book was Robert W. Johnson, with significant contributions by Steven W. Rudy and Amy J. Sato of Battelle's Process Safety and Risk Management group. Grateful acknowledgement is given to Caroline J. Cadwell for compiling the survey results of Appendix B and to Vicki G. Paddock for her careful editing.

We gratefully acknowledge the comments and suggestions submitted by the following companies and peer reviewers:

Mr. A. Sumner West,	CCPS Staff Consultant
Dr. Daniel A. Crowl,	Michigan Technological University
Mr. Thomas O. Gibson,	The Dow Chemical Company
Mr. John A. Hoffmeister,	Martin Marietta Energy Systems
Mr. Gregory Keeports,	Rohm and Haas Company
Mr. Peter N. Lodal,	Eastman Chemical Company
Mr. John D. Snell,	Occidental Chemical Corporation
Mr. R. Scott Strickoff,	Arthur D. Little, Inc.
Mr. Anthony A. Thompson,	Monsanto Company
Ms. Nita Marie Tosic,	Bayer Corporation.

Reviews and comments from Harold G. Fisher and Jonathan Kurland of Union Carbide Corporation are gratefully recognized. We also express our appreciation to Thomas W. Carmody, former director of CCPS; Bob G. Perry, AIChE Managing Director, Technical Activities; and Jack Weaver, Director of CCPS, for their support and guidance.

Acronyms

Acronyms used only in a particular section of this book are defined where they are used in the book. Acronyms that are used more prevalently are listed and defined here.

AIChE	American Institute of Chemical Engineers
AIT	Autoignition temperature
ARC	Accelerating Rate Calorimeter (Columbia Scientific Instrument Company)
ASTM	American Society for Testing and Materials
CCPS	Center for Chemical Process Safety
CHETAH	Chemical Thermodynamic and Energy Release Program
DIERS	Design Institute for Emergency Relief Systems
DOT	U.S. Department of Transportation
DSC	Differential scanning calorimeter; differential scanning calorimetry
DTA	Differential thermal analysis
ESCA	Electron scanning chemical analysis
HAZOP	Hazard and Operability [Study]
LFL	Lower flammable limit
LOC	Limiting oxidant concentration
MSDS	Material safety data sheet
NFPA	National Fire Protection Association
P&ID	Piping and instrumentation diagram
PSM	Process safety management
SADT	Self accelerating decomposition temperature
TGA	Thermogravimetric analysis
UFL	Upper flammable limit

Introduction

What is a *reactive material*? It is a substance that can liberate sufficient energy for the occurrence of a hazardous event by readily polymerizing, decomposing, rearranging, oxidizing in air without an ignition source, and/or reacting with water. Some commercially produced reactive materials are listed in Table 1.

Thus, reactive materials are not a homogeneous group; this definition can include such diverse substances as monomers, explosives, organic peroxides, pyrophorics, and water-reactive materials. Likewise, initiation of a hazardous reaction can be spontaneous, by heat input, by mechanical shock or friction, or by catalytic activity. Nevertheless, there is much in common among the various reactive materials with respect to their safe storage and handling.

This book addresses the on-site storage and handling of reactive materials. Off-site transportation, laboratory handling, and general warehousing requirements are not covered. Operations other than storage and handling, such as chemical processing, mixing, and blending are likewise not addressed. The scope of this book does not include commercial explosives or materials that are only flammable or combustible.

This book contains guidelines. These guidelines are intended to provide engineers, managers, and operations personnel with a technical overview of current good industry practice. They can, if prudently employed, significantly reduce the likelihood and severity of accidents associated with storing and handling reactive materials.

To store and handle reactive materials safely, the following questions must be addressed:

What kind of reactivity hazards are posed?

What is the magnitude of the reactivity hazards?

How can we design and operate our facility to store and handle safely the reactive materials?

TABLE 1
High-Volume Commercial Reactive Materials (see Note below for explanation)

N_r	*Material*	*1993 Volume*	*Self-Reactive*	*Reactive with Other Materials*	*Oxidizer*
3	Ammonium nitrate	16,790	shock-sensitive		yes
3	Ethylene oxide	5,684	polymerizing, decomposing		
3	Hydrogen peroxide, 100% (1991 data)	500	decomposing		yes
2	Sulfuric acid	80,306		water-reactive	
2	Ethylene	41,244	polymerizing, decomposing		
2	Vinyl chloride	13,746	polymerizing		
2	Styrene	10,063	polymerizing		
2	Propylene oxide	3,300	polymerizing		
2	1,3-Butadiene	3,092	polymerizing	peroxide-forming	
2	Vinyl acetate	2,827	polymerizing		
2	Acrylonitrile	2,508	polymerizing		
2	Methyl methacrylate	1,088	polymerizing		
2	Phosphorus	534		pyrophoric	
2	Lithium	6		water-reactive	

Note: U.S. production volumes in millions of pounds (*Chemical & Engineering News,* July 4, 1994). Only the highest-volume materials with N_r of 2 or higher in the categories of inorganic chemicals, organic chemicals, and minerals are listed. The N_r numbers are the NFPA reactivity ratings for each material from NFPA 49 (*Hazardous Chemical Data*, NFPA, Quincy, Mass., 1994) or NFPA 325M (*Fire Hazard Properties of Flammable Liquids, Gases, and Volatile Solids,* NFPA, Quincy, Mass., 1994). Only pyrophoric, peroxide-forming, and water-reactive characteristics are considered under "Reactive with Other Materials."

The first question is addressed in Chapters 1 and 2, which describe the several kinds of reactive chemical hazards and how they have been classified. The third question is addressed in Chapters 3 and 4, which summarize methods to conduct reactivity testing and calculate the severity of consequences of a reactive chemical incident. The last question is addressed in Chapters 5 through 7, which give both general and chemical-specific design considerations and operating practices.

1

Chemical Reactivity Hazards

This chapter gives a systematic overview of chemical reactivity hazards. It will enable the user to answer the questions

What kind of reactivity hazards are posed?

The chemical process industry by nature involves chemical reactions, the production of reactive chemicals and intermediates, and the handling of reactive materials. Most chemicals handled in the industry are not unstable or reactive under normal storage conditions without a strong initiator; however, the reaction of some materials is easily initiated with only a slight deviation from normal conditions, releasing sufficient energy to cause a hazardous event. Reactive chemicals and uncontrolled chemical reactions are often described using various descriptive adjectives such as *unstable, shock-sensitive, vigorous, violent, runaway,* and *explosive.*

Accident and Postaccident Concerns

Potential reactive chemical accidents include fires, explosions, and the generation and release of toxic materials. Reactive chemical incidents have resulted in the loss of hundreds of lives and many millions of dollars in property. Perhaps the most notable reactive chemical incidents are those that are now known merely by the location of their occurrence; namely, the Bhopal methyl isocyanate release and the Seveso dioxin release (documented in Marshall, 1987 and elsewhere).

Reactivity hazards may continue to exist after an incident has occurred and mitigation efforts are underway. Water-reactive materials such as aluminum alkyls, for example, can pose particularly difficult fire-fighting problems. Reactive metals such as sodium and metal hydrides also preclude the use of carbon dioxide or halogenated extinguishing agents. Many reactive materials are thermally unstable and can decompose rapidly if involved in a fire situation. Some reactive chemicals can cause spontaneous combustion in absorbents used for spill control. These examples illustrate the necessity for thorough analysis and careful design of systems to identify, contain, and control reactive chemicals and respond to reactive chemical incidents.

1.1. Framework for Understanding Reactivity Hazards

In order to identify reactive chemical hazards in a storage or handling facility systematically, a structured understanding of reactivity hazards is important. To this end, an overall framework for identifying reactive chemical hazards is presented in Section 1.2.1, along with brief descriptions of the types of hazards encountered within the given framework. In Section 1.2.2, some fundamentals of chemical reactivity are reviewed in the context of how both thermodynamic and kinetic factors affect reactive chemical systems. The common concept of "runaway reactions," which cuts across many types of reactivity hazards, is discussed in Section 1.2.3. Initiators of reactive chemical incidents are examined in Section 1.2.4.

1.1.1. Grouping of Reactivity Hazards into General Categories

Reactive materials can be grouped into several general categories, as shown in Table 1.1 and described below. While there is some overlap between the categories and subcategories presented here, they nevertheless can serve as a useful framework for understanding the range of reactivity hazards presented by industrially important chemicals.

Table 1.1 divides reactive chemicals into two major groups; namely, those that **"self-react"** and those that **react with other materials**. Each of the common types of reactive materials, such as pyrophoric and shock-sensitive materials, are discussed below within this framework. Those items discussed in detail in this book are shaded in Table 1.1.

Self-Reactive Materials

Reactive materials that are capable of self-reaction will react in one or more of three ways: they will *polymerize*, or form more complex molecules by polymerization-type mechanisms; *decompose*, or break down into simpler molecules such as water and nitrogen; and/or *rearrange* to form variants on the same basic chemical structures or formulas.

Polymerizing compounds are often monomers that self-react, often in the presence of a catalyst, to form polymers or other similar larger, more complex molecular structures by chaining, crosslinking, or similar reactions. Polymerization reactions are generally self-sustaining once initiated, and often highly exothermic. In addition to the heat of reaction, off-gases from the reaction can also pose a significant overpressurization hazard.

Decomposing materials have chemical structures that are relatively unstable and break down easily. Decomposing materials include shock-sensitive and thermally decomposing compounds. The decomposition of a *shock-sensitive*

TABLE 1.1
Reactivity Hazard Types

REACTIVE MATERIALS					
Self-Reactive (Unstable)			**Reactive with Other Materials**		
Polymerizing	*Decomposing*	*Rearranging*	*Oxygen-reactive*	*Water-reactive*	Nitrogen-Reactive
	Shock-sensitive	Isomerizing	Pyrophoric	Class A	Reactive with Metals
	Thermally decomposing	Disproportionating	Flammable	Class B	Oxidizing/Reducing
			Combustible		Acidic/Basic
			Peroxide former		Toxics; Others
Section 1.2	Section 1.3	Section 1.4	Section 1.5	Section 1.6	Sections 1.7, 1.8

NOTES: Only the shaded categories are treated in detail in these guidelines. Section numbers indicate text sections where categories are discussed. Many reactive materials such as 1,3-butadiene fall into two or more categories. Subcategories within the categories of Decomposing, Rearranging, Oxygen-reactive, and Water-reactive are listed in approximate order of decreasing reactivity.

material can be initiated by a sudden input of mechanical energy. This "shock" can be generated by a number of different mechanisms, such as by the impact of a dropped weight or by hydraulic shock. The decomposition reaction for shock-sensitive materials generally has a relatively small activation energy (discussed in Section 1.2.2), such that the mechanical energy input is sufficient to initiate the reaction, and the reaction is exothermic enough to be readily self-sustaining once initiated.

Thermally decomposing materials require a minimum thermal input before a significant decomposition reaction occurs; however, once initiated, the material may decompose at an accelerating rate until it proceeds at an uncontrollably high rate of reaction ("runaway" decomposition reaction).

Peroxides are a subset of decomposing materials that deserve special mention because of their industrial importance. Peroxides are chemical compounds that contain the peroxy (–O–O–) group. Peroxides can be considered as derivatives of hydrogen peroxide (HOOH), with organic and/or inorganic substituents replacing one or both hydrogens. Some peroxide formulations are shock-sensitive, but most are thermally decomposing. Many organic peroxides have particular stability problems that make them among the most hazardous of industrial chemicals.

Rearranging materials may undergo reactions in which their chemical bonds or chemical structure is simply rearranged. Isomerizing and disproportionating chemicals are part of this group.

Reactive with Other Materials

Substances may be stable by themselves, but will readily react with one or more common materials such as atmospheric oxygen, water, or metals. While quantitative chemical reactions such as oxidation–reduction and acid–base reactions, as well as biological reactivity (toxicity), are also in this category, they will not be treated in detail in this book. Likewise, materials that are only flammable or combustible are not given detailed treatment. They are not generally considered "reactive" chemicals, and the storage and handling of flammable and combustible materials are covered extensively in publications by such organizations as the National Fire Protection Association and the American Petroleum Institute.

Oxygen-reactive materials may be further broken down into pyrophoric, low-temperature autoignition, flammable, combustible, and peroxide-forming substances.

Pyrophoric materials are highly reactive with atmospheric oxygen and/or humidity. The energy released by the oxidation and/or hydrolysis reaction is great enough to cause ignition of the material after only a brief delay.

Materials exhibiting *low-temperature autoignition* require an above-ambient temperature but well below the normal autoignition temperature (AIT) range for self-sustained combustion in air to be initiated. A notable example is carbon disulfide, which has an AIT around 212°F (100°C).

Flammable and *combustible* materials will burn in air at normal or elevated temperatures but require an ignition source to start the oxidation reaction. "Combustible" is the more general of the two terms, and can refer to any solid, liquid, or gaseous substance that will burn in air. When applied to liquids, it generally refers to those liquids having a closed-cup flash point of 100°F (37.8°C) or greater. Flammable liquids are those having a closed-cup flash point below 100°F (i.e., that can be easily ignited at normal ambient temperatures). NFPA 321 (1991) gives more specific information on the classification of combustible and flammable liquids.

A *peroxide former* is a material that slowly reacts with air without an ignition source ("autoxidation") to form a peroxidic compound. Peroxide formers pose longer-term hazards; nevertheless, these hazards are significant in that the reaction products can include highly unstable organic peroxides. A few inorganic compounds, such as potassium and the higher alkali metals and sodium amide, can autoxidize and form peroxides or similarly hazardous reaction products.

Water-reactive materials are another category of reactive materials that will react with water, more or less violently. In addition to the problems surrounding the exclusion of all water in storage and handling operations, water-reactive materials also pose obvious fire-fighting difficulties.

1.1.2. Key Parameters That Drive Reactions

The reactions associated with the types of reactive materials outlined above have several governing principles in common. Thermodynamic, kinetic, and physical parameters are important in determining the potential for, and nature of, uncontrolled reactions. Table 1.2 summarizes these parameters.

Smith (1982) provides a good summary of both the objective and the difficulties of obtaining the proper thermodynamic and kinetic data:

> The primary objective of thermokinetic studies is to determine a temperature ceiling below which one can safely work. In principle, it is not possible to state such a temperature because the reaction-rate curve does not simply decrease to zero as temperature decreases. In fact, there is no [fundamental] physical quantity such as the decomposition or onset temperature, except for decompositions that start at melting points.
>
> The heat generation rates of specific samples depend on temperature, degree of conversion, and often, previous thermal history. The onset of a particular heat release rate will be detected at widely different temperatures, depending on the sensitivity of the instrument used.

To be able to obtain and interpret the necessary thermokinetic data properly, a basic understanding of reactivity parameters is necessary. Stepwise assessment of reactivity hazards by theoretical calculations and physical testing is detailed in Chapters 3 and 4.

TABLE 1.2
Parameters of Exothermic and Runaway Reactions[a]

THERMODYNAMIC PARAMETERS
Reaction energy
Adiabatic temperature increase
Specific quantity of gas generated
Maximum pressure in a closed vessel
KINETIC PARAMETERS
Reaction rate
Rate of heat production
Rate of pressure increase in a closed vessel
Adiabatic time to maximum rate
Apparent activation energy
Initial temperature of detectable exothermic reaction
PHYSICAL PARAMETERS
Heat capacity
Thermal conductivity
Surface-to-volume ratio

[a]After Smith, 1982.

Thermodynamic Parameters

One of the key measures of the magnitude of a reactive chemical hazard is the overall energy that could be released in the event that a reaction does take place. This potential energy release is known by various terms, depending on the type of reactive system. For self-reactive chemicals, it is the *heat of polymerization, heat of decomposition,* or *heat of rearrangement.* For systems with more than one reactant, the potential energy release is the *heat of reaction.* (For combustion reactions, the heat of reaction is further specified as the *heat of combustion.*)

The potential energy release is calculated as the difference between the total heat of formation of the product(s) and the total heat of formation of the reactant(s). Heats of formation for many individual chemicals can be obtained from standard chemical engineering and thermodynamics references (e.g., Perry and Green, 1984, 3-147ff). Most reactive chemical systems of concern for safe storage and handling considerations have a greater total chemical energy "content" in the initial reactant(s) than in the products; consequently, energy is released when the reaction occurs, the reaction is termed *exothermic,* and the reaction energy such as the heat of decomposition or the heat of combustion has a negative value. (The international convention of positive values for energy absorption and negative values for energy release is used here.)

The liberated thermal energy can cause pressure generation by vaporization and/or gas generation, ignition of nearby materials, acceleration of chemical reactions, burns to nearby personnel, etc., and thus is the major concern in safely storing and handling reactive chemicals. This reaction energy parameter can be used, for example, to calculate the adiabatic temperature rise for a reaction, which can be combined with the specific volume of the gas generated by the reaction to calculate a maximum internal pressure that can be developed inside a storage tank or other containment.

A highly exothermic reaction usually indicates a very energetic and reactive material or combination of materials. For example, as a rule of thumb, an individual compound is apt to be "explosive" if its heat of decomposition is greater than about 100 cal/g (420 kJ/kg). However, the spontaneity or irreversibility of a reaction is determined by both the reaction energy (enthalpy) and the tendency of a system to go from an ordered state to a more disordered state (increased entropy). A measure that combines enthalpy and entropy is the *Gibbs free energy,* calculated as follows for a compound:

$$\Delta G_f = \Delta H_f - T\Delta S_f$$

where ΔG_f is the Gibbs free energy of formation of the compound in J/mol, DHf is the heat of formation of the compound in J/mol, T is the absolute temperature in Kelvin, and ΔS_f is the entropy of formation of the compound in J/mol·K. The more negative the Gibbs free energy of reaction, the greater the tendency of the material(s) to react spontaneously and irreversibly at the conditions of interest (such as standard state). Stull (1977, 10–13) gives a basic discussion of entropy and Gibbs free energy. The heat of reaction is likely used more than the Gibbs free energy in thermodynamic evaluations because it is a measure of the total energy available if the reaction occurs.

One other commonly used thermodynamic term used in identifying reactive materials needs to be noted. Self-reactive materials that have a significantly positive heat of formation are often called *endothermic compounds,* since these substances require energy input for their formation from constituent elements. For example, acetylene gas has a heat of formation ΔH_f of +54.2 kcal/g-mol. This 54.2 kcal/g-mol would be released to the environment as thermal energy upon decomposition of acetylene to its elements. Consequently, endothermic compounds tend to be relatively unstable, unless the entropy change for the decomposition reaction is significantly positive. This terminology can lead to confusion, since the decomposition of an endothermic compound is an *exothermic* or heat-releasing reaction.

Kinetic Parameters

The basic kinetic variable that must be considered in reactive chemical systems is the **reaction rate**. The reaction energy (e.g., heat of decomposition) and the reaction rate together determine the rate of heat release that must be dealt with in the control of reactive chemical systems. The reaction rate, in turn, is a function of both temperature and reactant concentrations:

$$\text{Rate} = k_T \cdot [\text{concentration-dependent term}]$$

where k_T is the temperature-dependent "rate constant." In most chemical reactions, the temperature dependence of the reaction rate is of the form shown in Figure 1.1. This particular temperature dependence is commonly characterized using the empirical Arrhenius relationship:

$$k_T = e^{-e_a/RT}$$

where A is the Arrhenius frequency factor (usually assumed to be independent of temperature), E_a is the activation energy of the reaction, R is the gas constant, and T is the temperature. More detailed treatment of reaction rate parameters can be found in texts such as Levenspiel (1972) and Carberry (1976).

The activation energy can be considered as the energy input or "barrier" required by a reactive system to initiate a particular chemical reaction, as illustrated in the energy diagram of Figure 1.2. This figure also shows the interrelationship between the kinetic parameter of activation energy and the thermodynamic factor of reaction energy (heat of reaction). The most hazardous reactive systems have low activation energies (and therefore easily initiated) and highly negative heats of reaction (and therefore capable of releasing large amounts of energy). Reactive chemical types are qualitatively tabulated as to their expected activation energy and reaction energy in Table 1.3 for self-reactive chemicals and Table 1.4 for systems involving more than one reactant.

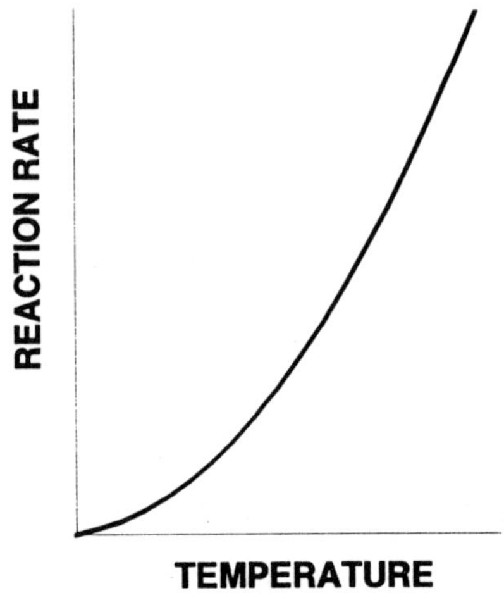

FIGURE 1.1. **Basic Temperature Dependence of Reaction Rate**

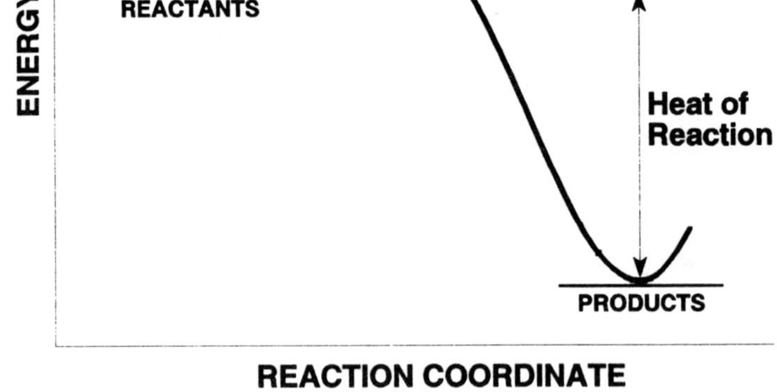

FIGURE 1.2. **Activation Energy and Heat of Reaction.**

TABLE 1.3
Thermokinetic Matrix: Self-Reactive Chemicals

Heat of Polymerization, Decomposition, or Rearrangement	*Activation Energy (can be reduced by presence of catalyst)*		
	Near Zero	*Low to Medium*	*High*
Highly Negative (exothermic)	Unstable intermediates and transition forms (spontaneously decompose)	Explosively decompose; initiation by shock or small thermal energy input	Explosively decompose or polymerize; initiation requires a greater shock or thermal energy input
Slightly Negative		Decompose, polymerize, or rearrange with moderate thermal energy input	Decompose or polymerize with fairly high thermal energy input
Zero	Most stable elemental states at standard conditions		
Slightly Positive	Readily decompose or rearrange; proportional to energy input	Decompose or rearrange with moderate thermal energy input	Decompose with considerable energy input

TABLE 1.4
Thermokinetic Matrix: Reactive Mixtures

Heat of Reaction	*Activation Energy (can be reduced by presence of catalyst)*		
	Near Zero	*Low to Medium*	*High*
Highly Negative (exothermic)	Hypergolic mixtures; pyrophoric; Class A water-reactive	Shock-sensitive explosive mixtures; highly flammable	Reactive/combustible/polymerizing mixtures; initiation requires greater energy input
Slightly Negative		Autoxidizing; Class B water-reactive; slightly flammable	Reactive with significant energy input
Zero	Sum of heats of formation of reactant(s) equal to sum of heats of formation of product(s)		
Slightly Positive	Readily reactive; proportional to energy input	Reactivity takes moderate thermal energy input	Reactivity would take considerable energy input
Highly Positive	Reactivity takes moderate thermal energy input	Reactivity would take considerable energy input	Reactivity would take extreme conditions

NOTE: Qualitative trends rather than absolute categories are indicated. Reactivity will depend on entropy change as well as heat of reaction, and outcome will depend on physical parameters such as heat dissipation to surroundings.

The thermodynamic parameters such as heat of decomposition and heat of reaction can be calculated from literature values for common compounds. However, the kinetic parameters for a given reaction must generally be obtained from physical testing. Measurement of the adiabatic temperature rise as a function of time can yield the activation energy, adiabatic time to a runaway reaction, and maximum rate of pressure rise for a given system; other parameters such as decomposition stoichiometry and vessel fill rate may also need to be addressed. The initial temperature of a reaction system is another important parameter. The initial temperature determines where the system begins along the curve of reaction rate versus temperature. Test methods for determining kinetic parameters are discussed in Chapters 3 and 4.

The reaction rate can be a complex function of the reactant and product concentrations. However, Smith (1982, 81) notes that:

> The assumption of a zero-order kinetic model produces a good approximation of dynamic behavior for most systems. Figure 1.3 shows a first-order decomposition reaction. Note that the self-heat rate starts out as a straight line, but falls off as reactant concentration decreases. We also note from Figure 1.3 that the apparent curvature is drastically reduced when the end-point of the study is defined as maximum rate. With this restriction, we can see why an *n*th-order reaction approaches a zero-order reaction in studies mostly concerned with the reaction up through the maximum rate.

Physical Parameters

The third group of parameters which, in combination with thermodynamic and kinetic parameters, largely determine the future of a reaction system, are the physical parameters: heat capacities of the reactants and products, heats of vaporization, the overall thermal conductivity of the reacting volume and of the containment vessel, and the surface-to-volume ratio of the containment. Based on relationships between these parameters, critical radii or volumes can be determined for various vessel configurations.

How These Parameters Affect Thermal Stability

The thermal stability of a chemical system is the ability of the system to safely remove the heat of an exothermic reaction (either self-reaction or reaction with other materials). This stability is determined by a combination of the thermodynamic, kinetic, and physical parameters, with the kinetic and physical parameters being of most importance. Two outcomes are possible for a system involving an exothermic reaction: the system will either come to a thermal equilibrium between heat generation and heat removal, or the system will spiral upward in temperature. Even very slow heat-evolution rates may lead to dangerous situations if they occur under heat-accumulation conditions. On the other hand, highly exothermic reactions can be safely handled with adequate heat transfer (Smith, 1982).

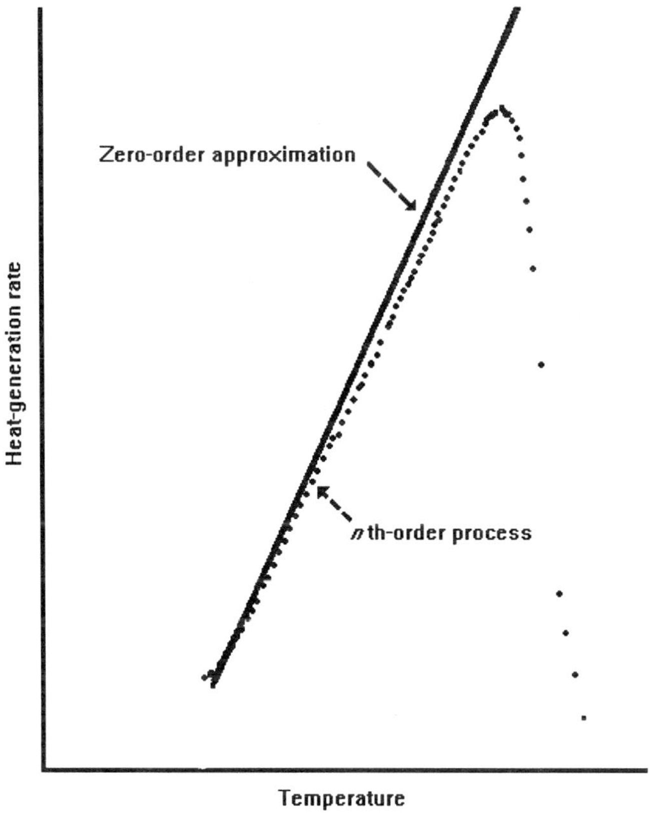

FIGURE 1.3. **Zero-Order Kinetic Model versus Dynamic Behavior of a System** (Smith, 1982)

As seen in the preceding section, the reaction rate, and thus the rate of heat generation, is an exponential function of temperature. The practical rule of thumb that the reaction rate often doubles or even triples for each increase of 10°C in reaction temperature illustrates this strong dependence of reaction rate on temperature. On the other hand, the heat transfer of a system (which is determined by the physical parameters) is a linear or nearly linear function of temperature difference. Hence, as the temperature of a reaction system increases, the heat-generation rate becomes more important than the heat-transfer ability of the system. At some point, called the "temperature of no return" (T_{NR}), the reaction heat can no longer be removed, the system temperature increases exponentially, and a thermal runaway ensues. These relationships are illustrated in the Semenov plot of Figure 1.4. (The heat generation curve flattens out due to consumption of reactant. The actual curve of temperature versus heat generation rate for a given system, such as one involving multiple reactions, may be significantly different than in Figure 1.4.)

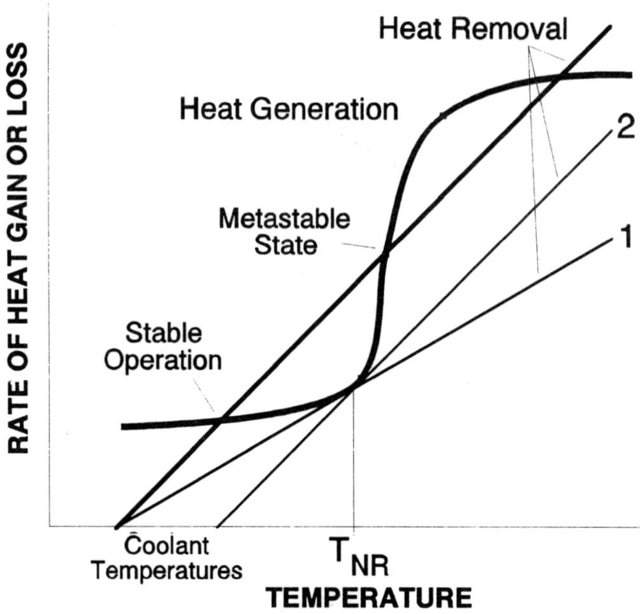

FIGURE 1.4. **Thermal Stability Determined by Heat Generation and Heat Removal**

Thermal runaways occur when the rate of heat generation from a process exceeds the rate of heat loss to the environment from the process container or vessel. The challenge is the determination of the rates of heat loss and heat generation as a function of the system variables. Having a thorough knowledge of the heat generating process, especially early in the reaction, can serve a useful part in helping to elucidate the potential thermal runaway.

The concepts represented in Figure 1.4 are critical to understanding the safe handling of reactive materials. Referring to this figure, there are two ways to affect the thermal stability, or equilibrium, of a system:

- Loss of heat-removal ability, such as by loss of cooling, loss of mixing, or similar deviations, will lower the slope of the heat-transfer line, although the stationary cooling temperature remains the same. If the slope decreases farther than is shown in curve 1, the heat-generation and heat-removal curves will no longer intersect. This is known as the hypercritical state, where heat removal is always less than heat generation. The reaction temperature will rise uncontrollably, creating a thermal runaway.
- The coolant temperature may increase while the heat-removal slope remains unchanged. If this is done, the heat-transfer line moves parallel to itself until it becomes tangential to, and then goes below, the heat-generation line. This situation is represented by curve 2. Even if the two

curves are tangential, a slight increase in the coolant temperature will cause loss of thermal equilibrium. Above this temperature T_{NR} (temperature of no return), thermal equilibrium is not possible, and a runaway reaction must occur.

1.1.3. Types of Runaway Reactions

Most industrially important chemicals that have the potential for a thermal runaway reaction are stored and handled in the liquid phase and have the potential for self-reaction by either decomposition or polymerization. However, autoaccelerating reactions can occur in gaseous, solid, or mixed phases. Reaction mechanisms such as autocatalysis can also lead to uncontrolled autoacceleration of a reaction.

Gas-Phase Runaways

A few storage/handling configurations for gases have the potential for gas-phase runaway reactions when the heat evolved from an exothermic reaction occurring at moderate rates cannot be dissipated sufficiently. For example, runaway has occurred in several instances of regeneration of ethylene dryers using hot ethylene at about 518°F (270°C) and 1000 psia (Britton, 1994). Exposing acetylene and other gases that will decompose at elevated temperatures to fire conditions can initiate a runaway decomposition reaction. Flammable gases mixed with air undergo a "runaway" combustion reaction when the *autoignition temperature* is reached for the particular containment configuration, gas concentration, and initial pressure.

Condensed-Phase Runaways

Runaway reactions involving a solid phase are possible. For example, adsorption of organic vapors in a bed of activated carbon is exothermic, and sufficient heat of adsorption can result in autoignition of the carbon bed if sufficient oxygen is present and heat dissipation is inadequate.

Importance of Coupled Reactions

Fully identifying reactive chemical hazards must include all pertinent reactions. For example, a polymerization reaction might generate sufficient heat to lead to another reaction, say a decomposition reaction, at a higher reaction temperature. Identifying and controlling only the polymerization reaction may lead to inadequate system safeguards against the decomposition reaction. This is particularly important to note when interpreting screening test data. Testing a reactive mixture in a differential scanning calorimeter (DSC) might show just one exotherm. Testing the same mixture in an accelerating rate calorimeter (ARC) might show two exotherms. A highly adiabatic apparatus might show four to five or more different exotherms for the same mixture. Often, the higher exotherm result is gas generation and not just heat evolution.

1.1.4. How Reactive Chemical Storage and Handling Accidents Are Initiated

Many of the underlying causes of incidents and accidents that have involved unexpected violent chemical reactions are related to a lack of appreciation of the effects of physical and chemical factors on the kinetics of practical reaction systems (Bretherick, 1986).

Accidents involving reactive chemicals are initiated by a number of factors arising from within the storage and handling containment, and from the environment outside the containment. In general, these factors affect reactivity by

- reducing the apparent activation energy E_a
- adding energy to the system (i.e., helping the system obtain the activation energy)
- changing the reaction path
- reducing heat loss to the surroundings

or a combination of the above.

How Storage and Handling Conditions Affect Reactivity Containment

Important factors in preventing the types of runaway reactions discussed previously are mainly related to the control of reaction velocity to as slow a rate as possible without changing the reaction path, as well as keeping the temperature within suitable limits. This may involve such considerations as adequate **cooling capacity** in both liquid and vapor phases of a reactive chemical storage and handling system. Cooling can be impacted by the use of **solvents** as diluents, changing the **viscosity** of the reactive medium, changing **agitation** in the storage vessel, and control of vessel **pressure** (Bretherick, 1990). Loss of agitation or loss of cooling has often been the main contributory factor in cases where inadequate temperature control has caused exothermic reactions (normal, polymerization, or decomposition) to run out of control. The converse, the addition of **heat** (energy) to a system, may either initiate or accelerate a chemical reaction by providing the energy input to overcome the apparent activation energy of the reactive system, thus increasing the reaction rate.

The **ratio of volume to surface** area for a system will impact the flow of excess energy out of the system. This in turn will affect the temperature of the system over time. As a result, some substances or mixtures that are not hazardous in small amounts may turn out to be hazardous when the quantity of material is increased. Accordingly, the scaling up of a storage system should take these factors into account.

To save some of the expense of heating or cooling, vessels are often **insulated,** particularly during long-term hot storage. Materials of limited thermal stability, or which possess self-heating capability, are potentially hazardous in such near-adiabatic systems if insulation is used (Bretherick, 1990, xxiv). Also, if a reactive material leaks into insulation, similar self-heating and ignition at the hot spot may occur.

Some reactions have a significant **induction time**, which may lead to a false sense of security. Even with the use of inhibitors, a self-accelerating reaction may still only be delayed and not prevented. In addition, some inhibitors change the path of the reaction to one that may be even more hazardous (Cardillo and Nebuloni, 1992).

Consequently, **prolonged storage** is of concern in any case. The same is true for pressurized systems or systems where **fresh metal surfaces** or metal powder may be produced. Catalysts, usually present in storage/handling situations as **contaminants**, effectively reduce the energy of activation, thus increasing the rate of reaction or changing the reaction path (Bretherick, 1990, xxii).

Mechanical shock can be sufficient to provide the activation energy for the decomposition of some highly unstable materials. **Friction** between materials in any phase, especially during transfer operations, may either increase the local temperature of the materials or lead to the development of a static electric charge. **Viscosity** can contribute via friction to static charge build-up and can also help produce hot spots in a reactive material by reducing convective heat transfer. **Accumulation of static electricity** and release of the energy as an electrostatic discharge is a well-documented ignition source.

External heat sources such as from **hot tapping** can initiate a thermal decomposition reaction. Sufficient heat can also be produced by process equipment, such as a **dead-headed pump** (particularly when blocked in on both the inlet and outlet of the pump) or a **hot seal**; even the pumping of reactive material around during standby operations, such as through a pump recirculation line, can gradually increase the temperature of the material.

Any significant **electromagnetic radiation** with such a wavelength distribution that it can be absorbed by the system, or can heat up the system indirectly via its containment, can pose an initiation hazard. Ultraviolet radiation is energetic enough to provide the activation energy for self-reaction of some substances.

Light can form free radicals that can initiate an uncontrolled polymerization reaction. **Direct sunlight** is capable of creating areas where temperature is significantly above ambient. For example, a fire occurred in a material whose autoignition temperature was much higher than ambient temperature. The material was stored on pallets outside on asphalt during summer. The material was heated by solar energy, then self-heated until it ignited the wooden pallets it was stored on (Englund, 1991).

For chemicals that are reactive with air, an **inert atmosphere** may be required to ensure stability. A loss of inerting, or a leak or vacuum break, may lead to an uncontrolled reaction in such reactive systems.

Hazards associated with certain **materials of construction,** such as the use of aluminum in an environment with halogenated hydrocarbons, have also been identified (Cardillo and Nebuloni, 1992).

Autoignition Temperature

In addition to the above considerations, the autoignition temperature (AIT) for many reactive chemicals is highly dependent upon pressure. Autoignition temperatures tabulated for many materials are only accurate for the pressure and configuration of the test conditions, but can be used as approximations for other configurations. Increased pressure will usually lower the autoignition temperature (Figure 1.5).

The hazards of **cool flames** should also be noted. *Cool-flame ignition* is a relatively slow, self-sustaining, barely luminous gas-phase reaction of a material or its decomposition products with air or another oxidant (NFPA 325M, 1991). Cool flames often cause temperature increases of about 150°K. Transition from the slow combustion of cool flames to hot-flame ignition may occur under certain conditions in process equipment.

Figure 1.5 shows the behavior of a material (e.g., an ether) that exhibits cool-flame ignition behavior. At 1 atm pressure, only slow combustion, with a cool flame, is found up to the autoignition temperature (AIT), at which point a rapid, hot-flame combustion reaction is initiated. In this case, the lowest *cool-flame reaction threshold* (CFT) is much lower than the AIT. At a higher initial pressure (such as 2 atm in Figure 1.5), it is possible to enter the ignition region at a temperature even lower than the ambient-pressure CFT (CCPS/AIChE,

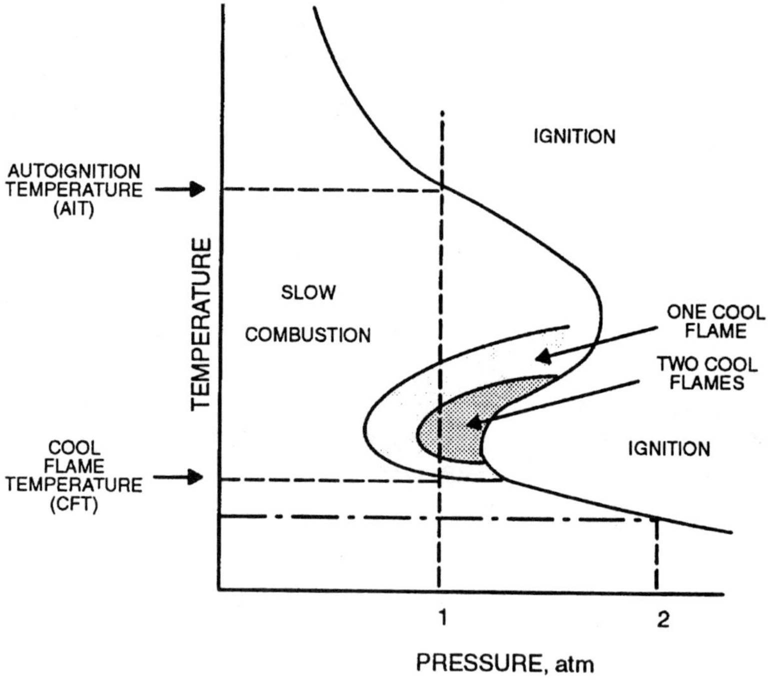

FIGURE 1.5. **Schematic Autoignition Temperature–Pressure Diagram**

1993, 323). With reference to Figure 1.5, if the cool flame region were entered in a closed vessel initially at 1 atm pressure, the increased temperature and pressure from the cool-flame reaction could shift the entire system into the hot-flame ignition region. Consequently, in a closed vessel, a cool flame could increase the pressure of the vessel to the point that the autoignition temperature could be attained.

1.2. Self-Reactive Polymerizing Chemicals

$$nM \rightarrow \sim MMM \sim + \Delta$$

Some chemicals present a hazard due to their ability to react to form larger molecules in a self-sustaining polymerization reaction. The reaction is exothermic, often producing significant quantities of off-gas, with the heat and/or gas generation being capable of overpressurizing storage and handling equipment and possibly resulting in a vessel rupture explosion. Some high-volume industrial chemicals that are self-polymerizing are listed in Table 1.5.

Undesired polymerization reactions can usually be prevented or controlled by the addition of reaction inhibitors, by controlling the bulk temperature of the material, or by controlling the pressure of the system. However, contamination of materials subject to polymerization, as well as external heating from fire exposure, can overwhelm these safeguards. Consequently, emergency response to fires and uncontrolled situations involving polymerizable chemicals must be planned carefully.

1.2.1. Thermal Instability

Materials that are *thermally unstable* will, at some specific temperature (usually above ambient), undergo some type of potentially hazardous reaction such as

TABLE 1.5
Some Self-Polymerizing Chemicals

Acrylonitrile	Propylene
1,3-Butadiene	Propylene Oxide
Ethylene	Styrene
Ethylene Oxide	Vinyl Acetate
Methacrylic Acid	Vinyl Chloride
Methyl Methacrylate	

polymerization, decomposition, or rearrangement, with consequent release of energy. In general, the *reaction onset temperature* is reported. The onset temperature is essentially the lowest temperature for which the thermal energy of the system is sufficient for the reaction to proceed at a measurable rate in a given experimental apparatus or process vessel. Tests designed to detect this onset temperature are described in Chapter 3. If known, the *self-accelerating decomposition temperature,* discussed in Section 1.4, is also reported.

Onset temperature is an important concept for reactive mixtures as well. The NFPA *Manual of Hazardous Chemical Reactions* (NFPA 491M, 1991), and *Bretherick's Handbook of Reactive Chemical Hazards* 1990 have information on reactions between specific chemicals, although onset temperatures are generally not reported. In general, onset temperatures are difficult to measure reliably and are frequently difficult to define precisely in complex systems.

1.2.2. Induction Time

Induction time or *induction period* is an important safety consideration in the storage and handling of polymerizable materials, in that inhibitors are often added to monomer storage to prevent the onset of a polymerization reaction. Bretherick (1990, 1636–1637) has a good description of induction period for hazardous reactions:

> In the absence of anything to prevent it, a chemical reaction will begin when the components and any necessary energy of activation are present in the reaction system. If an inhibitor (negative catalyst or chain-breaker) is present in the system, it will prevent the onset of normal reaction until the concentration of the inhibitor has been reduced by decomposition or side reactions to a sufficiently low level for reaction to begin. This [apparent] delay in onset of reaction is termed the *induction period.*

Chemicals and reactive systems exhibiting induction periods, such as Grignard reagents, are listed in the same reference. As illustrated in Figure 1.6, it is not necessary to have an inhibitor present to have an induction period for a given reaction. The induction time τ_2 in Figure 1.6 begins either at the start of the storage time for a material with no inhibitor present or at time τ_1 when all of the inhibitor has been used up.

The induction period can be affected by many variables, such as the presence of impurities, the temperature and pressure history of the contents, and the presence and concentration of inhibitors (Nicolson, 1991). Several means are available to monitor and ensure an adequate inhibitor concentration, such as periodic sampling, temperature or rate of temperature rise measurement, and inhibitor loss-of-flow alarms. Detailed design considerations for polymerizable compounds are given in Section 6.2.

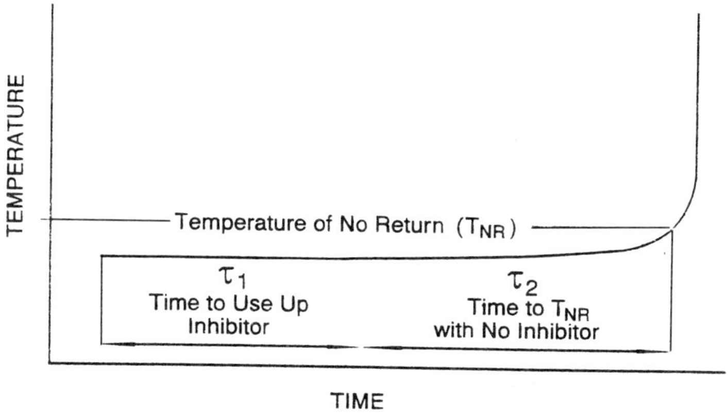

FIGURE 1.6. **Induction Time**

1.2.3. Example

Nicolson 1991 reports that hydroquinone and monomethyl ether of hydroquinone (MEHQ) can be used as inhibitors of methacrylic acid polymerization. However, both hydroquinone and MEHQ require the presence of oxygen to be effective as inhibitors. The oxygen concentration has a great impact on the stability of the monomer. At the right concentrations, oxygen increases the induction period of methacrylic acid. However, peroxide decomposition is associated with the onset of polymerization, with the potential consequence of a violently exothermic polymerization reaction initiated by the peroxide. Note also that needing oxygen to be present for inhibitor activation may be in conflict with inerting requirements, so that a controlled oxygen addition scheme may be necessary.

1.3. Self-Reactive Decomposing Chemicals

$$AB \rightarrow A + B + \Delta$$

A second major group of reactive chemicals are self-reactive by decomposing into smaller molecules, rather than combining together to form larger molecules as in polymerization. Decomposition reactions can liberate large amounts of energy, often with explosive violence. Some decomposition reactions can be initiated by mechanical shock alone, such as by a falling object striking the material; such materials are termed *shock sensitive*. At the other end of the spectrum of decomposing chemicals are those that will decompose after being exposed to an elevated temperature for a period of time; such materials are termed *thermally decomposing*. Shock and elevated temperature are not the only stimuli that can initiate decomposition reactions; other mechanisms include friction, light, trace contaminants such as rust or metal powders, and trace chemical contaminants such as oxidizing or reducing agents. Elevated pressure may also be important. For example, acetylene has decomposed explosively when faulty pressure control has allowed pressures in acetylene generators and distribution systems to approach 20.3 psig in the presence of moisture (Bretherick, 1990, 231).

1.3.1. Peroxides

One group of industrially important chemicals that has characteristics of both the shock-sensitive and thermally decomposing categories is the *peroxides*, characterized by the oxygen–oxygen single bond. Peroxides can be divided into inorganic peroxides, organic peroxides, and organomineral peroxides, depending on the substituents on either or both sides of the oxygen-oxygen bond.

Peroxides have a specific half-life, or rate of decomposition, under any given set of conditions. A low rate of decomposition may autoaccelerate and cause a violent explosion, especially in bulk quantities of peroxide. These compounds are sensitive to heat, friction, impact, and light, as well as to strong oxidizing and reducing agents.

Some peroxides such as many of the organic peroxides are extremely shock sensitive; other peroxides are quite shelf stable. As a class, organic peroxides are hazardous because of their extreme sensitivity to shock, sparks, heat, or other forms of accidental ignition. Some peroxides that are routinely handled in industry are more sensitive to shock than secondary explosives such as trinitrotoluene (TNT). For this reason, many peroxides are shipped in a diluent; ensuring that the diluent is always present is, therefore, a safety-critical design and operating consideration for such peroxides.

The highly reactive nature of many organic peroxides can be attributed to their having both oxidizing and combustibility properties. Specific recommendations regarding organic peroxides are given in NFPA 43B, *Code for the Storage of Organic Peroxide Formulations* (1993).

1.3.2. Self-Accelerating Decomposition Temperature

An important measurement of the storage stability for potentially unstable materials is the temperature at which an uncontrolled decomposition reaction can be initiated. The following basic description of the self-accelerating decomposition temperature (SADT) is taken from NFPA 49 (1994, 145):

> Certain compounds, such as organic peroxides and [some] swimming pool chemicals, when held at moderate ambient temperatures for an extended period of time, may undergo an exothermic reaction that accelerates with increase in

> temperature. If the heat liberated by this reaction is not lost to the environment, the bulk material increases in temperature, which leads to an increase in the rate of decomposition. Unchecked, the temperature grows exponentially to a point at which the decomposition cannot be stopped or slowed. The minimum temperature at which this exponential growth occurs in a material packed in its largest standard shipping container is defined as the *self-accelerating decomposition temperature.* Self-accelerating decomposition temperature is a measure of the ease in which decomposition occurs under normal storage [shipment] conditions. It is not an indicator of the violence of any decomposition reaction under conditions of fire exposure or contact with incompatible materials.

It should be noted that the SADT only applies to the container size and surface–volume configuration in which the material was tested. UN transportation requirements specify that a material to be shipped must be stable at 55°C for one week. U.S. DOT requires stability at 130°F (54.4°C) for the duration of the shipment. It can be argued that this "duration" could be anywhere from one week to six months. Methods for measuring the SADT are described in Section 3.4.

1.3.3. Predicting Instability Potential

A general characteristic of self-reactive decomposing chemicals is their inherent "instability" or propensity to decompose. Three indicators can be used to point to the likelihood of unstable behavior in a given chemical compound: (a) endothermicity, (b) presence of certain bonds or functional groups common to unstable materials, and (c) stoichiometry and oxygen balance. These three indicators are discussed in Section 3.2.

1.3.4. Deflagration and Detonation of Pure Material

A decomposition reaction can range from a slow, gas-evolving reaction to the detonation of a high explosive. Decomposition reactions, particularly in the liquid phase, generally occur as reactions in the bulk of the material rather than at a reaction front such as in a deflagration (although deflagrations can occur in the liquid phase). It is often possible to provide emergency relief protection for storage vessels handling such materials, although multiphase flow must usually be considered.

On the other hand, materials such as high explosives that can decompose at detonation velocities cannot generally be vented or contained. (See Glossary for definitions of *deflagration* and *detonation.*) Hence, safeguards and mitigation approaches will be different depending on the speed of the potential decomposition reaction.

1.3.5. Slow Gas-Forming Reactions

Numerous process incidents have occurred in closed systems where evolution of gas over time has caused pressure to build up in the system to the point of containment failure. The gas can evolve from a slow decomposition or hydrolysis reaction, an accelerating thermal decomposition, or a process upset or side reaction. Bretherick (1990, 1604–1607) provides an extensive list of chemicals and reactive systems that have been reported in the literature as causing gas-evolution incidents.

Slow gas-forming decomposition reactions can occur slowly enough that any heat of reaction can be dissipated to the environment, so that an accelerating thermal decomposition/runaway reaction does not occur. Another possible cause of such incidents is materials that have very small heats of decomposition, but the decomposition with gas evolution is caused by a catalytic impurity or other initiator being present that controls the rate of decomposition. Catalytic decomposition of hydrogen peroxide due to metal or metal oxide contaminants, liberating oxygen into the containment in gaseous form (with obvious amplification of any combustion–oxidation hazard if oxidizable material is present), is a common example of this category of reactions.

1.3.6. Heat of Compression

Thermal energy can be imparted to a reactive material by several mechanisms. In addition to thermal energy possibly coming from the surrounding environment (including external fires, steam leaks, etc.) and from slow reactions for which the heat of reaction is not dissipated, the mechanism of adiabatic compression can also generate thermal energy that is sometimes sufficient to initiate a decomposition reaction. Adiabatic compression arises when a compressible fluid is pressurized such that the pressure–volume (P-V) work done on the fluid is manifest in an increased fluid temperature. The fluid must have a positive Joule–Thompson coefficient for a temperature increase to occur.

As an example, the diesel engine works on the principle of autoignition by heat of compression. Hence, fuels that readily self-ignite are desirable for diesel engine fuels. (Converse behavior is found in the spark-ignition engine, where self-ignition causes engine "knock" and is therefore undesirable, so that fuels which resist pre-spark ignition reactions are favored for spark-ignition engines.)

Adiabatic compression has also been known to initiate decomposition of tetrafluoroethylene gas with atmospheric oxygen present. Adiabatic compression can be caused by such mechanisms as opening of a rupture disk below a relief valve, so that the high-pressure process gas compresses the gas in the space between the rupture disk and relief valve. Initiation of ethylene decomposition by compression mechanisms is discussed by Britton et al. (1986).

The autoignition behavior of gases and mists is highly complex, especially when related to the dynamic conditions in equipment such as an engine.

Similarly, dynamic conditions in process equipment containing possible sources of hot-spot autoignition are very difficult to quantitatively analyze (CCPS/AIChE, 1993), although they may be amenable to computer modeling.

1.3.7. Minimum Pressures for Vapor Decomposition

Decomposition of vapor-phase reactive compounds often requires an elevated pressure for the decomposition reaction to propagate. Adiabatic compression situations, as described in the preceding section, can provide both thermal energy for initiation and an elevated pressure if such a pressure is required for vapor-phase decomposition.

Table 1.6 lists available data on materials that may be stored in bulk quantities and that can decompose in the vapor phase without oxygen present, thus having effective limiting oxygen concentrations (LOCs) of zero when above the limiting pressures indicated. It should be noted that, even though the materials listed in Table 1.6 can indeed decompose in the absence of oxygen, having oxygen present will often result in a more easily initiated reaction, a lower pressure requirement for propagation, and/or a greater energy release rate due to a more energetic reaction.

1.3.8. Shock Sensitivity

The decomposition of *shock-sensitive* materials can be initiated simply by the mechanical energy input of a physical impact or a pressure (shock) wave. When

TABLE 1.6
Minimum Pressures for Vapor Decomposition

Gas or Vapor	*Limiting Pressure at Given Temperature*	*Reference*
Acetylene	~1 atm (101 kPa) at 25°C; depends on container diameter and ignition source	Medard, 1989; Bulletin 680
1,3-Butadiene	unknown; theoretically possible	Medard, 1989
Ethylene	~68 atm (6.9 MPa) at 25°C	Bulletin 680
Ethylene oxide	~300 mmHg (40 kPa) at ~30°C	Union Carbide, unpublished
Hydrazine	12 mmHg (1.6 kPa) at 25°C	Bulletin 680
Methyl acetylene	>4 atm (>400 kPa) at 25°C	Bulletin 680
Propadiene	>2.2 atm (>225 kPa)	Bulletin 680

NOTES. Ethylene oxide limiting pressure is reported as 525 mmHg (70 kPa) in Bureau of Mines Bulletin 680. Numbers in Bulletin 680 are order-of-magnitude accuracy. The T-P ranges for the "unknown" values are probably not industrially important except from the standpoint of fire exposure, plus possibly erroneous literature testaments to their stabilities.

some organic peroxides are said to be extremely sensitive to shock (as well as sparks, friction, heat, light, and other initiating mechanisms), this likely refers to such mechanisms as dropping a tool onto a surface where a shock-sensitive peroxide has been spilled. Table 1.7 gives a list of shock-sensitive materials compiled by the National Research Council.

TABLE 1.7
Shock-Sensitive Compounds[a]

Acetylenic compounds, especially polyacetylenes, haloacetylenes, and heavy metal salts of acetylenes (copper, silver, and mercury salts are particularly sensitive)
Acyl nitrates
Alkyl nitrates, particularly polyol nitrates such as nitrocellulose and nitroglycerine
Alkyl and acyl nitrites
Alkyl perchlorates
Amminemetal oxosalts; metal compounds with coordinated ammonia, hydrazine, or similar nitrogenous donors and ionic perchlorate, nitrate, permanganate, or other oxidizing group
Azides, including metal, nonmetal, and organic azides
Chlorite salts of metals, such as $AgClO_2$ and $Hg(ClO_2)_2$
Diazo compounds such as CH_2N_2
Diazonium salts, when dry
Nitrides (silver nitride, Ag_3N) can form in the reaction mixture from the Tollens' test for aldehydes if it is allowed to stand for some time; this can be prevented by adding dilute nitric acid to the test mixture as soon as the test has been completed)
Hydrogen peroxide becomes increasingly treacherous as the concentration rises, forming explosive mixtures with organic materials and decomposing violently in the presence of traces of transition metals.
***N*-Halogen compounds** such as difluoroamino compounds and halogen azides
***N*-Nitro compounds** such as *N*-nitromethylamine, nitrourea, nitroguanidine, nitric amide
Oxo salts of nitrogenous bases: perchlorates, dichromates, nitrates, iodates, chlorites, chlorates, and permanganates of ammonia, amines, hydroxylamine, guanidine, etc.
Perchlorate salts; most metal, nonmetal, and amine perchlorates can be detonated and may undergo violent reaction in contact with combustible materials
Peroxides and hydroperoxides, organic
Peroxides (solid) that crystallize from or are left from evaporation of peroxidizable solvents
Peroxides, transition-metal salts
Picrates, especially salts of transition and heavy metals, such as Ni, Pb, Hg, C, and Zn; picric acid is explosive but is less sensitive to shock or friction than its metal salts and is relatively safe as a water-wet paste
Polynitroalkyl compounds such as tetranitromethane and dinitroacetonitrile
Polynitroaromatic compounds, especially polynitro hydrocarbons, phenols, and amines

[a]Data from National Research Council, 1983; Shanley and Ennis, 1991.

The importance of shock sensitivity becomes especially acute when considering the possibility of an explosion at one location initiating a more severe decomposition reaction at some distance away from the original location. This potential consequence, as it relates to high explosives, is termed *sympathetic detonation*, and has been extensively studied. Safe separation distances for storage of high explosives in bulk have been determined, and are published as *quantity-distance relationships* in the American Table of Distances for Storage of Explosives (NFPA 495, 1992).

1.3.9. Examples of Shock Sensitivity

In addition to the well-recognized hazards of organic peroxides (NFPA 43B, 1993), other shock-sensitive materials pose hazards to storage and handling operations. For example, traces of unstable materials such as insoluble polyperoxides can accumulate gradually over time until a dangerous concentration is reached. The highly energetic, shock-sensitive compound nitrogen trichloride can be formed when excess chlorine or a chlorinating agent contacts aqueous ammonia, ammonium salts, or a compound containing a hydrolyzable amino-derivative (Bretherick, 1990, 1040). Nitromethane, while not very sensitive, can detonate if sufficient material and a sufficiently energetic stimulus are present.

1.4. Self-Reactive Rearranging Chemicals

$$R \rightarrow R' + \Delta$$

Chemicals that are self-reactive by internal molecular rearrangement include those which can liberate energy by isomerizing, disproportionating, or tautomerizing. Although the total risk associated with these types of reactivity is historically small, a few important chemical structures are discussed in this section.

1.4.1. Isomerization

Isomers are different compounds having the same empirical formula, such as ethyl alcohol (CH_3CH_2OH) and methyl ether (CH_3OCH_3) that both have the empirical formula C_2H_6O. Isomerization is the rearrangement of atoms in a molecule, such as a change from a straight chain to a branched molecule. It is employed, for example, in petroleum refining to convert straight-chain to branched-chain hydrocarbons or alicyclic to aromatic hydrocarbons. In some cases, the conversion from a more energetic to a less-energetic isomer can liberate sufficient energy to pose a significant hazard. For example, the conversion of the C_2H_4O isomers ethylene oxide (heat of formation –16.1 kcal/g-mol at 25°C) to acetaldehyde (heat of formation –39.72 kcal/g-mol at 25°C) will liberate 23.6 kcal for each gram mole converted, or 2.24 MJ for each kilogram converted.

One special kind of isomerization occurs with molecular hydrogen, which exists as two "spin isomers" known as *orthohydrogen* and *parahydrogen.* These two distinct forms are possible because two hydrogen atoms may combine with their nuclei spinning in the same (ortho) direction or in opposite (para) directions. These spin isomers are ordinarily fairly stable. However, they may be rapidly interconverted by using a catalyst such as activated charcoal or platinized asbestos. The ratio of the two isomers in an equilibrium mixture varies markedly with temperature. At temperatures approaching absolute zero, equilibrium hydrogen consists entirely of parahydrogen; at room temperature and above, parahydrogen constitutes only 25% of the mixture. Pure parahydrogen may be readily prepared by passing liquid hydrogen over activated charcoal. (Pure orthohydrogen has been prepared only in small quantities.) When ordinary hydrogen is liquefied, the heat evolved during the slow conversion of the equilibrium mixture to parahydrogen is responsible for the evaporation of large amounts of liquid hydrogen during storage (Kaplan, 1982, 6-743). Specific design criteria for storing and handling liquid hydrogen to avoid an explosion or release hazard associated with spin isomerization are given in Section 6.6.

1.4.2. Disproportionation

Disproportionation is a self-reaction in which a compound can form two similar products by transfer of electrons (reduction–oxidation). For example, upon heating, a hypochlorite can yield a chlorate and a chloride. Because most compounds require energy input for disproportionation to occur, such compounds generally do not pose a significant storage and handling hazard except for situations such as an external fire.

1.5. Reactivity with Oxygen

$$R\text{–}H + O_2 \rightarrow R\text{–}O\text{–}O\text{–}H$$

Oxygen-reactive materials have an obvious reactivity hazard due to the ubiquitous presence of atmospheric oxygen. The hazard can translate into a reactive chemical incident either by (a) air getting into the containment that stores or handles the oxygen-reactive material or (b) releasing the oxygen-reactive material into the environment, such as by an accidental spill or release.

If such an oxygen-reactive material does come into contact with air, the ensuing results will be determined by whether the material is *pyrophoric, flammable, combustible,* or *peroxide forming.* The basic characteristics of these and other oxygen-reactive categories are given in Table 1.8. Flammable and combustible materials are not treated in detail in these guidelines.

TABLE 1.8
Categories of Oxygen-Reactive Materials (in general order of decreasing reactivity)

Subset of Oxygen-Reactive Materials	*Basic Characteristics*	*Time Frame of Accident Potential*
Hypergolic	Immediate, spontaneous ignition on contact of fuel with oxidant at ambient conditions; large, extremely rapid energy release	Instantaneous ignition; no additional heat input required
Pyrophoric	Spontaneous ignition in air at or below about 130°F (54.4°C); may require moist air to ignite; large, very rapid energy release	Instantaneous or slightly delayed ignition; no additional heat input required
Low-Temperature Autoigniting	Autoignition temperature above ambient but relatively low (e.g., CS_2, AIT=90-125°C); large, rapid energy release	Immediate or delayed; requires some heat input for autoignition
Flammable	A gas, volatile liquid, or volatile solid with vapors that will burn in air at atmospheric pressure and below 100°F (37.8°C); large, rapid energy release	Immediate or delayed; requires ignition source (e.g., spark, flame, hot surface, or bulk heating to autoignition temperature)
Combustible	Oxygen-reactive gas that will ignite at elevated pressure (e.g., ammonia), liquid with flash point ≥100°F (37.8°C), or oxygen-reactive nonvolatile solid; large, more gradual energy release unless substance is finely divided and premixed with air prior to ignition	Immediate or delayed; requires heat input and ignition source
Autoxidizing	Propensity to slowly oxidize; relatively small energy release (Note: most autoxidation is via peroxide formation)	Ignition source not required for oxidation to occur; can lead to rapid combustion if evolved heat is not dissipated

1.5.1. Spontaneous Ignition and Pyrophoricity

Pyrophoric ("fire-bearing") compounds are highly reactive with atmospheric oxygen and/or humidity, to the extent that contact with the atmosphere causes oxidation and/or hydrolysis at a sufficiently high rate to cause ignition (Bretherick, 1986, 71–72). The criterion for pyrophoricity is not precise, as many materials show pyrophoric behavior under some atmospheric conditions but not others. A working definition is any material that will ignite spontaneously in air, sometimes with a short ignition delay, under normal conditions of ambient temperature and humidity. Sax and Lewis (1987, 985) give a more precise definition; namely, any liquid or solid that will ignite spontaneously in air at about 130°F (about 54°C). However, gases such as silane can also be pyrophoric.

Pyrophoricity is found in many different classes of compounds, but a few types of structures are notable for this behavior. Examples of pyrophoric materials are given in Table 1.9. A more extensive list that includes less common chemicals can be found in Bretherick (1990, 1778–1781).

Alkylaluminum derivatives (aluminum alkyls) are of particular industrial importance as powerful reducing agents. Literature on these compounds is reviewed by Bretherick (1990, 1492). Compounds with alkyl groups of C_4 and below ignite spontaneously and nearly immediately on exposure to air, unless diluted with a hydrocarbon solvent to 10–20% concentration. Even these dilute solutions may ignite on prolonged exposure to air due to exothermic autoxidation, which becomes rapid if solutions have a high surface-to-volume ratio such as after being spilled. Compounds with C_5 to C_{14} alkyl groups can be handled

TABLE 1.9
Some Pyrophoric Materials

Category	*Examples*
Finely divided metals (without oxide film)	Aluminum, calcium, cobalt, iron, magnesium, manganese, palladium, platinum, titanium, tin, zinc, zirconium
Many hydrogenation catalysts containing adsorbed hydrogen (before and after use)	Raney nickel catalyst with adsorbed hydrogen
Alkali metals	Potassium, sodium
Metal hydrides	Germane, lithium aluminum hydride, potassium hydride, silane, sodium hydride
Partially or fully alkylated metal hydrides	Butyllithium, diethylaluminium hydride, triethylbismuth, trimethylaluminum
Arylmetals	Phenylsodium
Alkylmetal derivatives	Diethylethoxyaluminium, dimethylbismuth chloride
Analogous derivatives of nonmetals	Diborane, dimethylphosphine, phosphine, triethylarsine
Carbonylmetals	Pentacarbonyliron, octacarbonyldicobalt
Grignard reagents (RMgX)	Ethylmagnesium chloride, methylmagnesium bromide
Miscellaneous	Phosphorus (white); titanium dichloride

References: Bretherick, 1986, 71-72; Britton, 1989; Cardillo and Nebuloni, 1992; National Research Council, 1983, 240-241; Sax and Lewis, 1987, 985.

safely at 20 to 30% concentrations. They smoke in air but do not burn unless ignited externally or unless the air is very moist.

Fairly extreme precautions must be taken when storing and handling a pyrophoric material, due to the need for a highly reliable means of keeping the material isolated from the atmosphere. Instead of a fire triangle requiring fuel, air, and an ignition source to have an accidental fire, only the contact of the fuel (pyrophoric material) with the atmosphere is necessary for a fire to occur. Typical handling procedures include storage in tightly closed containers under an inert atmosphere or immersed in a nonreacting liquid and carrying out all transfers and other operations under an inert atmosphere or liquid. Care must be taken to maintain the inerting or diluting material over time, since pyrophoric materials such as metal alkyls in solvents will become more and more reactive as the solvent evaporates. Special precautions must also be exercised in safely disposing of pyrophoric materials such as spent hydrogenation catalysts. Since a further ignition source is not required for combustion of pyrophoric materials, it is not necessary to use electrically classified equipment around such materials (NFPA 70, 1993).

"Low-autoignition materials" are closely related to pyrophoric materials, in that they require only a small amount of heat input to the material in contact with the atmosphere to ignite a combustion reaction. This heat input can come from such sources as heated process equipment or solar radiation. Specific design considerations for pyrophoric and low-autoignition materials are given in Section 6.8.

1.5.2. Pyrophoricity versus Hypergolic Properties

Hypergolic behavior is characterized by immediate, spontaneous ignition of an oxidation reaction upon mixing of two or more substances. Accordingly, many pyrophoric compounds can be considered as exhibiting hypergolic behavior. The term is most commonly used, however, to refer to liquid rocket fuels or propellants that consist of combinations of fuels and oxidizers that ignite spontaneously on contact (Sax and Lewis, 1987, 627). The oxidant in a hypergolic mixture does not need to be oxygen.

Presumably, hypergolic behavior is exhibited whenever an oxidation reaction has an extremely low activation energy, such that, at ambient temperatures, as soon as the materials contact each other, the oxidation reaction proceeds rapidly enough to liberate considerable thermal energy and thus propagate the reaction. It would be expected that the reaction rate is, therefore, only limited by the rate at which the reactants are mixed and the product gases can escape. The near-zero activation energy in a hypergolic mixture can be inherent in the particular combination of fuel and oxidizer, or it can be lowered due to catalytic effects such as is the case with hydrogenation catalysts containing adsorbed hydrogen.

The concepts of hypergolic behavior can be extended to other types of reactions as well, such as water-reactive materials. Yoshida (1987, 217–221) describes an apparatus to test for hypergolic ignition by breaking and mixing 500-g reagent bottles and estimating the violence of the ensuing reaction. The test results reported by Yoshida illustrate that some fuel–oxidant combinations may exhibit spontaneous ignition without exhibiting hypergolic behavior (immediate, spontaneous ignition). For example, smoking was observed 5 s after mixing and ignition 28 s after mixing when bleaching powder and ethylene glycol were combined. Ignition occurred 2 s after mixing with the combination of sodium chlorite, 60% nitric acid, and toluene.

1.5.3. Accumulation and Explosion of Pyrophoric Materials

Since pyrophoric materials ignite spontaneously in contact with the atmosphere, circumstances of an accidental event do not generally allow accumulation of a large quantity of spilled material prior to ignition. However, a few situations may allow such an occurrence, which could result in a more severe consequence. Among them are the following:

- Some materials are only marginally pyrophoric. For example, they may ignite spontaneously at 125°F with an optimum mixture, but not at normal ambient temperatures. Dichlorosilane is one example of a marginally pyrophoric material.
- Some materials are pyrophoric due to their extremely fast reaction with atmospheric moisture. If the humidity is extremely low, such as outdoors in a desert-type climate or in a heated indoor environment during very cold days, there may be insufficient atmospheric moisture to allow initiation of the combustion reaction.
- Loss of an inert gas purge may allow gradual accumulation of oxygen until a minimum oxygen concentration for ignition is achieved. Accumulation of sufficient humidity in systems that are highly water-reactive can have a similar effect.
- Likewise, loss of an inerting liquid by evaporation or other means may allow eventual exposure of the pyrophoric material to the atmosphere. An example is the handling of Grignard reagents, which are often dissolved in a highly volatile ether that can evaporate if not fully contained.
- Pyrophoric gases such as silane may have effective "flammable limits," such that a minimum concentration may need to accumulate before spontaneous ignition occurs.
- A chemical inhibitor may be present in the system that interferes with the reaction of concern. Depletion of the inhibitor can allow the reaction to begin after an induction period. Some pyrophoric materials may have an

induction period before spontaneous ignition due to other physical/chemical phenomena.

- Formation of a pyrophoric material due to an undesired reaction between incompatible materials can result in the accumulation of sufficient material to pose a significant hazard. Cardillo and Nebuloni (1992) have determined the conditions under which aluminum and halohydrocarbon mixtures will react, with the possible formation of pyrophoric alkylaluminums.
- Pyrophoric iron sulfides (FeS and Fe_2S_3) may form in anaerobic atmospheres in the presence of hydrogen sulfide. They can produce a hot-spot ignition source upon sudden exposure to air (CCPS/AIChE, 1993).
- Finely divided metals such as iron can also ignite upon sudden exposure to air as a result of their large reactive surface areas. Ignition of reduced oxides such as FeO may be possible (CCPS/AIChE, 1993). Other examples are given by Medard (1989).

In each case, it is essential to recognize the hazards of handling the particular pyrophoric material of concern in a specific storage and handling situation. Protection measures can include the detection and warning of (a) approach to an autoignition temperature, (b) presence of moisture, (c) loss of the inerting atmosphere or liquid, and/or (d) the onset of the undesired reaction. In any case, due to the unpredictable nature of pyrophoric behavior (as illustrated in the next section), pyrophoricity itself should not be relied on as a safety feature, such as for preventing the accumulation of hazardous quantities of a material before ignition occurs.

1.5.4. Competition between Air and Atmospheric Moisture

Pyrophoric materials such as some alkylaluminum derivatives react with both the oxygen and the water (humidity) in the atmosphere. The outcome of a spill of such materials will depend on the atmospheric humidity at the time of the spill. The spilled material may burn rapidly under low-humidity conditions, but the same material may smoke away and not ignite under high-humidity conditions. The consequences of such a spill will also depend on the solvent concentration and the volatility of the solvent.

1.5.5. Peroxide Formation

$$R\text{–}H + O_2 \rightarrow R\text{–}O\text{–}O\text{–}H$$

At the other end of the oxygen-reactive materials spectrum from pyrophoric compounds is a group of chemicals that react with atmospheric oxygen much more slowly, but with possibly very hazardous results. This group of chemicals consists of those which can gradually form peroxides. The resulting peroxides can accumulate over time, eventually building up a sufficient concentration to pose a peroxide decomposition hazard.

Peroxide formation, or *peroxidation,* usually takes place slowly when the liquid materials are stored with limited access to air and exposure to light. The peroxides initially formed may subsequently react to form polymeric or cyclic peroxides, many of which are dangerously unstable when concentrated.

The common structural feature in organic peroxidizable compounds is the presence of a hydrogen atom that is susceptible to autoxidative conversion to the hydroperoxy group –OOH. Some of the typical structures susceptible to peroxidation are shown in Table 1.10. (See also Table 2.6 for a list of specific peroxide-forming compounds.) Two or more of these structures within an organic molecule will generally make the compound significantly more susceptible to peroxidation.

Many commonly used chemicals can form peroxides upon exposure to air over a period of time, and some by only brief exposures such as opening a container. A notable example is isopropyl ether (diisopropyl ether), which has been involved in numerous explosion incidents. Storage/handling design and procedures must, therefore, provide highly reliable means to isolate the peroxide-forming material from atmospheric oxygen, detect any dangerous peroxides formed over time, and safely dispose of suspicious inventories. Upon formation, the peroxidized compounds can become particularly dangerous in one of two ways; namely, concentration of the peroxide and initiation of a polymerization reaction. The decomposition of some peroxides is easily initiated and violently explosive in concentrated solution or when dried to a solid; consequently, the evaporation of liquids possibly containing unstable peroxides must be avoided. Sometimes, peroxide-forming solvents, including diethyl ether, tetrahydrofuran, dioxan, 1,2-dimethoxyethane ("glyme"), and *bis*-2-methoxyethyl ether ("diglyme"), are often stored without inhibitors, and are therefore susceptible to peroxidation. Many accidents involving distillation in use of the peroxide-containing solvents have been reported (Bretherick, 1986, 72–73). It is essential to test these solvents for peroxide (such as with sodium or potassium iodide in glacial acetic acid) before use and, if present, peroxides must be eliminated by suitable means before proceeding. Procedures for testing for peroxides and for removing small amounts from laboratory quantities of chemicals are given by the National Research Council (1983, 73–76).

Peroxide formation can also occur in many polymerizable compounds, with the peroxide then capable of initiating an uncontrolled polymerization reaction. Many peroxides are intentionally used as polymerization initiators.

If allyl and, particularly, vinyl monomers become peroxidized, they are potentially dangerous for both of the reasons discussed above. The peroxides of some vinyl monomers such as 1,1-dichloroethylene and 1,3-butadiene separate from solution and are extremely explosive. Even when this does not happen, the peroxide present may initiate the exothermic and sometimes violent polymerization of any vinyl monomer during storage. Reactive monomers must therefore be inhibited against oxidation and stored cool, with regular checks for presence of peroxide. However, contained air should not be completely displaced by nitrogen, as some is essential for effective inhibition (Bretherick, 1986, 72–73).

TABLE 1.10
Some Chemical Structures Susceptible to Peroxide Formation

	ORGANIC SUBSTANCES
Structure	*Explanation (NOTE: not all bonds are shown)*
–O–CH	Ethers and acetals with α hydrogen atoms; especially cyclic ethers and those containing primary and secondary alcohol groups, form dangerously explosive peroxides on exposure to air and light
C=C–CH	Allyl compounds (olefins with allylic hydrogen atoms), including most alkenes
C=C–X	Chloroolefins, fluoroolefins
C=CH	Vinyl and vinylidene halides, esters, ethers, styrenes
C=C–C=C	Dienes (i.e., monomers)
C=CH–C≡CH	Vinylacetylenes with α hydrogen atoms
CH–C≡CH	Alkylacetylenes with α hydrogen atoms
CH–Ar	Tetrahydronaphthalenes, decahydronaphthalenes, alkylarenes with tertiary hydrogen atoms (such as cumene)
CH	Isopropyl compounds; alkanes and cycloalkanes with tertiary hydrogen atoms
C=C–CO_2R	Acrylates, methacrylates
CH–OH	Secondary alcohols
O=C–CH	Ketones with α hydrogen atoms
O=CH	Aldehydes
O=C–NH–CH	Ureas, amides, and lactams that have a hydrogen atom on a carbon atom attached to nitrogen
	INORGANIC SUBSTANCES
Alkali metals, especially potassium, rubidium, and cesium	
Metal amides	
Organometallic compounds with a metal atom bonded to carbon	
Metal alkoxides	

References: Bretherick, 1986, 72–73; National Research Council, 1981, 63–64; National Research Council, 1983, 244–245

1.6. Reactivity with Water

$$R + H_2O \rightarrow P + \Delta$$

Water reactivity is the key type of reactivity of concern for fire protection interests; indeed, a *reactive material* in NFPA 704 (1990) is defined for the purposes of that standard as "one that can enter into a violent chemical reaction with water." (The term *unstable material*, as discussed in Section 1.4, is also used in NFPA 704 with respect to self-reactivity.)

Water is, after air, likely the most common substance other than containment materials to come into contact with reactive materials, even if the contact is unintentional. Since water can have a dilution effect as a solvent and a heat sink, it is usually necessary for a relatively limited amount of water to come into contact with a water-reactive material to give a significant hazard.

Water reactivity can be hazardous by one or more of several mechanisms. The direct heat of reaction may be dangerous to persons or equipment nearby. The violence of the reaction may also cause dispersal of reactive materials. Toxic reaction products such as hydrogen chloride are often formed. Some water-reactive materials evolve flammable gases; for example, potassium, sodium, and many metal hydrides react with water to form hydrogen. The heat of the reaction can often be sufficient to ignite flammable/combustible materials or initiate another type of secondary chemical reaction; for example, the heat of reaction between solid sodium metal and water is sufficient to ignite the hydrogen evolved from the reaction (Yoshida, 1987, 2).

Some concentrated acids or bases can generate considerable heat of dilution when mixed with water. However, this is a physical effect rather than a chemical reaction.

1.6.1. Water Reactivity: Fast and Slow Reactions

Just as there is a broad range of oxygen reactivity (Section 1.6), the intensity with which materials react with water covers a broad range also. Water-reactive materials might be considered in two categories. Some materials react rapidly to violently with water and have an NFPA reactivity rating of 2 or higher based on water reactivity alone. Other materials react relatively slowly but can generate heat and/or gases that can result in elevated pressure if contained. These categories are discussed in more detail in Section 2.5.

1.6.2. Water-Reactive Structures

A list of some typical water-reactive chemical structures is given in Table 1.11. Many of these types of chemicals can react violently with water, especially if the amount of water present is limited. These compounds should always be stored and handled so that they do not come into contact with liquid water or water vapor.

A list of individual chemicals exhibiting water-reactive behavior is presented in Table 1.12. This list was generated by extracting all of the chemicals listed as water-reactive (symbol ₩) from both NFPA 49 (1994) and NFPA 325M (1994). Ethylene oxide is also included; even though it also is highly reactive and can explosively decompose, it may undergo runaway reaction with water (NFPA 49, 1994). Class B materials react slowly with water, but can build up pressure if confined, possibly rupturing their container.

TABLE 1.11
Some Chemical Categories Susceptible to Water Reactivity (Violent or Slow)

Category	*Examples*
Alkali and alkaline-earth metals	Calcium, potassium, sodium, lithium
Anhydrous metal halides	Aluminum tribromide, germanium tetrachloride, titanium tetrachloride
Anhydrous metal oxides	Calcium oxide
Grignard reagents	Ethylmagnesium chloride, methylmagnesium bromide
Metal alkyls	Aluminum alkyls, lithium alkyls
Metal amides	Lead amide, potassium amide, silver amide, sodium amide
Metal hydrides	Calcium hydride, lithium aluminum hydride, sodium borohydride, sodium hydride
Nonmetal halides	Boron trifluoride, phosphorus trichloride, silicon tetrachloride
Nonmetal halide oxides (inorganic acid halides)	Phosphoryl chloride, sulfuryl chloride, chlorosulfuric acid
Nonmetal oxides	Phosphorus pentoxide, sulfur trioxide
Low-molecular-weight organic acid halides and anhydrides	Acetic anhydride, acetyl chloride
Other	Calcium carbide

References: Bretherick, 1986, 72-73; National Research Council, 1983, 244-245

1.7. Reactivity with Other Common Substances

Some substances are reactive with other common substances besides atmospheric oxygen and water. The hazards of a substance that can react with anything commonly in close proximity to the substance must be identified and carefully controlled. Other common substances can include

- atmospheric nitrogen;
- biological materials;
- concrete, asphalt, wood, and other structural materials;
- metals and other containment materials.

Nitrogen is generally inert to most chemical substances. However, under the proper conditions, it can react violently with lithium, neodymium, or titanium. Extreme precautions must obviously be taken when handling such materials in their pure state. Table 1.13 lists some nitrogen-reactive materials.

TABLE 1.12
Some Materials That React with Water[a]

Water-reactive materials with NFPA reactivity hazard rating of 2, 3, or 4		
Acetyl chloride	Diethyl telluride	Phosphorus pentachloride
Alkylaluminums	Diethylzinc	Phosphorus pentasulfide
Allyl trichlorosilane	Diisobutylaluminum hydride	Phosphorus tribromide
Aluminum chloride, anhydrous	Dipropylaluminum hydride	Phosphorus trichloride
Aluminum phosphide	Ethyltrichlorosilane	Potassium, metal
Amyl trichlorosilane	Ethylaluminum dichloride	Potassium-sodium alloys
Benzoyl chloride	Ethylaluminum sesquichloride	Silicon tetrachloride
Boron tribromide	Fluorine	Silicon tetrafluoride
Bromine pentafluoride	Gallium arsenide	Sodium, metal
Bromine trifluoride	Gallium phosphide	Sodium hydride
Butylacrylate	Germane	Sulfuric acid
Butyllithium	Isophorone diisocyanate	Sulfuryl chloride
Calcium, metal	Lithium, metal	Thionyl chloride
Calcium carbide	Lithium aluminum hydride	Titanium tetrachloride
Chlorine trifluoride	Lithium hydride	Toluene diisocyanate
Chlorosilanes	Methylaluminum sesquibromide	Trichlorosilane
Chlorosulfonic acid	Methylaluminum sesquichloride	Triethylaluminum
Chromium oxychloride	Methyldichlorosilane	Triisobutylaluminum
Diborane	Methyl isocyanate	Trimethylaluminum
Dichloroacetyl chloride	Methyltrichlorosilane	Trimethylchlorosilane
Dichlorosilane	Monochloro-*s*-triazinetrione acid	Tripropyl aluminum
Diethylaluminum chloride	Mono-(trichloro)tetra-(monopotassium dichloro)-penta-*s*-triazinetrione	Vanadium tetrachloride
Diethylaluminum hydride		Vinyl trichlorosilane
Diethyl carbamyl chloride	Phosphorus oxychloride	Zirconium tetrachloride
Examples of materials that react slowly with water		
Acetic anhydride	Butyric anhydride; isobutyric anhydride	Sodium dichloro-*s*-triazinetrione dihydrate
Boron trifluoride	Methylene diisocyanate	Sulfur chlorides

[a]Sources: NFPA 49, 1994; NFPA 325M, 1994.

TABLE 1.13
Some Nitrogen-Reactive Materials[a]

Lithium	Neodymium	Titanium dust	Zirconium dust
Magnesium dust	Thorium dust	Uranium dust	

[a]Sources: NFPA 69, 1992; NFPA 491M, 1991.

Reactivity with surfacing and structural materials are generally well-recognized for a given reactive material. Design and construction considerations will depend on the actual reactive material of concern. Commonly considered here are such hazards as liquid oxygen spills reacting with asphalt to form impact-sensitive explosive mixtures, corrosive materials causing accelerated corrosion and weakening of structural steel, and hydrofluoric acid spills damaging concrete by reacting with the silicon dioxide in the concrete to form silicon tetrafluoride.

1.7.1. Reactions with Metals

Reactivity with metals and other containment materials can take the form of bulk, rapid reaction or slower, corrosive action. Materials of construction must be selected to avoid both immediate and long-term, corrosive reaction between a reactive material and its containment under all normal and foreseeable upset conditions. Few materials will react in bulk with metals used for process equipment; however, there are some notable exceptions. Dry chlorine gas can be handled in iron-based piping under normal conditions, but if a local hot spot exists due to an external fire or other cause, the iron can begin to actually burn due to chlorine being a strong oxidizer. Carbon steel ignites in chlorine near 483°F (251°C). Titanium reacts violently with dry chlorine (CGA, 1990, 314).

Cardillo and Nebuloni (1992) studied the reaction of halogenated hydrocarbons with light metals to form highly reactive metal alkyls. They concluded that there is little evidence of any hazard associated with short-term exposure of bulk metal to halogenated hydrocarbons at ambient temperatures and pressures. However, systems in which light metals such as aluminum are exposed for a prolonged period of time to halogenated hydrocarbons, particularly at elevated pressures and where fresh metal surfaces or metal powders are produced, are susceptible to initiating a violent reaction between the metal and the halogenated hydrocarbon. The referenced article thermochemically evaluated over 2000 binary mixtures of aluminum with several halogenated compounds to identify mixtures with high reactivity. Bretherick (1990, 24–26) reviews 25 other articles on the reactivity of aluminum with halogenated hydrocarbons.

Confinement may make localized elevated pressures possible in such systems. A series of explosions has occurred in recent years with aluminum paint sprayers, when the closing of valves isolated an aluminum part while still in contact with residual paint. Current procedures call for flushing and leaving sprayer valves open.

1.7.2. Surface Area Effects

Reactions with metals and other solids are generally controlled by surface transport phenomena such as diffusion of the reactant or reactants to the metal surface and diffusion of the products from the surface. For this reason, the amount of surface area exposed to the reactive material has a very strong effect on the outcome of the reaction.

For example, in half of the reported incidents involving reaction of aluminum with halogenated hydrocarbons, aluminum in the form of powder, dust, filings, and turnings was involved. It was concluded that along with temperature and pressure, the physical division of the metal increases its reactivity and any contact of finely divided aluminum with a halogenated hydrocarbon should be viewed with caution (Cardillo and Nebuloni, 1992).

Due to the large surface area and high reactivity of many finely divided metals, intense powdered metal fires, dust explosibility or even pyrophoric behavior can be exhibited (Section 1.6). Active metals with intentionally high surface areas used for such purposes as hydrogenation catalysts are particularly susceptible to high reactivity. Hydrogen adsorbed on the catalyst surface increases the potential severity of incidents involving hydrogenation catalysts.

1.7.3. Catalyst Deactivation and Surface Passivation

Metals having catalytic activity can be deactivated by various contaminants and reactants that either block active sites, alter the chemical or physical nature of the surface, or cause a surface coating to be formed. The surfaces of storage and handling equipment may be intentionally deactivated to prevent catalytic activity from occurring, such as the preparation of a vessel for use in peracetic acid service by passivation of the internal vessel surfaces. The electrochemical, corrosive, and/or catalytic activity of iron, chromium, and related metals can be passivated by treatment with a strong oxidizing agent such as nitric acid or by evolving oxygen onto the surface by electrolysis to form an oxide coating (Sax and Lewis, 1987, 876).

1.8. Reactive with Other Chemicals: Incompatibility

$$A + B \rightarrow P + \Delta$$

There are a virtually unlimited number of possible chemical reactions. However, the combinations that have been involved in hazardous storage and handling incidents are mostly limited to those that produce or lead to an exothermic reaction too large or too fast for effective dissipation under the particular facility conditions (Bretherick, 1986). Besides the self-reactive, oxygen-reactive, and water-reactive chemicals that are emphasized in the preceding sections, quantitative chemical reactions can also result in storage and handling accidents. For example, if a tank truck load of acid is inadvertently unloaded to a tank containing a cyanidic waste, an energetic reaction will ensue and toxic hydrogen cyanide gas will be produced.

Several resources are available to aid in the identification of incompatible materials. The U.S. Environmental Protection Agency has published an extensive list of chemicals in 41 reactivity groups, with expected outcomes for interactions between each group. The document is aimed at incompatibilities related to hazardous wastes (Hatayama et al., 1980). A computerized version of the EPA interaction data is available from Wertz (1989). Bretherick (1990) gives an extensive compilation of incompatibility incidents reported in the literature. An electronic database of Bretherick's data is available from the same publisher. NFPA 491M (1991) is a manual of hazardous chemical interactions based on an extensive literature survey. The use of a *chemical interaction matrix* to systematically identify potentially incompatible mixtures and trace contaminants is detailed in Section 4.1.

Two groups of reactive substances are particularly common: those with oxidizing or reducing properties, and those with acidic or basic properties. These and other important incompatibility potentials are discussed below.

1.8.1. Oxidizing and Reducing Properties

Oxidation–reduction reactions involve the transfer of electrons from one molecule or atom (the one being oxidized) to another (the one being reduced). Oxidizers are electron receptors, and reducing agents (or "fuels") are electron donors. Oxidizers can be considered as solids, liquids, or gases that yield oxygen or other oxidizing gas during the course of chemical reaction or that readily react to oxidize combustible materials. Accordingly, oxidizers include chlorates, perchlorates, bromates, peroxides, nitrates, nitrites, nitric acid, and permanganates. The halogens (fluorine, chlorine, bromine, and iodine), which react similarly to oxygen under certain conditions are, therefore, classed as oxidizers (NFPA 49, 1994). Stull (1977, 5ff) cites specific examples and gives warning that some oxidizers and fuels are less easily recognizable as such, but their combination, nevertheless, results in an energetic reaction.

The rates of oxidation–reduction reactions vary considerably and may be rapid enough to be quite violent. The hazard greatly increases with increasing temperature, and explosive deflagrations are possible under fire exposure conditions. Further, combustible materials that are contaminated or mixed with oxidizers may become sensitive to heat, shock, friction, or impact.

While most inorganic oxidizers such as chlorine gas are not themselves combustible, they can greatly intensify a fire in which they become involved by speeding the oxidation of the burning material. They may also play a role, through their increased concentration, in aiding ignition in air. Small traces of solid oxidant can, in some cases, cause fuels to ignite at room temperature. For example, a mixture of glycerol with potassium permanganate will ignite, though after some delay. Liquid oxygen (or air) can be condensed by very cold equipment such as in liquid nitrogen or helium service. This can be a hazard outside the equipment if liquid air condenses and mixes with a combustible material. Inside equipment, a hazard can result from traces of oxygen in an inert stream should this condense out (CCPS/AIChE, 1993).

Most of the oxidation–reduction reactions involve a compound that is capable of being oxidized plus air as the most likely oxidizer (see Section 1.6), although many other oxidizers are possible. For example, glycerol reacts vigorously with solid potassium permanganate as the oxidizer. Also, ethanol undergoes violent or explosive oxidation in excess concentrated nitric acid; this reaction involves formation of the unstable compound fulminic acid.

Other oxidation–reduction reactions involve a recognized reducing agent in contact with a material capable of being reduced. For example, a heated mixture of aluminum powder and a metal sulfate will explosively decompose, and sodium in contact with chlorinated solvents is shock-sensitive. Stull (1977, 5) gives examples of non-obvious fuel–oxidant mixtures such as lead chromate, a yellow pigment, and ferric ferrocyanide, a blue pigment.

The most hazardous interactions are between a strong oxidizer and a strong reducing agent. Bretherick cites a severe explosion upon attempted reduction of dibenzoyl peroxide by lithium tetrahydroaluminate. Hydrazine decomposes explosively in contact with chromium trioxide. Space propulsion and power technology also involve examples of strong oxidation–reduction systems capable of extreme energy release (Bretherick, 1986).

1.8.2. Acidic and Basic Properties

Acids are ionic or molecular species that donate protons (hydrogen ions) in solution. Bases, likewise, are proton receptors. Most acid–base reactions are simply characterized as causing heat generation and gas evolution. Acidic and basic properties of materials are generally easier to identify than oxidizing and reducing properties.

1.8.3. Formation of Unstable Materials

One of the most important types of reactions that do not pose a direct hazard, but rather form unstable, and therefore hazardous, substances is that of peroxide formation, which is discussed in Section 1.6.5. Reactions between incompatible materials often form other unstable materials that are either shock-sensitive or can thermally decompose. For example, the highly energetic, shock-sensitive compound nitrogen trichloride can be formed when excess chlorine or a chlorinating agent contacts aqueous ammonia, ammonium salts, or a compound containing a hydrolyzable amino-derivative (Bretherick, 1990, 1040).

1.8.4. Thermite-Type Reactions

Thermite is a mixture of ferric oxide and powdered aluminum. On ignition by a ribbon of magnesium, the reaction produces a temperature of 2200°C which is sufficient to melt steel. This is typical of some oxide–metal reactions that provide their own oxygen supply and thus are very difficult to stop (Sax and Lewis, 1987).

Mechanical sparks usually require high impact velocities in order to become incendiary. Exceptions include sparks of special materials used as flints in gas lighters, where the hot fragments become incandescent upon reaction with air. Another exception occurs when contacting surfaces react together. The most common of these is reaction between aluminum and oxides such as iron rust or red lead, where the thermite-type reaction may occur. In this reaction, aluminum is oxidized and the iron oxide is reduced, releasing a significant amount of heat. Analogous reactions can occur in other systems of metals and oxides. It is important to recognize that thermite reactions can be hazardous ignition sources even if the metal oxide is present only superficially. The ignition of methane by the impact of an aluminum alloy with a rusty surface has been studied by the Bureau of Mines (CCPS/AIChE, 1993). The use of aluminum with transition metal oxides such as manganese dioxide may lead to explosion (Bretherick, 1990). Thermite-type reactions of aluminum with various oxides of chromium, cobalt, copper, iron, manganese, and nickel are possible.

1.8.5. Incompatibility with Heat Transfer Fluids and Refrigerants

The storage and handling of reactive materials often involves the use of heat transfer fluids for either heating or refrigeration of the material. Heat exchanger tube leaks and other means of mixing heat transfer fluids with reactive materials can lead to serious incidents if the substances are incompatible. Some potential incompatibilities include:

- water with water-reactive materials
- ammonia with acids, halogens, oxidizing materials, etc.
- biphenyl–biphenyl oxide heat transfer fluids in the presence of sulfuric acid or air (sulfonation or oxidation possible)
- chlorofluorocarbon refrigerants with alkali- and alkaline-earth metals; for example, dichlorodifluoroethane (refrigerant-12) with aluminum (Schwab, 1971)
- Coastal Chemical HITEC® oxidizing heat transfer salt mixture containing potassium nitrite, sodium nitrite and sodium nitrate, with fuels and other materials capable of being reduced.

Incompatibilities with alternative refrigerants, particularly various hydrochlorofluorocarbons (HCFCs) and hydrofluorocarbons (HFCs), will need to be examined closely as the use of these materials is expected to increase.

1.8.6. Adsorbents

Adsorption is the physical or chemical adherence of atoms, molecules, or ions in any phase to the surface of a solid or liquid called the *adsorbent*. Finely divided or microporous materials, which, therefore, present a large area of active surface, are strong adsorbents; examples include activated carbon, activated alumina, and silica gel.

Adsorbents may be used in storage and handling operations for such purposes as drying or collection of off-gases. Adsorbents can pose subtle but significant hazards when used in operations handling reactive materials. For example, an adsorbent may serve to catalyze the decomposition of an unstable material in the liquid or gaseous state. Also, the heat of adsorption in a dryer or other unit operation can be high enough to pose a thermal hazard by heating the bulk of the reactive material or by reaching the autoignition temperature of the off-gas vapors. This may be particularly true during process upsets that may, for example, admit impurities into the adsorption unit or cause a high flow rate of organic vapors to pass through the unit.

It can be seen that temperature monitoring of the adsorption unit is an important provision for safe operation. To be effective, such monitoring must measure the actual bed temperature and not just the wall temperature. Britton et al. (1986) and Britton (1994) discuss bed temperature monitoring, as well as the initiation of ethylene decomposition by the heat of adsorption in an adsorbent bed. Purging and pre-loading procedures are also discussed.

Bibliography

L. Bretherick, *Bretherick's Handbook of Reactive Chemical Hazards,* 4th Ed., Butterworths, London, xxi-xxvi and 1477-1824, 1990.

F. P. Lees, *Loss Prevention in the Process Industries,* Butterworths, London, 1980.National Research Council, *Prudent Practices for Handling Hazardous Chemicals in Laboratories,* Committee on Hazardous Substances in the Laboratory, Assembly of Mathematical and Physical Sciences, National Research Council, National Academy Press, Washington, D.C., 1981.

D. R. Stull, *Fundamentals of Fire and Explosion,* AIChE Monograph Series No. 10, Volume 73, American Institute of Chemical Engineers, New York, 1-44, 1977.

T. Yoshida, *Safety of Reactive Chemicals,* Industrial Safety Series Volume 1, Elsevier Science Publishers B.V., Amsterdam, 1987.

References

CCPS/AIChE (1988). *Guidelines for Storage and Handling of High Toxic Hazard Materials.* American Institute of Chemical Engineers, New York.

CCPS/AIChE (1993). *Guidelines for Engineering Design for Process Safety.* American Institute of Chemical Engineers, New York.

Bretherick, L. (1986). *Reactive Chemical Hazards.* Butterworths, London.

Bretherick, L. (1990). *Bretherick's Handbook of Reactive Chemical Hazards,* 4th ed. Butterworths, London.

Britton, L. G., Taylor, D. A., and Wobser, D. C. (1986). "Thermal Stability of Ethylene at Elevated Pressures." *Plant/Oper. Prog.,* 5(4, October), 238–251.

Britton, L. G. (1989). "Combustion Hazards of Silane and Its Chlorides." *Plant/Oper. Prog.,* 9(1, January).

Britton, L. G. (1994). "Loss Case Histories in Pressurized Ethylene Systems." *Proc. Safety Prog. 13*(3, July).

Bulletin 680. Bureau of Mines Bulletin 680, U.S. Bureau of Mines, Pittsburgh, Pennsylvania.

Carberry, J. J. (1976). *Chemical and Catalytic Reaction Engineering.* McGraw-Hill, New York.

Cardillo, P., and Nebuloni, M. (1992). "Reactivity Limits of Aluminium and Halohydrocarbon Mixtures: Determination by the ASTM CHETAH Program." *J. Loss Prev. Proc. Ind.* 5(2), 81-88.

Englund, S. M. (1991). "Selected Case Histories from the Manufacturing Chemists Association, 1951-1973." The Dow Chemical Company, Midland, Michigan, June.

Hatayama, H. K., et al. (1980). "A Method for Determining the Compatibility of Hazardous Wastes." EPA-600/2-80-076. U.S. Environmental Protection Agency, Washington, D.C., April. Available through NTIS.

CGA (1990). *Handbook of Compressed Gases,* 3rd ed. Compressed Gas Association, Arlington, Virginia. Published by Van Nostrand Reinhold, New York.

Kaplan, L. (1982). "Hydrogen." *McGraw-Hill Encyclopedia of Science and Technology.* McGraw-Hill, New York.

Levenspiel, O. (1972). *Chemical Reaction Engineering,* 2nd ed. John Wiley & Sons, New York.

Marshall, V. C. (1987). *Major Chemical Hazards.* Ellis Horwood Ltd., Chichester, England.

Medard, L. A. (1989). *Accidental Explosions.* Volume 2: Types of Explosive Substances. John Wiley & Sons, New York.

National Research Council (1981). *Prudent Practices for Handling Hazardous Chemicals in Laboratories.* Committee on Hazardous Substances in the Laboratory, Assembly of Mathematical and Physical Sciences, National Research Council. National Academy Press, Washington, D.C.

National Research Council (1983). *Prudent Practices for Disposal of Chemicals from Laboratories.* Committee on Hazardous Substances in the Laboratory, Assembly of Mathematical and Physical Sciences, National Research Council. National Academy Press, Washington, D.C.

NFPA 321 (1991). *Basic Classification of Flammable and Combustible Liquids.* National Fire Protection Association, Quincy, Massachusetts.

NFPA 325M (1994). *Fire Hazard Properties of Flammable Liquids, Gases, and Volatile Solids.* National Fire Protection Association, Quincy, Massachusetts.

NFPA 43B (1993). *Storage of Organic Peroxide Formulations.* National Fire Protection Association, Quincy, Massachusetts.

NFPA 49 (1994). *Hazardous Chemical Data.* National Fire Protection Association, Quincy, Massachusetts.

NFPA 491M (1991). *Manual of Hazardous Chemical Reactions.* National Fire Protection Association, Quincy, Massachusetts.

NFPA 495 (1992). *Explosive Materials Code.* National Fire Protection Association, Quincy, Massachusetts.

NFPA 69 (1992). *Explosion Prevention Systems.* National Fire Protection Association, Quincy, Massachusetts.

NFPA 70 (1993). *National Electric Code®*. National Fire Protection Association, Quincy, Massachusetts.

NFPA 704 (1990). *Standard System for the Identification of the Fire Hazards of Materials.* National Fire Protection Association, Quincy, Massachusetts.

Nicolson, A. (1991). "The Effect of O_2 Concentration on Methacrylic Acid Stability." *Plant/Oper. Prog. 10*(3, July), 171-183.

Perry, R. H. and Green, D. W. (1984). *Perry's Chemical Engineers' Handbook,* 6th ed. McGraw-Hill, New York.

Sax, N. I. and Lewis, R. J. (1989). *Dangerous Properties of Industrial Materials* (3 volumes), 7th ed. Van Nostrand Reinhold, New York.

Schwab (1971). "Chlorofluorohydrocarbon Reaction with Aluminum Rotor." *Loss Prev. 5.* American Institute of Chemical Engineers, New York, 111.

Shanley, E. S., and Ennis, J. L. (1991). "The Chemistry and Free Energy of Formation of Silver Nitride." *Ind. Eng. Chem. Res. 30*(11), 2503-2506.

Smith, D. W. (1982). "Runaway Reactions and Thermal Explosion." *Chem. Eng. 89*(25), 79.

Stull, D. R. (1977). *Fundamentals of Fire and Explosion.* AIChE Monograph Series No. 10, Vol. 73. American Institute of Chemical Engineers, New York.

Wertz, D. (1989). "REACT Hazardous Chemical Reactivity Code." Uniformed Services University of the Health Sciences, Bethesda, Maryland 20814-4799, (301) 295-3323.

Yoshida, T. (1987). *Safety of Reactive Chemicals.* Industrial Safety Series Volume 1. Elsevier Science Publishers B.V., Amsterdam.

2

Chemical Reactivity Classifications

The breakdown of reactivity hazard types that was presented in Chapter 1 is not the only method to classify reactivity hazards. Several other classification systems are currently in use. Each system was developed for a specific purpose, and some indicate the magnitude of reactivity hazard as well. This chapter will review other classification systems that are in use or have been proposed for use.

2.1. NFPA Reactivity Hazard Signal

The NFPA *Recommended System for the Identification of the Fire Hazards of Materials* (NFPA 704, 1990) gives specifications for a standard system for the assigning and identification of the health, flammability, and reactivity (instability and water reactivity) hazards of materials under emergency conditions such as spills, leaks, and fires. It is not intended to identify the nonemergency health hazards of chemicals. As stated in NFPA 704:

> The system provides a simple, readily recognized, easily understood system of markings that provide a general idea of the hazards of a material and the severity of these hazards as they relate to handling, fire prevention, exposure and control.

This system was initially developed in 1957. The 1990 revision has been approved by the American National Standards Institute (ANSI).

NFPA 704 also specifies a standard way to display the ratings for health, flammability, and reactivity hazards. It basically comprises three adjacent diamonds, at the 9, 12, and 3 o'clock positions, each containing one of the hazard ratings. The fourth diamond position is reserved for special symbols indicating reactivity with water, or, if not water-reactive, then some other unusual hazard, such as a strong oxidizer. The most commonly used configuration for displaying the NFPA hazard identification signal is shown in Figure 2.1.

At present, definitions are in place for ratings from 0 to 4 for each of the three types of hazards (health, flammability, reactivity). The **reactivity** rating

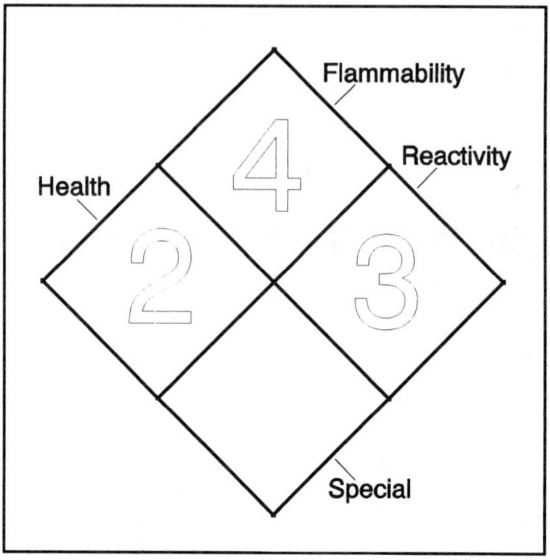

FIGURE 2.1 **NFPA Hazard Signal**

system is of primary interest to reactive materials storage and handling, and will be the only hazard type discussed.

2.1.1. NFPA 704 Rating System for Overall Reactivity

The NFPA reactivity rating is on an integer scale from 0 to 4, with 4 being the most severe hazard. These degrees of reactivity are intended to indicate the susceptibility of a material to release energy, either by itself or in combination with water. Fire exposure is one of the factors considered, along with conditions of shock and pressure (NFPA 704, 1990). The same 0-to-4 rating scale is also used for toxicity and flammability hazards.

2.1.2. Definitions for Reactivity Signal Ratings

Definitions for each integer value in the NFPA reactivity rating system are given in Table 2.1 (NFPA 704, 1990). It is evident from these definitions that the primary concern for the NFPA reactivity classification system is for fire-fighting situations. Nevertheless, the classification provides a widely used summary of most reactivity hazards associated with chemicals of industrial importance.

An alternate proposed test method and criteria for determining NFPA reactivity ratings are also shown in Table 2.1. This approach uses the *instantaneous power density* (IPD), which is the product of the heat of reaction and the

TABLE 2.1

NFPA Hazard Identification System for Reactivity Hazards (NFPA 704, 1990)

Identification of Reactivity (Stability) Color Code: YELLOW				
Susceptibility to Release of Energy				
Signal	*Stability*	*This Usually Includes:*	*Water Reactivity*	*Proposed Criteria*
4	Materials that, in themselves, are readily capable of detonation, explosive decomposition, or explosive reaction at normal temperatures and pressures	Materials that are sensitive to localized thermal or mechanical shock at normal temperatures and pressures		Instantaneous Power Density ≥1000 W/mL (see text)
3	Materials that, in themselves, are capable of detonation, explosive decomposition, or explosive reaction, but require a strong initiating source or heating under confinement	Materials that are sensitive to thermal or mechanical shock at elevated temperatures and pressures	Materials that react explosively with water without heat or confinement	IPD ≥100 but <1000 W/ml *or* explosive water reactivity ($-\Delta H_r \geq 600$ cal/g)
2	Materials that, in themselves, are normally unstable and readily undergo violent chemical change at elevated temperatures and pressures but do not detonate	Materials that can undergo chemical change with rapid release of energy at normal temperatures and pressures; Materials that exhibit an exotherm at temperatures ≤150°C when tested by differential scanning calorimetry	Materials that react violently with water or form potentially explosive mixtures with water	IPD ≥10 but <100 W/ml *or* violent water reactivity ($-\Delta H_r \geq 100$ but <600 cal/g)
1	Materials that, in themselves, are normally stable but can become unstable at elevated temperatures and pressures	Materials that change or decompose on exposure to air, light, or moisture; Materials that exhibit an exotherm at temperatures >150°C but ≤300°C when tested by differential scanning calorimetry	Materials that will react with water with some energy release but not violently	IPD ≥0.01 but <10 W/ml *or* vigorous water reactivity ($-\Delta H_r \geq 30$ but <100 cal/g)
0	Materials that, in themselves, are normally stable, even under fire exposure conditions	Materials that exhibit an exotherm at temperatures >300°C (if any) when tested by differential scanning calorimetry	Materials that are not reactive with water	IPD <0.01 W/mL *and* not water-reactive ($-\Delta H_r$ <30 cal/g)

reaction rate at 482°F (250°C), to categorize reactive materials by quantity and rate of energy release. These criteria, if approved as changes to NFPA 704 in 1996, will replace the DSC testing criteria. For water reactivity, the proposal specifies the use of a Two Drop Mixing Calorimeter (Hofelich, 1994) or equivalent technique using a 1 : 1 ratio of material to water to determine the heat of mixing.

2.1.3. Reactivity Hazards Not Identified by NFPA 704

It is important to note that, due to its primary emphasis on fire-fighting situations, not all reactivity hazards may be identified by the NFPA classification system. For example, rearranging materials and materials that could cause slow, pressure-accumulating reactions would not necessarily be identified as reactive materials using the NFPA classification system. Consequently, a material having an NFPA reactivity rating of 0 or 1 should not automatically be considered as a "nonreactive" substance.

2.1.4. NFPA Reactivity Ratings for Specific Chemicals

Reactivity ratings for many pure chemicals are given in NFPA 49, *Hazardous Chemical Data* (1994) and NFPA 325M, *Fire Hazard Properties of Flammable Liquids, Gases, and Volatile Solids* (1994). Reactivity ratings can be obtained for mixtures by applying the same definitions as are in Table 2.1. For example, if a mixture is not water-reactive or shock-sensitive, and differential scanning calorimetry tests on the mixture show an exotherm at 200°C, then a reactivity rating of 1 can be assigned to the mixture.

NFPA 49 contains health, fire, and reactivity hazard data and hazard ratings on 325 chemicals of commercial significance that have an NFPA 704 health hazard rating of 2 or higher, that have a reactivity rating of 1 or higher, that present unusual spill control or fire-fighting problems, or that present unusual storage problems. Not addressed in NFPA 49 are explosives and blasting agents and those chemicals that only pose flammability, mild oxidizing ability, mild irritation, or chronic health hazards.

NFPA 325M contains fire hazard properties for about 1,500 flammable liquids, flammable gases, and volatile flammable solids. Health, flammability, and reactivity hazard ratings are provided for most chemicals, in addition to other typical fire hazard properties that are directed at emergency response (flash point, ignition temperature, flammability limits, liquid/vapor density, boiling point, water solubility, and extinguishing methods).

It should be noted that the reactivity ratings in NFPA 49 and NFPA 325M are not always consistent. A notable example is methyl isocyanate. It has a reactivity rating of 2 in NFPA 49, whereas in NFPA 325M it has a rating of 3 and is identified as water-reactive (see Section 2.5 below).

2.2. NPCA Hazardous Materials Identification System

The Hazardous Materials Identification System (HMIS) published by the National Paint & Coatings Association is a comprehensive hazard communication program involving hazard assessment, labeling, material safety data sheets, and employee training (NPCA, 1985). It is broader than the NFPA 704 system in that the chemical identity, personal protective equipment requirements, and chronic health hazards are included, in addition to the acute health, flammability, reactivity, and special hazards identified by the NFPA 704 system.

The HMIS label uses a five-part rectangle rather than a four-part diamond to summarize the chemical identity, acute hazard ratings, chronic hazard information, and personal protective equipment designations. The HMIS reactivity hazard rating criteria are given in Table 2.2. A comparison of Tables 2.1 and 2.2 shows that the HMIS labeling uses essentially the same reactivity hazard ratings as the NFPA 704 system.

2.3. Classifications of Organic Peroxides

Two systems that are used for classifying organic peroxides by severity of hazard are summarized in this section. The system proposed by The Society of the Plastics Industry (SPI) combines the results of several different tests to classify an organic peroxide into Classes I through IV, with a material in Class I posing the greatest hazard. A system developed by the National Fire Protection Association for organic peroxides (NFPA 43B, 1993) uses more qualitative definitions for its Classes I through V, with Class I again posing the greatest hazard. The documents describing the two classification systems also give the classifications of several specific organic peroxides according to their respective systems.

2.3.1. SPI 19A Classification of Organic Peroxides

The Organic Peroxide Producers Safety Division of the Society of the Plastics Industry gives a suggested classification system for organic peroxides in the SPI Bulletin 19A, *Suggested Relative Hazard Classification of Organic Peroxides* (SPI 19A, 1975). Organic peroxide formulations are classified based on the seven test results listed in Table 2.3. Details of each test method are also presented in the SPI 19A bulletin.

Each of the seven test results is compared to the criteria in Table 2.3 and assigned a rating of 1 to 3. These ratings are then totaled over all seven test results to obtain an overall score between 7 and 21, inclusive. This overall score is used to classify the organic peroxide formulation in either Class I (Danger), Class II (Warning), Class III (Caution), or Class IV (Unregulated) using the ranges shown in Table 2.4. Example formulations in each class are also given in Table 2.4.A similar classification system for commercial organic peroxides was proposed by

TABLE 2.2
HMIS Reactivity Hazard Ratings (NPCA, 1985)

Hazard Index	*Reactivity Rating Summary*	*Reactivity Rating Criteria*
4 Severe Hazard	Materials that are readily capable of detonation or explosive decomposition at normal temperatures and pressures	Materials thst, in themselves, are readily capable of detonation or of explosive decomposition or explosive reaction at normal temperatures and pressures. This degree should include materials which are sensitive to mechanical or localized thermal shock at normal temperatures and pressures.
3 Serious Hazard	Materials that are capable of detonation or explosive reaction but require a strong initiating source or must be heated under confinement before initiation; or materials that react explosively with water	Materials that, in themselves, are capable of detonation or of explosive decomposition or explosive reaction but which require a strong initiating source or which must be heated under confinement before initiation. This degree should include materials which are sensitive to thermal or mechanical shock at elevated temperatures and pressures or which react explosively with water without requiring heat or confinement.
2 Moderate Hazard	Materials that, in themselves, are normally unstable and will readily undergo violent chemical change but will not detonate. These materials may also react violently with water	Materials that, in themselves, are normally unstable and readily undergo violent chemical change but do not detonate. This degree should include materials which can undergo chemical change with rapid release of energy at normal temperatures and pressures or which can undergo violent chemical change at elevated temperatures and pressures. It should also include those materials which may react violently with water or which may form potentially explosive mixtures with water.
1 Slight Hazard	Materials that are normally stable but can become unstable at high temperatures and pressures. These materials may react with water but they will not release energy violently	Materials that, in themselves, are normally stable, but which can become unstable at elevated temperatures and pressures or which may react with water with some release of energy but not violently.
0 Minimal Hazard	Materials that are normally stable, even under fire conditions, and will not react with water	Materials which, in themselves, are normally stable, even under fire exposure conditions and which are not reactive with water.

Noller et al. (1964). This classification system, also summarized by Stull (1977, 23–24), has five hazard classes based on the results of drop weight, lead pipe deformation, and pressure vessel tests.

TABLE 2.3
Test Ratings for SPI Classification of Organic Peroxides (SPI 19A, 1975)

Test	*Parameter*	*Test Result*	*Rating*
Pressure Vessel Test (PVT)	Vent diameter	20–14 mm	3
		<14–1.0 mm	2
		<1.0 mm	1
Rapid Heat Test	Force of decomposition	Rapid (equal to or greater than *t*-butyl peracetate in benzene)	3
		Moderate	2
		Mild	1
Self-Accelerating Decomposition Temperature (SADT)	Force of decomposition	Rapid decomposition (considerable damage to test oven)	3
		Moderate decomposition (considerable damage to container and possible slight damage to test oven)	2
		Mild decomposition (no damage to oven and some or no damage to container)	1
Self-Accelerating Decomposition Temperature (SADT)	Temperature	<10°C	3
		10–49°C	2
		>49°C	1
Impact Sensitivity (solids and pastes only)	Height	<10 cm	3
		10–25 cm	2
		>25 cm	1
Modified Trauzl	Expansion	60-35 ml	3
		<35–15 ml	2
		<15 ml	1
Burning Test	Flame height	>152 cm or very vigorous burning	3
		152–91 cm	2
		<91 cm	1

2.3.2 NFPA 43B Classification of Organic Peroxides

The NFPA Code *Storage of Organic Peroxide Formulations* (NFPA 43B, 1993) applies to commercially available organic peroxide formulations in approved packages. It does not apply to formulations capable of detonation in normal shipping containers under conditions of fire exposure; such formulations are to

TABLE 2.4
SPI 19A Classification of Organic Peroxides (SPI 19A, 1975)

	Code Word	*Numerical Range (Total of 7 Tests)*	*Examples*
Class I	Danger	15–21	Dibenzoyl peroxide, dry *t*-Butyl peracetate, 75% in benzene
Class II	Warning	11–14	*t*-Butyl peracetate, 75% in odorless mineral spirits *t*-Butyl perbenzoate Acetyl peroxide, 25% in dimethyl phthalate
Class III	Caution	8–10	Dilauroyl peroxide Dibenzoyl peroxide, 70% (wet) Di-*t*-butyl peroxide
Class IV	No code word Unregulated	7	*t*-Butyl hydroperoxide 70 (with ~30% water) Dibenzoyl peroxide, 50%, fire retardant paste

be handled and stored as Class A explosives. NFPA 43B classifies organic peroxide formulations using the system shown in Table 2.5. The definitions for the five NFPA 43B classes are more qualitative and subjective than the SPI rating system. However, NFPA 43B does emphasize the need to take into account not only the composition of each organic peroxide formulation (concentration of peroxide, active oxygen, and diluent) but also the strength and size of the shipping container.

2.4. Classification of Materials That Form Peroxides

The common grouping of peroxide formers into the three "lists" shown in Table 2.6 can be considered as a classification system for peroxide formers, based on severity of hazard. List A contains materials that pose a peroxidation hazard even without further concentration. List B contains materials that pose a peroxidation hazard only upon concentration. List C contains both liquid and gaseous peroxide-forming monomers that by themselves may not be particularly hazardous but that on decomposition may initiate uncontrolled polymerization of the bulk monomer (Trommsdorf effect).

This particular list has been widely used since its original publication in a summary of a Du Pont safety study on peroxide formers (Jackson et al., 1970). Both the list and the safety study are discussed in Bretherick (1990, 1747–1748).

TABLE 2.5
NFPA 43B Classification of Organic Peroxides (NFPA 43B, 1993)

	Definition	*Examples*	*Maximum container size*
Class I	Those formulations that are capable of deflagration but not detonation	Dibenzoyl peroxide, 98 wt%	1 lb
		Diisopropyl peroxydicarbonate, 99 wt%	10 lb
Class II	Those formulations that burn very rapidly and that present a severe reactivity hazard	*t*-Butyl hydroperoxide, 70 wt% in di-*t*-butyl peroxide and *t*-butanol	55 gal
		t-Butyl peroxypivalate, 75 wt% in odorless mineral spirits	5 gal
Class III	Those formulations that burn rapidly and that present a moderate reactivity hazard	Dibenzoyl peroxide, 75 wt% in water	25 lb
		Methyl ethyl ketone peroxide, 9 wt% active oxygen in dimethyl phthalate	5 gal
Class IV	Those formulations that burn in the same manner as ordinary combustibles and that present a minimal reactivity hazard	Dibenzoyl peroxide, 70 wt% in water	25 lb
		Dilauroyl peroxide, 98 wt%	110 lb
Class V	Those formulations that burn with less intensity than ordinary combustibles or do not sustain combustion and that present no reactivity hazard	Dibenzoyl peroxide, 35 wt% in dicalcium phosphate dihydrate or calcium sulfate dihydrate	100 lb
		2,4-Pentanedione peroxide, 4 wt% active oxygen in water and solvent	5 gal

The above system is based on the behavior of certain specific **formulations** in DOT-approved shipping containers and under conditions of fire exposure. Other typical organic peroxide formulations in each of the above classes are given in Appendix B of NFPA 43B.

This categorization has been useful for labeling purposes, where the maximum storage time after opening of a container of the material may range from three to twelve months. It should be noted that this classification was primarily developed for laboratory chemicals, which may be more likely to contact atmospheric oxygen than bulk chemicals in storage; hence, the retention times listed in Table 2.6 may not apply to bulk storage. Likewise, such monomers as styrene and vinyl acetate only form peroxides when an inhibitor is consumed and additional oxygen is present, so that bulk commercial materials can be stored for longer periods of time as long as inhibitor is present.

TABLE 2.6
Common Peroxide-Forming Chemicals (Jackson et al., 1970)

LIST A	LIST B	LIST C
Severe peroxide hazard on storage with exposure to air	*Peroxide hazard on concentration; do not distill or evaporate without first testing for the presence of peroxides*	*Hazard of rapid polymerization initiated by internally formed peroxides (liquids)*[b]
Discard within 3 months	*Discard or test for peroxides after 6 months*	*Discard or test for peroxides after 6 months*
• Diisopropyl ether (isopropyl ether)	• Acetaldehyde diethyl acetal (acetal)	• Chloroprene[a,c]
• Divinylacetylene[a]	• Cumene (isopropylbenzene)	• Styrene[a]
• Potassium metal	• Cyclohexene	• Vinyl acetate[a]
• Potassium amide	• Cyclopentene	• Vinylpyridine[a]
• Sodium amide (sodamide)	• Decalin (decahydronaphthalene)	
• Vinylidene chloride (1,1,-dichloroethylene)[a]	• Diacetylene (1,3-butadiyne)	*Hazard of rapid polymerization initiated by internally formed peroxides (gases)*
	• Dicyclopentadiene	*Discard after 12 months*[d]
	• Diethyl ether (ether)	• Butadiene[a,c]
	• Diethylene glycol dimethyl ether (diglyme)	• Tetrafluoroethylene[a,c]
	• *p*-Dioxane	• Vinylacetylene[a,c]
	• Ethylene glycol dimethyl ether (glyme)	• Vinyl chloride[a]
	• Ethylene glycol ether acetates	
	• Ethylene glycol monoethers (cellosolves)	
	• Furan	
	• Methylacetylene	
	• Methylcyclopentane	
	• Methyl isobutyl ketone	
	• Tetrahydrofuran	
	• Tetralin (tetrahydronaphthalene)	
	• Vinyl ethers[a]	

[a]Polymerizable monomers should be stored with a polymerization inhibitor from which the monomer can be separated by distillation just before use. (See discussion in text on bulk inhibited monomers.)

[b]Although common acrylic monomers such as acrylonitrile, acrylic acid, ethyl acrylate, and methyl methacrylate can form peroxides, they have not been reported to develop hazardous levels in normal use and storage.

[c]The hazard from peroxides in these compounds is substantially greater when they are stored in the liquid phase, and if so stored without an inhibitor they should be considered as in LIST A.

[d]Although air will not enter a gas cylinder in which gases are stored under pressure, these gases are sometimes transferred from the original cylinder to another in the laboratory, and it is difficult to be sure that there is no residual air in the receiving cylinder. An inhibitor should be put into any such secondary cylinder before one of these gases is transferred into it; the supplier can suggest inhibitors to be used. The hazard posed by these gases is much greater if there is a liquid phase in the secondary container, and even inhibited gases that have been put into a secondary container under conditions that create a liquid phase should be discarded within 12 months.

2.5. Classification of Water-Reactive Materials

Just as there is a broad range of oxygen reactivity, the intensity with which materials react with water covers a broad range also. No particular classification system is in use for water-reactive materials. Water-reactive materials can be considered within the framework of two broad categories, as described below.

2.5.1. *Materials That React Violently with Water*

Many water-reactive substances react violently and immediately on contact with water. These substances are of particular concern in fire-fighting situations. Note that chemicals in both NFPA reactivity hazard category 2 (chemicals that may react violently with water or form potentially explosive mixtures with water) and NFPA reactivity hazard category 3 (chemicals that react explosively with water without requiring heat or confinement) with respect to water reactivity would fall into this group. The NFPA ₩ symbol is always used in the bottom special-hazards quadrant of the NFPA hazard signal for these water-reactive substances.

2.5.2. *Materials That React Slowly with Water*

Some water-reactive substances react slowly upon contact with water, in a way that tends to build pressure in any containment that may be present. Thus, the greater hazard for this category of water-reactive materials is generally the potential to rupture a container or vessel due to internal overpressurization. A water-reactive material of this type would generally be in NFPA reactivity hazard category 1, unless the material also posed a significant instability hazard.

Use of the NFPA ₩ symbol for water-reactive materials has been inconsistent. This is due to changes in NFPA 704, which defines the hazard rating methodology, causing NFPA 49 and NFPA 325M, which are being updated at different times, to currently have some discrepancies. Some facilities mark storage tanks with the NFPA W symbol not only for violently water-reactive materials, but also materials such as acetic anhydride that react more slowly with water and have the potential for overpressuring a container or vessel. It may also be used if the reaction with water generates toxic gases.

Bibliography

NFPA 43B, *Storage of Organic Peroxide Formulations*, Quincy, Massachusetts: National Fire Protection Association, 1993.

NFPA 704, *Recommended System for the Identification of the Fire Hazards of Materials*, Quincy, Massachusetts: National Fire Protection Association, 1990.

NPCA, *Hazardous Materials Identification System Revised Implementation Manual,* Washington, D.C.: National Paint & Coatings Association, January 1985.

SPI 19A, *Suggested Relative Hazard Classification of Organic Peroxides,* Washington, D.C.: The Society of the Plastics Industry, Inc., January 1975.

References

Bretherick, L. (1990). *Bretherick's Handbook of Reactive Chemical Hazards,* 4th ed. Butterworths, London.

Hofelich, T. C., Frurip, D. J., and Powers, J. B. (1994). "The Determination of Compatibility via Thermal Analysis and Mathematical Modeling." *Proc. Safety Prog. 13*(4, October), 227-233.

Jackson, H. L., McCormack, W. B., Rondesvedt, C. S., Smeltz, K. C., and Viele, I. E. (1970). "Control of Peroxidizable Compounds." *J. Chem. Ed. 47*(3), March.

NFPA 325M (1994). *Fire Hazard Properties of Flammable Liquids, Gases, and Volatile Solids.* National Fire Protection Association, Quincy, Massachusetts.

NFPA 43B (1993). *Storage of Organic Peroxide Formulations.* National Fire Protection Association, Quincy, Massachusetts.

NFPA 49 (1994). *Hazardous Chemical Data.* National Fire Protection Association, Quincy, Massachusetts.

NFPA 704 (1990). *Standard System for the Identification of the Fire Hazards of Materials.* National Fire Protection Association, Quincy, Massachusetts.

Noller, D. C., Mazurowski, S. J., Linden, G. F., DeLeeuw, F. J. G., and Mageli, O. L. (1964). *Ind. Eng. Chem. 56*(12), 18-27.

NPCA (1985). *Hazardous Materials Identification System Revised Implementation Manual.* National Paint & Coatings Association, Washington, D.C., January.

SPI 19A (1975). *Suggested Relative Hazard Classification of Organic Peroxides.* The Society of the Plastics Industry, Inc., Washington, D.C., January.

Stull, D. R. (1977). *Fundamentals of Fire and Explosion.* AIChE Monograph Series No. 10, Vol. 73. American Institute of Chemical Engineers, New York.

3

Materials Assessment

This chapter presents guidelines for the evaluation of chemical reactivity. The main question to be answered in this chapter is

Are we or will we be storing/handling reactive material(s) that can release sufficient energy rapidly enough to pose a threat to health and property?

Answering this question may require the application of various recognition aids, thermodynamic calculations and tests, as well as the judgment and experience of persons knowledgeable in the field of reaction chemistry. Figure 3.1 gives a "how-to" flow chart outlining this initial hazard assessment of reactive materials. It can be used as a stepwise program to evaluate the reactivity hazards of substances for which the molecular structure is known and/or for which sample material is available. This figure does not represent the only method for materials assessment; many companies handling reactive materials have somewhat different approaches. Nevertheless, most of these same elements are involved in nearly all programs having a thorough, systematic approach to identifying reactive chemical hazards.

It is generally not necessary to perform *screening tests* on all potentially reactive materials to decide whether a significant reactivity hazard exists and whether consequence analysis is warranted. *Prior experience* with the same or similar materials (Section 3.1), *theoretical evaluations* (Section 3.2), and *expert determination* (Section 3.3) can be employed prior to performing screening tests such as shock sensitivity tests and adiabatic calorimetry measurements (Section 3.4). The location in this chapter of each of these steps in the materials assessment strategy of Figure 3.1 is indicated by the section numbers next to each shaded block in the figure. Flow sheets showing more details, particularly related to reaction hazard evaluation testing strategies, are presented in CCPS/AIChE (1995).

Once a substance is determined to pose a significant reactivity hazard, severity testing and/or consequence analysis will need to be performed, in order

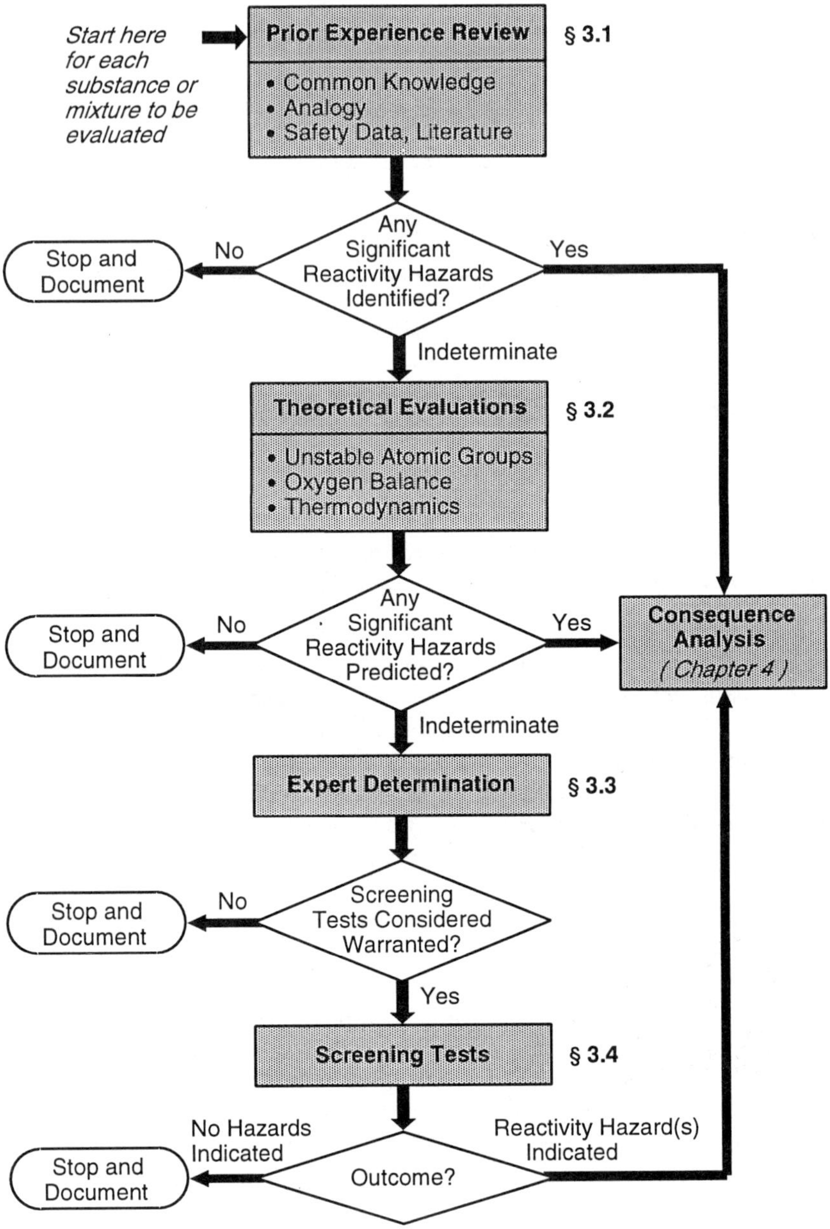

FIGURE 3.1. **Materials Assessment Strategy.**

to specify the design and operating criteria for safe handling of the material. Guidelines for conducting a consequence analysis are given in Chapter 4.

3.1. Prior Experience Review

Prior experience may be able to be employed to determine whether significant reactivity hazards exist for a given facility. This is suggested as the first step in the materials assessment strategy of Figure 3.1, since it will often be the easiest and least costly approach. The objective of using prior experience is to judge, based on prior knowledge and practice, whether materials to be handled can release sufficient energy rapidly enough to pose a threat to health and property. Three outcomes are possible from this determination:

1. It is judged with certainty, on the basis of prior experience, that **no** significant reactivity hazard exists. If this is the case, then no further assessment is required besides documenting this decision and its rationale (and any other legal or company requirements that may be applicable). Of course, the material(s) may still pose toxicity and/or flammability hazards that will warrant taking other appropriate safety precautions.
2. It is judged with certainty, on the basis of prior experience, that one or more significant reactivity hazards **do** exist. If this is the case, then further hazards identification and screening tests are not needed, and the program can proceed directly to assessing the magnitude of the potential hazard(s), as described in Chapter 4.
3. It can **not** be judged with certainty, based on prior experience alone, whether or not a significant reactivity hazard exists. Further determination by theoretical evaluations is therefore warranted (Section 3.2).

If more than one material is being stored and handled in a facility, then this judgment must be made and documented for **each** substance or mixture. As a result, it is likely that different outcomes will be obtained for different materials. For example, it may be judged based on prior experience that storing aluminum alkyls in ton quantities does pose a significant reactivity hazard, but prior experience is not sufficient to determine whether or not handling trifluorochloroethylene in ton quantities is a significant hazard. The subsequent steps taken to evaluate and deal with each material will then be different, and will need to be carefully tracked to ensure all materials are fully assessed. A form such as is presented in Figure 3.2 can be used for this purpose.

Prior experience can be found in different forms and different places. Judgment concerning some materials can be made solely on the basis of *common knowledge*. For others, *analogy* can be drawn with other materials known to pose (or known not to pose) significant hazards. The *safety literature* can also be

MATERIALS ASSESSMENT

Complete for Each Substance or Mixture to be Stored/Handled

Facility______________________________

Material______________________________

Quantity______________________________

Equipment______________________________

Conditions______________________________

Basic Controls_______________________________

____

Can this quantity of material, stored and/or handled in this equipment under these conditions, release sufficient energy rapidly enough to threaten health and property?

Prior experience indicates:

☐ YES [stop; document]
☐ NO [stop; document]
☐ INDETERMINATE [continue]

Theoretical evaluations indicate:

☐ YES [stop; document]
☐ NO [stop; document]
☐ INDETERMINATE [continue]

Expert determination indicates:

☐ SCREENING TESTS NOT WARRANTED [stop; document]
☐ SCREENING TESTS WARRANTED [continue]

Screening tests indicate:

☐ YES [document]
☐ NO [document]

Basis for final determination:______________________________

____

____

____

____

Completed by__________________

Approved by__________________ Date___________

FIGURE 3.2. **Suggested Form for Materials Assessment Tracking.**

effectively used to find summaries of existing knowledge and past incidents involving a given material. Guidelines for each of these approaches are given below.

3.1.1. Common Knowledge

Common knowledge and common sense can quite often be used to make a definitive judgment as to whether a given material poses a significant reactivity hazard. Based on prior industry experience with methyl isocyanate, for example, common knowledge would indicate that this material can pose a significant toxicity hazard, although its water reactivity may be less well-known. Based on personal experience with the materials, the noxious properties of ammonia and chlorine are common knowledge to most people. Routine handling of various materials on-site over time will generally lead to an appreciation of the hazards of specific reactive materials, particularly where incidents and near misses are taken seriously and investigated. Thus, common knowledge usually involves materials that are frequently handled and materials that have been involved in past serious incidents.

3.1.2. Analogy

Even if a person does not have experience with a specific material, the hazards of the material may be able to be assessed by analogy with another substance or group of substances for which the hazards are known. For example, metal complexes involving different metals in the same periodic table group with similar electronegativities may be expected to have similar properties; likewise, organic compounds with the same basic structure, functional groups and substituents. If the hazards of pimelic acid (1,7-heptanedioic acid) are being assessed, an analogy to adipic acid (1,6-hexanedioic acid) can be drawn, although the latter is more reactive. Specialty chemical manufacturers often make slight variants within well-defined classes of organic compounds, and it is often known what the most hazardous compound within a class is expected to be. Care must be exercised, however, when using analogy to infer the significance of reactivity hazards. Many examples can be cited of two apparently similar compounds that have vastly different hazard properties, such as tetrachloroethylene (noncombustible and very stable under normal conditions) and tetrafluoroethylene (highly flammable and reactive).

3.1.3. Safety Data and Literature

If available, material safety data sheets (MSDSs) will likely be the first source of hazard information for a given material. Additional information may be found in hazardous materials databases such as CHRIS (DOT, 1978); hazardous property references such as NFPA 49 (1994), NFPA 325M (1994), and Sax and

Lewis (1989); chemical interaction surveys such as NFPA 491M (1991) and Bretherick (1990); and case studies such as in the periodic *Loss Prevention Bulletin* published by the Institution of Chemical Engineers (Rugby, England). Articles, studies, symposia, and research reports on the hazards of specific compounds may also be found in the literature, and on-line databases are being used with greater frequency. An annotated listing of many sources of safety data is presented in **Appendix A.**

Full advantage should be taken of this published safety information to establish whether a reactivity *hazard* exists; however, as with the theoretical calculations detailed in the next section, the facility-specific parameters such as the quantity of material handled, the storage temperatures and pressures, the heat transfer characteristics of the system, and the location of operating personnel will often be the determining factors as to whether an incident involving the material of concern can release sufficient energy rapidly enough to threaten health and property. Consequently, great care must be taken to ensure that literature data are interpreted and applied correctly to a given situation, if it is to be used for any other purpose than establishing that a reactivity hazard is likely to exist.

3.2. Theoretical Evaluations

If prior experience is insufficient to judge definitively whether a material poses a significant reactivity hazard, the next techniques available to aid this determination are *theoretical evaluations,* or methods to predict reactivity hazards based on the molecular structure and/or thermodynamic properties of the material. This implies, of course, that the molecular structure of the material (or components of a mixture) is known. If the molecular structure is not known (such as may be the case for a hazardous waste stream of uncertain origin, for example), then theoretical evaluations are not possible, and this step in the materials assessment strategy of Figure 3.1 must be considered as indeterminate.

The three possible outcomes of the theoretical evaluations are the same as for the prior experience review:

1. It is judged with certainty that **no** significant reactivity hazard exists.
2. It is judged with certainty that one or more significant reactivity hazards **do** exist. If this is the case, then further hazards identification and screening tests are not needed, and the program should proceed directly to assessing the magnitude of the potential hazard(s), as described in Chapter 4.
3. It can **not** be judged with certainty, based on theoretical evaluations, whether or not a significant reactivity hazard exists. Further assessment by expert determination is therefore warranted (Section 3.3).

As before, the rationale for each determination should be documented, and, if more than one material is being stored and handled in a facility, then this judgment must be made for each substance or mixture. A form similar to Figure 3.2 can be used to document the rationale.

Theoretical evaluations generally address the **thermodynamic** aspects of a material's reactivity. As such, these evaluations can generally predict the maximum amount of energy that can be released if, for example, the material decomposes at a given temperature. However, the kinetic dimension of the rate of energy release is not predicted by thermodynamic calculations. Kinetic parameters such as the energy of activation must be determined experimentally.

Nevertheless, theoretical evaluations can look at several aspects of a material's potential for reactivity, such as *unstable atomic groups, oxygen balance, heat of formation, heat of decomposition, heat of reaction,* and *equilibrium considerations.* Many of the theoretical evaluations have been incorporated into computer codes that predict reactivity. One important code, CHETAH, will be discussed in this Section, and an example determination will be given to illustrate the theoretical evaluations.

3.2.1. Unstable Atomic Groups

One means of predicting the reactivity of a material is through analysis of its chemical structure. Experience has shown that instability in single compounds, or high reactivity in combinations of different materials, is often associated with certain bond structures or atomic groupings. This is particularly true of weak bonds, strained molecular configurations, and high proportions or local concentrations of oxygen or nitrogen. Some typical structures in reactive molecules that influence their stability include double or triple bonds, nitro groups, peroxide groups, nitrogen-to-halogen or oxygen-to-halogen bonds, ring strain, and steric hindrance. Examples of such substances are acetylenes, azo and diazo compounds, nitro compounds, organic and inorganic peroxides, and halites; a more complete list is presented in Table 3.1. Each of the energetic groups listed in Table 3.1 have relatively weak bonds that tend to release a significant amount of energy upon cleavage.

The compilation of Table 3.1 is not exhaustive; however, it can assist in screening materials that are likely to pose significant reactivity hazards. It should be noted that some of the classes of materials in Table 3.1 are difficult to represent in line diagrams. For example, the amminemetal oxosalts are compounds containing ammonia or an organic base coordinated to a metal, with coordinated or ionic chlorate, nitrate, nitrite, nitro, perchlorate, permanganate or other oxidizing groups also present (Bretherick, 1986, 70–71).

The presence of one of the mentioned groups in a molecule does not necessarily imply that the substance is hazardous. For instance, a molecule consisting of a nitro group attached to a long aliphatic chain shows no significant instability. Trinitromethane, on the other hand, is highly energetic and thermally

TABLE 3.1
Bonds and Functional Groups Conferring Instability

Structural Feature*	Class
C≡C	Acetylenes
C≡C—M	Metal acetylides
C≡C—X	Haloacetylene derivatives
—N_3	Azides (acyl, amminecobalt(III), halogen, metal, nonmetal, and organic azides; azide complexes of cobalt(III); 2-azidocarbonyl compounds)
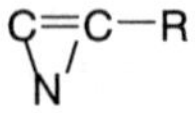	Aziridines
C—N=N—C	Azo compounds
Ar—N=N—O—C	Arenediazoates
Ar—N=N—O—N=N—Ar	Bis(arenediazo) oxides
Ar—N=N—S—Ar	Arenediazo aryl sulfides
Ar—N=N—S—N=N—Ar	Bis(arenediazo) sulfides
CN_2	Diazo compounds, diazoazoles
CN_2^+	Diazonium salts (carboxylates, perchlorates, sulfates, sulfides and derivatives, tetrahaloborates, and triiodides)
$N_2^+ArO^-$	Arenediazonium oxides
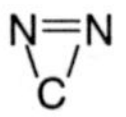	Diazirines
N=N—N—H	Triazenes
N=N—N—CN	
N=N—N—OH	
N=N—N—NO	
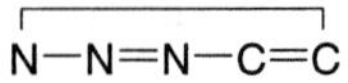	Triazoles
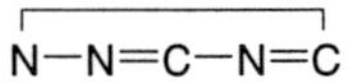	
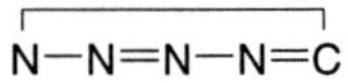	Tetrazoles
$N^+ EO_n^-$	Oxosalts of nitrogenous bases
$H_5N_2^+Z^-$	Hydrazinium salts
$H_4ON^+ Z^-$	Hydroxylaminium salts
$H_3N \rightarrow M^+ EO_n^-$	Amminemetal oxosalts
N—M	*N*-metal derivatives (especially heavy metals)
N—X	*N*-halogen compounds, including *N*-haloimides
N—F_2	Difluoroamino compounds

TABLE 3.1 (CONTINUED)

Structural Feature*	Class
C—NO_2	Nitroalkanes, *C*-nitro compounds
C=C—NO_2	Nitroalkenes
Ar—NO_2	Nitroaryl compounds
N—NO_2	*N*-Nitro compounds
$C(NO_2)_n$	Polynitroalkyl compounds
O_2N—C—C—NO_2	
$Ar(NO_2)_n$	Polynitroaryl compounds
C—N=O	Nitroso compounds
Ar—N=O	Nitroso arenes
N—N=O	*N*-Nitroso compounds
C=N—O	Oximes
MC≡N→O	Metal fulminates
C—O—N—O	Alkyl or acyl nitrites
C—O—NO_2	Alkyl or acyl nitrates
—O—O—	Peroxides (inorganic, organic, organomineral)
$(O_2)^{--}$	Inorganic peroxides
—O—O—H	Hydroperoxides (including alkyl hydroperoxides, peroxoacids, peroxyacids)
O_3	Ozonides
C—C \ / O	1,2-Epoxides
XO_n	Halogen oxides
O—X	Hypohalites, acyl hypohalites
O—X=O	Halites, halite salts
C—O—Cl=O	Alkyl chlorites
ClO_2^-	Chlorite salts
O—X—O_3	Perhalates
O—Cl—O_3	Perchlorates, alkyl perchlorates, perchloric acid
C—Cl—O_3	Perchloryl compounds
$(NH—Cl—O_3)^-$	Perchlorylamide salts
Ar—M—X	Halo-arylmetals
X Ar M	
Xe O_n	Xenon-oxygen compounds

*Abbreviations: Ar=aromatic; E=nonmetal; M=metal; R=organic; X=halogen; Z=anion

References: Bretherick, 1986, 70-71; Bretherick, 1990, 1477-1824 sensitive. Increasing the molecular weight of an unstable compound tends to make it more stable by, in effect, "diluting" the active groups. A similar effect is obtained by mixing (diluting) unstable molecules with more stable molecules.

Conversely, the absence of unstable groups does not guarantee a low reactivity hazard for a compound. For example, many ethers are easily oxidized to more unstable peroxides, and polymerizable chemicals may have relatively high resistance to decomposition. The unstable groups of Table 3.1 are primarily associated with self-reactive decomposition hazards.

Predictions of instability from chemical structure have been incorporated into computer programs such as CHETAH, as detailed in Section 3.2.6. Whether the prediction is made by computer or by similarity to, or extrapolation from, chemical structures already known to be unstable, this prediction is rarely used alone as a criterion to decide whether or not a significant reactivity hazard exists. It is generally used in combination with other theoretical evaluations and/or screening tests.

Example: Organic Peroxides

One functional group listed in Table 3.1 as conferring instability to organic molecules (as well as inorganics and organominerals) is the peroxide group, characterized by an oxygen-oxygen bond. Organic peroxides are widely used as initiators and cross-linking agents for polymerization reactions, since they typically form free radicals upon decomposition. This free radical formation can induce further decomposition by autocatalysis, and thus a runaway reaction hazard can exist with insufficient heat removal. The runaway decomposition reaction may eventually be followed by autoignition or a thermal explosion. The pressure has a marked accelerating effect on the reaction rate also, as do trace contaminants such as redox agents, strong ionizing acids, strong oxidizing and reducing agents, and heavy metals (Grewer, 1988) that can serve as additional catalysts for the decomposition reaction. The exothermic decomposition of organic peroxides can propagate in the absence of oxygen, and is often accompanied by spraying of liquid and formation of off-gas (McCloskey, 1989).

A wide range of stabilities exists within the organic peroxides, from quite stable to highly shock-sensitive. Nevertheless, as the above discussion indicates, any unfamiliar substance containing one or more peroxide groups should be suspected of potentially unstable behavior until demonstrated otherwise.

3.2.2. Oxygen Balance

Oxidation reactions can be extremely rapid and energetic. The reacting oxidant may be supplied by atmospheric oxygen (such as in a combustion reaction), from a liquid or solid oxidizer mixed with the substance (such as chlorine or hydrogen peroxide), or from within the molecule itself. When the oxidant is supplied within the molecule itself, then the resulting self-reaction mechanisms are generally decomposition reactions, exemplified by many high explosives containing nitro groups that provide the internal oxygen content. Hence, another means of predicting the reactivity of a material is by calculating and interpreting the internal oxidant content or "oxygen balance" of the material (Frurip, 1992, 226).

Oxygen Balance Definition

The *oxygen balance* is the difference between the available oxidizing atoms and the available reducing atoms within the molecular structure of a substance. Thus, it indicates the extent to which the substance, if it self-reacts, may fully oxidize its components. Since oxygen is the most common oxidizer, the oxygen balance generally indicates the difference between the oxygen content of a compound or mixture and that required to fully oxidize the carbon, hydrogen, metals, and other oxidizable elements present to their lowest states of valency (i.e., most stable oxidized forms).

How to Calculate Oxygen Balance

The following are general guidelines for calculating oxygen balance:

- Carbon is oxidized to carbon dioxide
- Hydrogen is oxidized to halides (preferentially) or water
- Noninert metals are oxidized to halides (preferentially) or metal oxides
- Halogens are reduced to halides if metal or hydrogen atoms are present
- Sulfur is oxidized to sulfur dioxide
- Nitrogen is assumed not to oxidize, so it does not enter into the oxygen balance calculation.

The oxygen balance is calculated such that a deficiency of oxygen gives a negative oxygen balance and an excess of oxygen gives a positive oxygen balance. The result is usually expressed as a positive or negative number in weight percent or in grams of oxygen per 100 grams of material. The formula to calculate oxygen balance in weight percent for a pure substance having the composition $C_iH_jN_kO_n$ is:

$$\text{Oxygen Balance} = \frac{-1600(2i + j/2 - n)}{\text{molecular weight}} \text{ wt\%} \tag{3-1}$$

Equation 3-2 below can be used to calculate the overall oxygen balance OB_{mixture} for mixtures of reducing and oxidizing substances:

$$OB_{\text{mixture}} = \frac{\sum_n m_n \cdot (OB)_n}{\sum_n m_n} \tag{3-2}$$

in which m_n refers to the mass and $(OB)_n$ to the oxygen balance of each component n. Fully oxidized substances such as water should not be included in the calculation.

Interpreting the Oxygen Balance

If the oxygen balance of a substance or mixture is near zero, then all reducing atoms can be completely oxidized without a source of external oxygen or other oxidizer. Hence, a near-zero oxygen balance indicates that the *potential* for significant self-reactivity exists, and in general, the closer the oxygen balance is to zero, the more energetic the reaction potential. However, since all energetic reactions are not self-decomposition reduction/oxidation reactions, then a material may still be reactive by another mechanism even if its oxygen balance is far from zero. Hence, oxygen balance should not be considered alone when predicting the reactivity of a material.

Empirical evidence confirms that compounds such as high explosives are most energetic as the oxygen balance approaches zero (Stull, 1970). For example, nitroglycerin ($C_3H_5N_3O_9$) can be shown by Equation (3-1) to have an oxygen balance of +4%. By contrast, the oxygen balance of glycerin ($C_3H_8O_3$) is –122% and that of propane (C_3H_8) is –363%.

In reality, however, most industrial high explosives have a significantly negative oxygen balance. Some examples are given in Table 3.2. It should be noted that if an oxidant is present when decomposition of a compound with negative oxygen balance (i.e., an oxygen-deficient compound) is initiated, then the oxygen balance will be effectively made more positive; i.e., closer to zero. For example, if the decomposition of trinitrotoluene (TNT) with an oxygen

TABLE 3.2
Oxygen Balance Examples (Bretherick, 1986, 69)

Compound	*Decomposition Reaction*	*Oxygen Balance*
Compounds of negative oxygen balance		
Trinitrotoluene	$C_7H_5N_3O_6 + 10.5\,O \rightarrow 7\,CO_2 + 2.5\,H_2O + 1.5\,N_2$	—74%
Peracetic acid	$C_2H_4O_3 + 3\,O \rightarrow 2\,CO_2 + 2\,H_2O$	—63%
Compounds of zero oxygen balance (maximum potential energy release)		
Performic acid	$CH_2O_3 \rightarrow CO_2 + H_2O$	0%
Ammonium dichromate	$Cr_2H_8N_2O_7 \rightarrow Cr_2O_3 + 4\,H_20 + N_2$	0%
Compounds of positive oxygen balance		
Glyceryl nitrate	$C_3H_5N_3O_9 \rightarrow 3\,CO_2 + 2.5\,H_2O + 1.5\,N_2 + 0.5\,O$	+4%
Ammonium nitrate	$H_4N_2O_3 \rightarrow 2\,H_2O + N_2 + O$	+20%
Dimanganese heptoxide	$Mn_2O_7 \rightarrow Mn_2O_3 + 4\,O$	+29%

balance of –74% is initiated in the presence of air, then the oxygen in the surrounding air will make up some of the oxygen lacking within the TNT molecule itself, making the decomposition reaction more energetic than if the air were not present. Likewise, if a fuel or reducing agent is present when a compound with positive oxygen balance is initiated, the oxygen balance will be effectively reduced; i.e., made closer to zero.

Three additional, important **cautions** should be noted when interpreting oxygen balance values:

- Oxygen balance is not relevant if the compound is already in its most stable oxidized form. This is obvious once it is noted that both water and carbon dioxide have oxygen balance values of zero.
- The relevance of the rating also depends on how oxygen is bonded within a molecule. For example, acetic acid ($C_2H_4O_2$) has an oxygen balance of –107%, which would be classified as having "high" potential using the CHETAH criteria described in the next paragraph. However, acetic acid is certainly not an explosive substance, indicating that the structure of the molecule precludes the ready oxidation of the carbon and hydrogen in the molecule by the oxygen of the organic acid group.
- The oxygen balance will not identify unstable compounds such as acetylides, explosive nitrides, and azides that do not have oxygen as part of the molecular structure (Shanley and Melhem, 1995).

Thus, it can be seen that the evaluation of self-reactivity potential based on oxygen balance alone is not an absolute rating; however, it can point to where a hazard may exist.

The CHETAH Criteria

The CHETAH program (Frurip, 1992) uses the following explosion hazard criteria with respect to oxygen balance: compounds with less than –240% or greater than +160% oxygen balance have low explosion potential, compounds with an oxygen balance between –240% and –120% or between +80% and +160% have medium explosion potential, and compounds with an oxygen balance between –120% and +80% have high explosion potential. These criteria are illustrated in Figure 3.3.

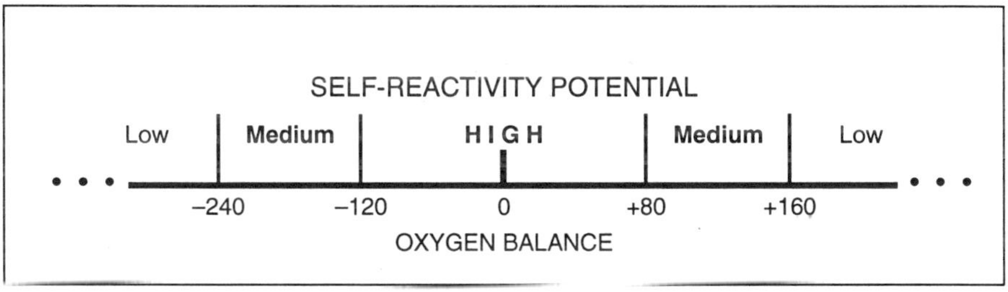

FIGURE 3.3. **CHETAH Oxygen Balance Criteria**

It should be noted that the oxygen balance criteria are only one of four sets of criteria for indicating explosion hazard used by the CHETAH program. The overall methodology used by CHETAH is discussed in more detail in Section 3.2.6.

Extension to Other Self-Reactions

The concept of oxygen balance for a given molecular structure is an extension of the principles of stoichiometry to reactive mixtures. The ease of ignition and the rate of reaction of a mixture of air and a flammable gas, for example, will be greatest near the stoichiometric concentration, and will continue to decrease the farther the mixture is on either the rich or the lean side of stoichiometric. With this in mind, the concept of oxygen balance may be expanded to more general oxidation–reduction reactions and other self-reactive features within molecules, particularly when, for example, the reducing and oxidizing parts of the same molecule are in close proximity. Bretherick (1986, 71) indicates that salts of reducing bases and oxidizing acids would fall into this category, such as hydroxylaminium nitrate, hydrazinium chlorite or chlorate, or double salts such as potassium cyanide–potassium nitrite.

3.2.3 Thermodynamics: Heat of Formation

Thermodynamics can be used to predict the reactivity of a material by providing a means to calculate the energy content of a substance or a reactive system and, in some cases, to predict equilibrium products of reactions. Thermodynamic measures of energy content include the *heat of formation*, *heat of polymerization*, *heat of decomposition*, *heat of reaction*, and *heat of combustion*. These five measures are compared in Table 3.3. The first of these measures, the heat of formation, is discussed in this section. The remaining measures, which are closely related, are discussed in Section 3.2.4.

Heat of Formation Definition

The *heat of formation* or *enthalpy of formation* (ΔH_f) is the basic measure of the chemical energy content of a substance. The heat of formation is the energy required to form a substance from its constituent elements at a standard condition of 25°C and 1 bar pressure. It is usually reported in units of kilocalories per gram mole (kcal/g-mol) or kilojoules per gram mole (kJ/g-mol).

The heat of formation is based on the assigning of a value of zero enthalpy to each of the elements in its most stable form at 25°C and 1 bar pressure. This most stable form, or standard state, is either gaseous (hydrogen, nitrogen, oxygen, fluorine, etc.), liquid (mercury, bromine), solid, or crystalline. The standard state for carbon is graphite.

TABLE 3.3
Comparison of Thermodynamic Heat Differences

Symbol	*Heat of*	*Starting Energy Level*	*Ending Energy Level*	*Energetic Sign*[a]
ΔH_f	Formation	Constituent elements in most stable forms at 25°C and 1 bar pressure (77°F and 14.5 psia); enthalpies=0 by definition	Material at 25°C and 1 bar pressure (77°F and 14.5 psia)	+
ΔH_{polym}	Polymerization	Polymerizable monomer	Polymer + by-products	–
ΔH_d	Decomposition	Shock-sensitive or thermally unstable compound	Decomposition products	–
ΔH_{rxn}	Reaction	Reactants	Reaction products and by-products	–
ΔH_c	Combustion	Flammable/combustible material [+ atmospheric oxygen implied]	Combustion products	–

[a]Indication of whether positive or negative heat values have the greatest energy potential; e.g., materials with a positive heat of formation are the most energetic, whereas materials having a negative heat of combustion have a greater energy release potential upon combustion than materials having a positive heat of combustion

How to Determine the Heat of Formation

Several approaches exist to determine the heat of formation of a substance. These approaches are summarized in Table 3.4 and discussed below as follows: experimental measurement, literature values, bond contribution and group contribution methods, and computer codes.

Experimental Measurement. It is possible to calorimetrically determine the heat of formation of a material by synthesizing it from its elements or from materials with known heats of formation. However, other methods are generally faster and less expensive, and adequately accurate for identifying reactive chemical hazards. The heat of formation of a material can also be determined from its heat of combustion, as long as the combustion products and their proportions are known or can be measured or predicted with confidence. The heat of combustion is merely the sum of the heats of formation of the products minus the heat of formation of the starting material (and oxygen, which has a heat of formation of zero); consequently, measurement of the heat of combustion of a combustible material, analysis of its combustion products, and measurement

TABLE 3.4
Methods for Determining Heats of Formation (CCPS/AIChE, 1993)

Experimental Measurement
- Direct measurement of heat of formation
- Measurement of heat of combustion and analysis of combustion products

Literature Values
- Thermodynamic data tables having heats of formation
- Thermodynamic data tables having heats of combustion

Bond Contribution Methods
- Average bond energy summation method of Craven
- Compilation of chemical bond strengths by Kerr, Parsonage and Trotman-Dickenson

Group Contribution Methods
- Benson method
- Andersen, Beyer and Watson method

Computer Codes
- CHETAH
- DIPPR (as source of thermal data)

of the final temperature will allow back-calculation of the heat of formation of the material. (The calculation must take any latent and sensible heats into account.)

Literature Values. The most common methods of determining a heat of formation are by finding a tabulated value or by using an estimating technique when a tabulated value cannot be found. Heats of formation for common, relatively simple organic and inorganic chemicals can be found in many standard handbooks such as Pedley et al. (1986), Perry and Green (1984, 3-147), and Weast (1985). Other literature sources are also available, such as Benson (1982), Domalski (1972), and Stull et al. (1969). Heats of formation for many less-common substances can be found in the data sources listed in Appendix A. As described for experimental determination, many heat of combustion values can also be found in standard handbooks such as Perry and Green (1984, 3-155) and Weast (1985) and can be used to back-calculate the heat of formation.

Bond Contribution and Group Contribution Methods. When ΔH_f cannot be determined directly, an attempt can be made to estimate it directly by means of a similar "model" reaction that is known. Estimation of a heat of formation is obviously less accurate than using an experimentally determined value; however, an estimated value is likely to be sufficiently accurate for determining whether or not a reactivity hazard exists. The theoretical methods can be characterized as those that use bond contribution and the ones that use group contribution.

Bond contribution methods have been developed by Craven (1987) and by Kerr, Parsonage and Trotman-Dickenson (Weast, 1985). Group contribution methods include the Benson method (1969) and the Andersen, Beyer, and Watson method (Houghan et al., 1959).

Computer Codes. ASTM's CHETAH computer program, discussed in Section 3.2.6, uses the Benson method of group contributions (Benson, 1969) to estimate ΔH_f. Computerized compilations of chemical data such as the one developed by AIChE's Design Institute for Physical Property Data (DIPPR, 1993) may also have heat of formation or heat of combustion data.

Comparison of Methods

CCPS/AIChE (1995) gives the result of a comparison between five different estimating methods and experimental results for ten chemicals. From this comparison, it was concluded that the method of Kerr, Parsonage and Trotman-Dickenson (Weast, 1985) shows large deviations in the calculated heats of formation and is only reliable in the case of simple linear aliphatic molecules. The method of Craven (1987) shows better results and has a wider scope of use. The largest errors occur when steric hindrance, such as in trinitrotoluene, or ring stress, such as in cyclopropane, is involved. The methods of Benson (1969) and of Andersen, Beyer, and Watson (1944; also in Houghan et al., 1959) that are based on group contribution show the best results, both in the scope of use and the accuracy in predicting the heat of formation. As would be expected, results from the CHETAH code were very similar to the results from the Benson method. Benson compensates for ring stress but does not take into account steric hindrance. Anderson et al. account for steric hindrance but not ring stress. Remarkably, none of the methods can handle heterocyclic pyrimidine because of its aromatic (double) bond between nitrogen and carbon. CCPS/AIChE (1995) provides a short description of each method and a comparison of calculated and experimental values for several substances.

Interpreting the Heat of Formation

The heat of formation is a measure of the net energy needed to form all of the chemical bonds in a given molecular structure. A positive heat of formation indicates that the substance required a net energy input for it to be formed from its elements; this implies that there is a degree of chemical energy "stored" in the molecular structure. Because of this net energy input required for producing such substances, they are often termed *endothermic compounds*. For example, acetylene ($H_2C{\equiv}CH_2$) has a highly positive heat of formation of +54.2 kcal/g-mol (+227 kJ/g-mol), indicating that absorption of energy is required to produce acetylene from its constituent elements in their standard states; i.e., carbon (graphite) and hydrogen (gas). Acetylene is therefore an endothermic compound.

When an endothermic compound decomposes, energy will be generally released during the decomposition reaction, and the more positive the heat of formation, the greater the potential energy release. Hence, another means of predicting the reactivity of a substance is by looking for a positive heat of formation. (As mentioned earlier, this terminology can lead to some confusion, since the decomposition of an *endothermic* compound is a heat-releasing, or *exothermic*, reaction.)

It should be remembered that the heat of formation is not a direct measure of the quantity of chemical energy that can potentially be released upon decomposition, since compounds do not generally decompose to their elements but to decomposition products such as water, carbon dioxide, hydrogen chloride, and sulfur dioxide. These decomposition products generally have negative heats of formation, and thus the heat of decomposition of a substance generally has a greater absolute value than the heat of formation of the substance.

Likewise, if a material has a negative heat of formation, this does not mean it cannot decompose or otherwise self-react with the liberation of energy, as long as the decomposition products or other self-reaction products have a more negative heat of formation than the starting material. Stull (1977, 9) gives the example of barium azide (BaN_6), which has a heat of formation of –5.3 kcal/g-mol (–22 kJ/g-mol) and is thus not an endothermic compound. Nevertheless, the decomposition of barium azide to barium nitride (Ba_3N_2) and nitrogen liberates 23.6 kcal/g-mol (99 kJ/g-mol) of barium azide, which is sufficient to raise the temperature of the products several hundred degrees Celsius. This is possible because the product barium nitride has a much more negative heat of formation (–86.9 kcal/g-mol or –364 kJ/g-mol) than the starting material barium azide. In summary, then:

- *A compound having a positive heat of formation (endothermic compound) has the potential to decompose with the liberation of a significant amount of energy. The more positive the heat of formation, the greater the potential energy release.*
- *A compound with a near-zero or negative heat of formation may or may not decompose with significant energy release, depending on the energy level (heats of formation) of the decomposition products.*

Identifying Endothermic Compounds

Some typical structural similarities in endothermic substances are a relatively high degree of unsaturation (double and triple bonds), a high proportion or high local concentration of nitrogen in the molecular structure, nitrogen-to-halogen bonds, and peroxide groups. Ring deformation and steric hindrance also influence the stability of a molecule. The bonds and functional groups that were presented in Table 3.1 are representative of structures likely to be found in endothermic compounds.

A few examples of endothermic substances having these structural features are shown in Table 3.5. Note that the heat of formation in kJ/g gives a significantly different relative indication of potential energy release than the heat of formation in kJ/g-mol.

3.2.4. Thermodynamics: Heats of Reaction and Self-Reaction

When a material polymerizes, decomposes, rearranges, or reacts with another material, some of its bonds may be broken and others created. If the products are at a lower energy state than the reactants, then thermal energy (heat) must have been liberated by the reaction, which is termed an *exothermic* reaction. Likewise, if the products are at a higher energy state than the reactants, then heat must have been absorbed by the reaction, which is termed an *endothermic*

TABLE 3.5
Structures of Some Endothermic Compounds

		ΔH_f°	
Compound	*Structure*	*kJ/g*	*kJ/g-mol*
acetylene	HC≡CH	+8.7	+227
cyanogen	N≡C–C≡N	+5.9	+308
allene	$H_2C{=}C{=}CH_2$	+4.8	+192
hydrogen cyanide	HC≡N	+4.8	+130
diazomethane	H_2CN_2	+4.6	+192
benzothiazole	$\overline{C_6H_4NHN{=}N}$	+2.1	+250
1,3-butadiene	$H_2C{=}CH{-}CH{=}CH_2$	+2.1	+112
nitrogen trichloride	NCl_3	+1.9	+230
ethylene	$H_2C{=}CH_2$	+1.9	+52
propylene	$H_2C{=}CH{-}CH_3$	+0.5	+20

reaction (not to be confused with an *endothermic compound,* as discussed in the previous section). From a safety standpoint, an exothermic reaction is generally the most hazardous, since the liberated heat energy may raise the temperature of the products hundreds of degrees, accelerate the reaction rate, cause phase changes, result in gas expansion and pressure effects, and/or initiate other side reactions or decomposition reactions. This is often the case during fast decompositions and complete oxidations. An endothermic reaction may be hazardous by generating more-energetic materials that may be susceptible to instability.

Calculating the Heat of Reaction

If the end products are known, then the heat of reaction or self-reaction can be determined by the difference between the chemical energy content of the products and that of the reactants. This difference can be calculated by using the heats of formation of the reactants and products, as in Equation (3-3):

$$\Delta H_{rxn} = \Delta H_f(\text{products}) - \Delta H_f(\text{reactants}) \qquad (3\text{-}3)$$

where ΔH_{rxn} is the specific heat of reaction in kJ/g, ΔH_f(products) is the sum of the heats of formation of the products in kJ/g, and ΔH_f(reactants) is the sum of the heats of formation of the reactants in kJ/g. (The same equation applies when calculating the heat of reaction in kJ/g-mol; however, the heat of reaction on a weight basis is more commonly used when evaluating reactivity.)

The term *heat of reaction* is the most general term used for various kinds of reactions; more specific terminology is often used when a particular type of reaction is involved: *heat of polymerization* for polymerization self-reactions; *heat of decomposition* for explosive or thermal decomposition self-reactions; *heat of combustion* for all combustion reactions. The reactants and products for these various kinds of reactions were summarized in Table 3.3.

Measuring the Heat of Reaction

Adiabatic calorimetry is employed to measure the heat of reaction for a given reactive material or mixture. However, measuring the heat of reaction is generally not performed as part of a theoretical evaluation. The heat of reaction might be determined as part of screening tests (Section 3.4) or severity testing (Section 4.2).

Interpreting the Heat of Reaction

One of the hazard criteria used by the CHETAH code (Section 3.2.6) is the maximum heat of reaction or decomposition on a weight basis. Substances having a $-\Delta H_{rxn}$ or $-\Delta H_d$ less than 720 cal/g (3.0 kJ/g) are indicated to pose less of a reactivity hazard, and materials having $-\Delta H_{rxn}$ or $-\Delta H_d$ values greater than 720 cal/g (3.0 kJ/g) are considered to have a greater reactivity hazard with respect to reaction or decomposition. The United Nations Orange Book, by contrast, uses a heat of reaction ($-\Delta H_{rxn}$ or $-\Delta H_d$) of 75 cal/g (0.3 kJ/g) as the criterion for determining a self-reactive substance and the need for obtaining a self-accelerating decomposition temperature (SADT) for shipment purposes (UN, 1991).

Example: Polymerization

Polymerization is an exothermic process that poses a thermal runaway hazard, where loss of control can develop into a thermal explosion. The reaction rate depends partly on the mechanism of polymerization (radical, ionic), the concentration of monomer and catalyst, and the temperature. Polymerization does not require the presence of catalysts (e.g., radicals), although they strongly increase the polymerization rate. Quite often a combination of processes can occur. For instance, in the case of styrene monomer, a high rate of polymerization in an isolated container can result in evaporation of the remaining monomer. If a container with styrene monomer is subjected to a large heat flux (fire, steam), the polymerization of the monomer causes the temperature to rise. At a certain elevated temperature, spontaneous decomposition of the styrene monomer and/or its polymers start. This secondary decomposition process generates twice as much energy as the polymerization process itself.

Heat of polymerization values are usually around 20 kcal/g-mol. Styrene, for example, has a heat of polymerization of 17 kcal/g-mol, which translates to 160 cal/g (0.7 kJ/g). This falls into the "low" hazard degree category, according to the CHETAH criterion cited above that is related to maximum heat of reaction. It should be kept in mind that the **rate** of polymerization must also be considered when evaluating the hazard.

3.2.5. Thermodynamics: Equilibrium Considerations

According to thermodynamics, the most stable equilibrium state of a system at constant pressure and temperature is the one with the minimum Gibbs free energy, G, defined as:

$$G = U - T \cdot S + p \cdot V \tag{3-4}$$

in which U is the internal energy of the system, T the absolute temperature, S the entropy, p the pressure and V the volume. The change in the Gibbs free energy ΔG during a process occurring at constant temperature and pressure is

$$\Delta G = \Delta U - T \cdot \Delta S + p \cdot \Delta V \tag{3-5}$$

The internal energy term can be replaced by enthalpy ΔH by noting that, at constant pressure,

$$\Delta H = \Delta U + p \cdot \Delta V \tag{3-6}$$

resulting in the relation

$$\Delta \mathrm{G} = \Delta H - T \cdot \Delta S \tag{3-7}$$

Thus, the Gibbs free energy is a function of both the enthalpy (heat content) and the entropy (randomness) of a system.

It has been found useful to define a Gibbs free energy of formation ΔG_f for substances, in order to perform calculations similar to those for heats of formation (Section 3.2.3). The Gibbs free energy of formation for a material in its standard state, denoted by a degree symbol, is a function of the standard-state enthalpy and entropy of formation of the material:

$$\Delta G_f^\circ = \Delta H_f^\circ - T \cdot \Delta S_f^\circ \tag{3-8}$$

Values for the standard-state Gibbs free energy of formation ΔG_f° of some substances can be found in standard handbooks such as Perry and Green (1984, 3-147) and Weast (1985).

The Gibbs free energy can be used in more than one way during the course of a theoretical evaluation. Three uses for the thermodynamic concepts surrounding Gibbs free energy are discussed below, followed by a summary of activation energy considerations that come into play in the interpretation of free energy calculations. A fuller discussion of Gibbs free energy is given in Stull (1977, 10–13).

Using Gibbs Free Energy to Determine Gaseous Equilibrium

The difference between the total free energies of formation of a reaction's products and that of the reactants is the Gibbs free energy of reaction ΔG_r. This value is related to the equilibrium constant K_p of a gas-phase reaction by the equation

$$\Delta G_r = -RT \log_e K_p \tag{3-9}$$

using ideal gas law assumptions, where R is the ideal gas constant and T is the absolute temperature. K_p is related to the partial pressures P of the gases and the reaction stoichiometry according to the expression

$$K_p = \frac{(P_C)^c (P_D)^d}{(P_A)^a (P_B)^b} \tag{3-10}$$

for a typical reaction

$$aA + bB \leftrightarrow cC + dD \tag{3-11}$$

where the number of moles of gas does not change. [Otherwise, a bulk pressure term needs to be added to Eq. (3-10).] Hence, if the Gibbs free energy of reaction is known, then the concentrations of the gases at equilibrium can be predicted.

Using Gibbs Free Energy to Determine Reaction Products

Since the final state of a system is that which has the lowest Gibbs free energy, this fact is used to determine the expected products of a reaction such as the decomposition of a shock-sensitive material. Because of the many reaction products that are possible, the calculations to determine what products are formed and in what ratios to minimize the Gibbs free energy of the products are often done by computer. For example, such a calculation indicated that the decomposition of ethylene at a final total pressure of one atmosphere would yield the following products (Cruise, 1964) :

$$C_2H_4 \rightarrow 0.146\ CH_4 + 1.854\ C + 1.708\ H_2 + 63\ \text{kJ/g-mol} \qquad (3\text{-}12)$$

Using Gibbs Free Energy to Determine Reaction Direction

Since a reaction will proceed spontaneously in the direction of minimum Gibbs free energy, this fact can be used to determine the expected course of a given reaction. The Gibbs free energies of formation on both sides of a chemical equation can be totaled, and the side of the equation with the lower total value will be the direction to which the reaction will tend to move.

Activation Energy Barrier

It should not be concluded that metastable (i.e., higher Gibbs free energy) thermodynamic states of a chemical system are not of practical importance, despite the fact that a system has a tendency to move spontaneously to its minimum Gibbs free energy state. In fact, an energy barrier, or *activation energy* E_a, as was shown in Figure 1.2, may slow to an imperceptible pace the spontaneous transition of a system to minimum Gibbs free energy, so that reaction rate may be considered to be negligible at ambient conditions. However, adding thermal energy to a system, such as by an external fire, can accelerate the reaction rate by providing the energy input needed to overcome the activation energy barrier. Once an exothermic reaction is initiated, the heat of reaction may be sufficient to maintain the higher temperature and thus cause the reaction to continue until all material is converted or until the reaction is stopped by forced cooling.

3.2.6. CHETAH

Due to the complexity of some thermodynamic calculations and the necessity to either look up or predict such quantities as heats of formation, heats of reaction, and decomposition products, the prediction of reactivity hazards is often done by computer. Several codes developed for reactivity hazard prediction, some of which may not be readily available, are reviewed in CCPS/AIChE (1995). Some of the more important considerations regarding computer codes currently available for thermodynamic evaluations include the following:

- The codes all require heats of formation of the substance being evaluated and its reaction products. This information must be input by the user, included in a database, or predicted by one of the methods referenced in Section 3.2.3.

- Most of the codes were developed to run on mainframe computers. The CHETAH program is also available in a PC version.
- In general, none of the programs take into account the complete molecular structure of the reacting molecules. The principle in prediction of the reaction products is based on the complete disintegration of the reactant molecules to atoms and subsequent rearrangement of the atoms into small molecules. With some programs, the choice of reaction products is left to the user.
- None of the codes can predict the kinetic parameters of reaction rate, activation energy or reaction order. These parameters can only be determined experimentally.
- The primary use of the codes is to compute the heats of decomposition and combustion. Acid–base neutralization, exothermic dilution, partial oxidation, nitration, halogenation and other synthesis reactions are not included in the codes.

The features of the widely used CHETAH code, developed by the American Society for Testing and Materials (ASTM), are summarized below as an example of the predictions that can be performed by computer.

General Description of the CHETAH Code

CHETAH stands for Chemical Thermodynamic and Energy Release Program. First introduced in 1974, CHETAH has been developed under the control of Subcommittee E27.07 of the American Society of Testing and Materials (ASTM). In addition to its primary objective of predicting reaction potential from molecular structure, it is also employed for estimating heats of reaction and combustion, heat capacities, and entropies. One large U. S. company uses CHETAH most frequently for the estimation of heats of reaction, since this can be done as part of a presynthesis hazard evaluation when only the chemical structure is known (Frurip, 1992).

The latest version of CHETAH, Version 7.0, differs dramatically from the previous Version 4.4 that was offered beginning in 1989 (Davies et al., 1990). Its user-friendliness has been increased, the data files have been expanded to include inorganic compounds, and a new hazard evaluation criterion (plosive density, described below) has been added. Lower flammable limits in air can be estimated using Version 7.0.

Heat of Reaction Estimation

CHETAH estimates thermodynamic properties (enthalpy, entropy, free energy) of organic and organometallic compounds from the group additivity method of Benson (1982) between 300 and 1500K, then combines the thermodynamic properties to obtain heats and free energies of reaction. The program includes thermodynamic data for about 500 group contributions and about 400 chemical compounds (Davies et al., 1985, 84). CHETAH uses a linear programming technique to select the reaction products that give the maximum energy of decomposition. The temperature can be chosen by the user, but the pressure cannot be varied.

The CHETAH program calculates the heat of reaction assuming all species are in the gaseous phase. If all reactants and products are in the condensed phase, then the error may be small (Frurip, 1992). However, corrections for heats of vaporization may need to be used when dealing with multiphase systems.

Hazard Evaluation Criteria

A unique and useful feature of the CHETAH program is its use of four hazard evaluation criteria for determining energy-release potential. Specifically, the program predicts the ability of a material to decompose with violence when subjected to a severe impact. A pattern-recognition technique is now employed with the program to combine the results of the four criteria and give an unambiguous determination. It is claimed that the program will make the correct hazard evaluation classification 95% of the time, an erroneous but conservative classification 4% of the time, and a nonconservative erroneous classification 1% of the time (Davies et al., 1985, 86). The four hazard evaluation criteria are summarized below.

Maximum Heat of Reaction or Decomposition. The first hazard evaluation criterion is the maximum heat of reaction or decomposition. The ranges for this criterion, from very low to high hazard potential, were given in Section 3.2.4.

Probability Correlation. The second criterion is based on the difference between the heat of combustion in excess oxygen and the maximum heat of decomposition. This second criterion follows the assumption that a substance that comprises sufficient oxidizer within its own structure to convert itself to normal oxidation products presents a larger energy hazard than one that does not. If the absolute difference between the heat of combustion and the maximum heat of decomposition is 3.0 kcal/g or less, then the substance is considered to have a high hazard potential (Yoshida, 1987, 80). A difference greater than 5.0 kcal/g indicates a low hazard potential and 3.0 to 5.0 kcal/g indicates a medium potential.

Oxygen Balance. The calculation of oxygen balance and the CHETAH criteria for oxygen balance were described in Section 3.2.2. An oxygen balance between −120 and +80 indicates a high hazard potential, and an oxygen balance lower than −240 or higher than +160 indicates a low hazard potential.

The Y Criterion. The fourth criterion applied in the CHETAH code takes into account the number of atoms in a molecule of the substance involved. It is defined as:

$$Y = 10 \cdot Q^2 \cdot MW/N \tag{3-13}$$

where Q is the decomposition energy in kcal/g, MW is the molecular weight of the substance, and N is the number of atoms in a molecule of the substance. A Y value less than 30 is classified as having low hazard potential, between 30 and 110 is assigned a medium hazard potential, and above 110 gives a high hazard potential classification.

Plosive Density. The new "plosive density" criterion, as described by Seaton (1989), is a group contribution method for predicting the "plosive" tendency of a chemical composition; i.e., whether or not it has the potential to cause an explosion. Although it has no rigorous theoretical basis, error data shows it to be at least 94% accurate in determining whether or not a composition has explosion potential.

3.2.7. Example Evaluation

The following situation will be used to illustrate the theoretical calculations of Section 3.2:

Example Evaluation

Charbroiled Chemicals' production facility is in the initial design stages for a *t*-butyl hydroperoxide (2-hydroperoxy-2-methylpropane) bulk storage system. A review of prior experience (Section 3.1) with this material was inconclusive as to whether a significant reactivity hazard existed. No computer codes for hazard prediction are available within the organization. A theoretical evaluation of reactivity hazards is to be performed before proceeding with screening tests.

Unstable Atomic Groups

The structure of *t*-butyl hydroperoxide is shown in Figure 3.4. Comparison of this structure with bonds and functional groups that generally confer instability (Table 3.1) indicates that the compound may be unstable and, hence, susceptible to shock sensitivity and/or thermal decomposition, because of the hydroperoxide group. However, it is not known whether the rest of the molecule has a tendency

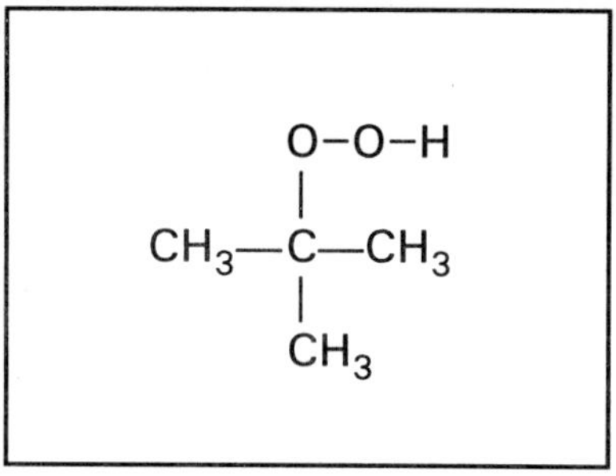

FIGURE 3.4. ***t*-Butyl Hydroperoxide Structure.**

to stabilize or destabilize the oxygen–oxygen bond. No other suspicious bonds or functional groups are present.

Oxygen Balance

The molecular formula for *t*-butyl hydroperoxide is $C_4H_{10}O_2$. It is not in its most stable oxidized form, so the oxygen balance can be calculated meaningfully. Substituting the numbers 4, 10, and 2 for the variables *i*, *j* and *n* in Equation (3-1) and using a molecular weight of 90.12 gives

$$\text{Oxygen Balance} = \frac{-1600(2 \cdot 4 + 10/2 - 2)}{90.12} = -195 \text{ wt\%} \qquad (3\text{-}14)$$

Comparing this oxygen balance of –195% to the criteria in Figure 3.3 indicates it is in the "medium" category with respect to self-reactivity potential based on oxygen balance alone. Since the oxygen balance is negative (i.e., indicating an oxygen deficiency), the presence of atmospheric oxygen could possibly make the decomposition of *t*-butyl hydroperoxide even more energetic.

Heat of Formation

The heat of formation is predicted by the group contribution method of the CHETAH program (Version 7.0) to be –57.61 kcal/g-mol (241 kJ/g-mol) for *t*-butyl hydroperoxide in the gaseous state. This is close to the measured value of –58.8 kcal/g-mol (–245.9 ± 5 kJ/g-mol) reported in the literature (Pedley et al., 1986). The literature value for the heat of formation in the condensed (liquid) state is –70.2 kcal/g-mol (–293.6 ± 5 kJ/g-mol). These values indicate that *t*-butyl hydroperoxide is not an endothermic compound.

Plosive Density

This criterion was added to Version 7.0 of the CHETAH program as an additional predictor of explosion potential. It applies a group contribution method similar to that used for predicting heat of formation, and then compares the sum of "plosophoric" groups (those that contribute instability to a molecule) and "auxoplosive" (those that alter the stability of a molecule such as by diluting the effect of plosophoric groups). Using the method in Seaton (1989), as illustrated in Table 3.6, the sum of plosive and auxoplosive weights divided by the molecular weight of *t* butyl hydroperoxide is 0.444. Since this is higher than the minimum value of 0.265 for classifying a composition as plosive, this criterion would indicate that *t*-butyl hydroperoxide does pose an explosion hazard.

Heat of Decomposition

The heat of decomposition ΔH_d is calculated by subtracting the heats of formation of the material from those of its expected decomposition products.

TABLE 3.6
Plosive Density Calculations for *t*-Butyl Hydroperoxide

Second-Order Groups in This Molecule	*Number of This Group in the Molecule*	*Plosophoric Weight of This Group*	*Auxoplosive Weight of This Group*	*Total Plosophoric and Auxoplosive Weight of This Group*
$C-(C)(H)_3$	3	0.00	0.00	0.00
$C-(C)_3(O)$	1	0.00	–0.52	–0.52
O–(C)(O)	1	19.82	0.00	19.82
O–(H)(O)	1	20.68	0.00	20.68
Sum of plosophoric and auxoplosive weights for this molecule =				**39.98**
Ratio of this sum to the molecular weight = 39.98/90.12 = **0.444**				

Without additional oxygen, predicted decomposition products and the resulting heat of decomposition are:

$$C_4H_{10}O_{2(\text{liquid})} \rightarrow 1.5\ CH_{4(\text{gas})} + 2.5\ C_{(\text{graphite})} + 2\ H_2O_{(\text{gas})}$$

$$\Delta H_d = -(-70.2) + 1.5(-17.89) + 2.5(0) + 2(-57.80) = -72.2 \text{ kcal/g-mol}$$

The negative ΔH_d indicates that self-reaction of *t*-butyl hydroperoxide to its decomposition products can occur exothermically. Using the molecular weight of *t*-butyl hydroperoxide of 90.12 g/g-mol, $-\Delta H_d$ is calculated on a weight basis to be 800 cal/g (3.4 kJ/g). This puts the compound into the higher-hazard category by a small margin, using the CHETAH heat of reaction criterion discussed in Section 3.2.3.

Heat of Combustion

The heat of combustion is calculated from the heats of formation of the material and its combustion products. Complete combustion to CO_2 and H_2O liberates

$$C_4H_{10}O_{2(\text{liquid})} + 5.5\ O_{2(\text{gas})} \rightarrow 4\ CO_{2(\text{gas})} + 5\ H_2O_{(\text{gas})}$$

$$\Delta H_c = -(-70.2) - 5.5\ (0) + 4\ (-94.05) + 5\ (-57.80) = -595 \text{ kcal/g-mol}$$

As expected for a compound with a negative oxygen balance, the heat of combustion is significantly more negative than the heat of decomposition; i.e., significantly more energy is released per mole of *t*-butyl hydroperoxide in combustion than in decomposition. The difference on a weight basis between $-\Delta H_c$ (6600 cal/g) and $-\Delta H_d$ (800 cal/g) indicates a "low" hazard potential using the CHETAH "probability correlation."

Equilibrium Considerations

The Gibbs free energy of formation of *t*-butyl hydroperoxide must be known to perform free energy calculations. Gas-phase equilibrium calculations are not applicable to *t*-butyl hydroperoxide, which is a liquid at ambient conditions.

Theoretical Evaluation Conclusions

The theoretical evaluations outlined above indicate that, although *t*-butyl hydroperoxide is not an endothermic compound, it nevertheless has medium to high potential for shock sensitivity and/or thermal instability. Since the results might be considered as borderline or inconclusive, a review by expert determination (Section 3.3) may be required. Such a review might bring to bear any prior experience with *t*-butyl hydroperoxide or with closely related compounds such as *t*-amyl hydroperoxide. By comparison, an NFPA reactivity rating of 4 for *t*-butyl hydroperoxide (NFPA 325M, 1994) signifies that it is classified in the highest reactivity category, indicating a material that in itself is readily capable of detonation or of explosive decomposition or reaction at normal temperatures and pressures (NFPA 704, 1990), so that the above indications of self-reaction potential are justified. However, by contrast, a commercial formulation of *t*-butyl hydroperoxide diluted with 30% water requires elevated temperatures and pressures for decomposition to occur, and is considered to burn in the same manner as ordinary combustibles (NFPA 43B, 1993).

3.3. Expert Determination

If prior experience (Section 3.1) and theoretical evaluations (Section 3.2) are borderline or otherwise indeterminate as to whether any significant reactivity hazards exist for a given storage and handling situation, then screening tests (Section 3.4) may be warranted. However, screening tests are often expensive and time-consuming. Many companies employ the judgment of reactive chemical experts before beginning a series of screening tests. The collective experience and knowledge of company and/or outside experts may point to a conclusion that screening tests are not warranted, averting the spending of unnecessary resources on a screening test program. However, a routine screening test, such as a DSC, might be considered for any new chemical substance as a place to start in the evaluation.

The three possible outcomes of the expert determination are:

1. It is judged with certainty that no significant reactivity hazards exist, and thus screening tests are not warranted.
2. It is judged that one or more significant reactivity hazards are likely to exist, and thus screening tests are warranted.
3. It can **not** be judged with certainty, based on the expert determination, whether or not screening tests are warranted. In this case, the conservative approach should be taken, and the screening tests should be conducted.

The expert determination might also bring to light additional information (incidents, test data, etc.) that was not part of the prior experience review or the theoretical evaluations; this additional information may indicate that a significant reactivity hazard does exist. In this case, the screening tests could be bypassed and a program of severity testing, as described in Chapter 4, could be started right away (see Figure 3.1). The rationale for each determination should be documented, and, if more than one material is being stored and handled in a facility, then this judgment should be made for each substance or mixture.

3.3.1. Expert Committees

How such an expert determination is made varies from company to company, depending on factors such as the size and organization of the company, the nature of the company's businesses, and the availability of research and testing personnel. One large company, for example, has established "Reactive Chemical Committees" on the division, site, location, or other organizational level. These committees typically consist of representatives from several disciplines, including research, engineering, production, safety/loss prevention, testing, and any other appropriate functions. Two of the responsibilities for these Reactive Chemical Committees are to provide data and consultation concerning hazard evaluation of materials and processes, and to coordinate programs for hazard review of existing materials and processes within organizational groups. Thus, these committees can serve to provide expert determination on whether screening tests are warranted in a given situation.

In most companies, many testing decisions need to be made that relate to small process changes, such as the use of a slightly different additive in a material to be stored on-site or the handling of a material at somewhat more severe conditions of temperature or pressure. Such decisions may not warrant as extensive of a Reactive Chemicals Committee as described above. A smaller committee or subcommittee can be designated as having responsibility for more day-to-day process change authorizations, with the ability to call together the full committee where it is deemed necessary.

3.3.2. Kinetics Determination Factors

Prime examples of factors that expert determination might consider are those related to reaction kinetics. The theoretical evaluations described in Section 3.2 can determine, by thermodynamics, the potential energy release in a given storage and handling arrangement, but they say nothing of the kinetic factors that determine how fast or slow that energy can be released. Since heat generation must be offset by heat removal in a stable reaction system, then the *rate* of heat generation may be more important than the overall thermodynamic potential energy release in the design of many storage and handling systems.

Several kinetics-related factors influence the conditions under which reaction or self-reaction can accelerate. The interaction between the rate of reaction or self-reaction and a number of these parameters is discussed below (Bretherick, 1987, 4.1).

Temperature

Temperature is the most important factor influencing the reaction rate in a reactive chemical system. This can be seen in the Arrhenius equation, which describes the relationship between reaction rate and temperature, where the reaction rate is an exponential function of absolute temperature. In practical terms, this means that an increase in temperature of 10°C can result in a two- to four-fold increase in reaction rate, depending on the energy of activation. At higher temperatures, side reactions and/or sequential reactions can be initiated, with a different set of kinetic parameters for each reaction.

In addition to possible dilution effects, low-boiling solvents can affect the temperature performance of a storage arrangement. If the reaction temperature exceeds the boiling point of the solvent, a refluxing effect may be achieved, particularly in the presence of a condenser. The refluxing effect prevents a further increase of the reaction rate by effectively limiting the temperature to the boiling point of the solvent system and removing the evolved heat of reaction as heat of vaporization. Water should also be considered as a solvent because its heat of vaporization (541 cal/g or 2.265 kJ/g) is higher than nearly all organic solvents.

Concentration

In general, the reaction rate is proportional to the concentration of the reactant(s) to the powers *a* and *b*, with *a* and *b* being the orders of the reaction:

$$r \approx c_A^{\,a} \qquad \text{or} \qquad r \approx c_A^{\,a} \cdot c_B^{\,b} \tag{3-15}$$

Dilution of a reactive mixture or a self-reactive substance with an inert solvent may be used to reduce the reactant concentration and even to minimize the temperature rise by evaporation of the solvent. In addition, the adiabatic temperature rise is decreased.

Impurities with Catalytic Effects

Small amounts of impurities that act as catalysts can increase the rate of reaction significantly. For example, sulfuric acid increases the rate of decomposition and decreases the observed decomposition onset temperature of various isomers of nitrobenzoic acid (Grewer, 1988). Other substances such as NaCl, $FeCl_3$, platinum, vanadium chloride and molybdenumchloride also show catalytic effects for nitrobenzoic acid decomposition. In the presence of molybdenum chloride, for instance, the decomposition temperature is lowered as much as 100°C. Although catalysts are not usually added intentionally to storage and handling operations, some substances (e.g., rust) may be present inadvertently. Also, some decomposition reactions are autocatalyzed, where one or more of the reaction or decomposition products will act as a catalyst to accelerate the rate of reaction. These potential catalytic effects cannot be ignored in a hazard evaluation.

Air

As described in Section 1.6, a number of substances are able to form unstable peroxides in contact with atmospheric oxygen. Inadvertent oxidation reactions can also cause quality losses, self-heating, and possible ignition and fire. On the other hand, oxygen can play a role in controlling the rate of some reactions, such as acting as an inhibitor in vinyl monomer systems.

Confinement

Confinement can increase the reaction or decomposition rate in many storage and handling arrangements. For example, ignition of a vapor–air deflagration at the bottom or at the center of a closed or semiclosed vessel may lead to an increase in the deflagration rate by a factor of more than 100 in comparison with top initiation. Another example of the effect of confinement is the enhancement of autocatalytic decomposition involving a volatile catalyst with confinement present. Decomposition rates in some substances such as azo compounds, peroxides, and lead(IV)oxide (Grewer, 1988) may accelerate with an increase in pressure, especially in the case where the governing decomposition reaction is gas-phase controlled.

3.4. Reactivity Screening Tests

Two situations where reactivity screening tests may be needed are when expert determination (Section 3.3) indicates that screening tests are warranted, and when a company submits all new substances to screening tests as a matter of routine procedure to be sure of the absence of instabilities and incompatibilities.

The flow chart in Figure 3.5 is a schematic strategy for conducting reactivity screening tests. This test strategy addresses both self-reactive materials (polymerizing, thermally decomposing, shock-sensitive, and rearranging) and those that are reactive with other materials (pyrophoric, water-reactive, peroxide forming, and reactive with other common materials). Flammability testing, although important, is not within the scope of this work.

Not all of the tests shown in Figure 3.5 will be conducted for every new substance to be characterized. The actual type of tests involved in the test strategy depends on the process development stage and prior experience. Selection of tests and determination of an appropriate test sequence should be decided by expert determination such as was discussed in Section 3.3. The flow chart of Figure 3.5 allows for the tests to be conducted either sequentially or in parallel (or a combination of sequential and parallel tests).

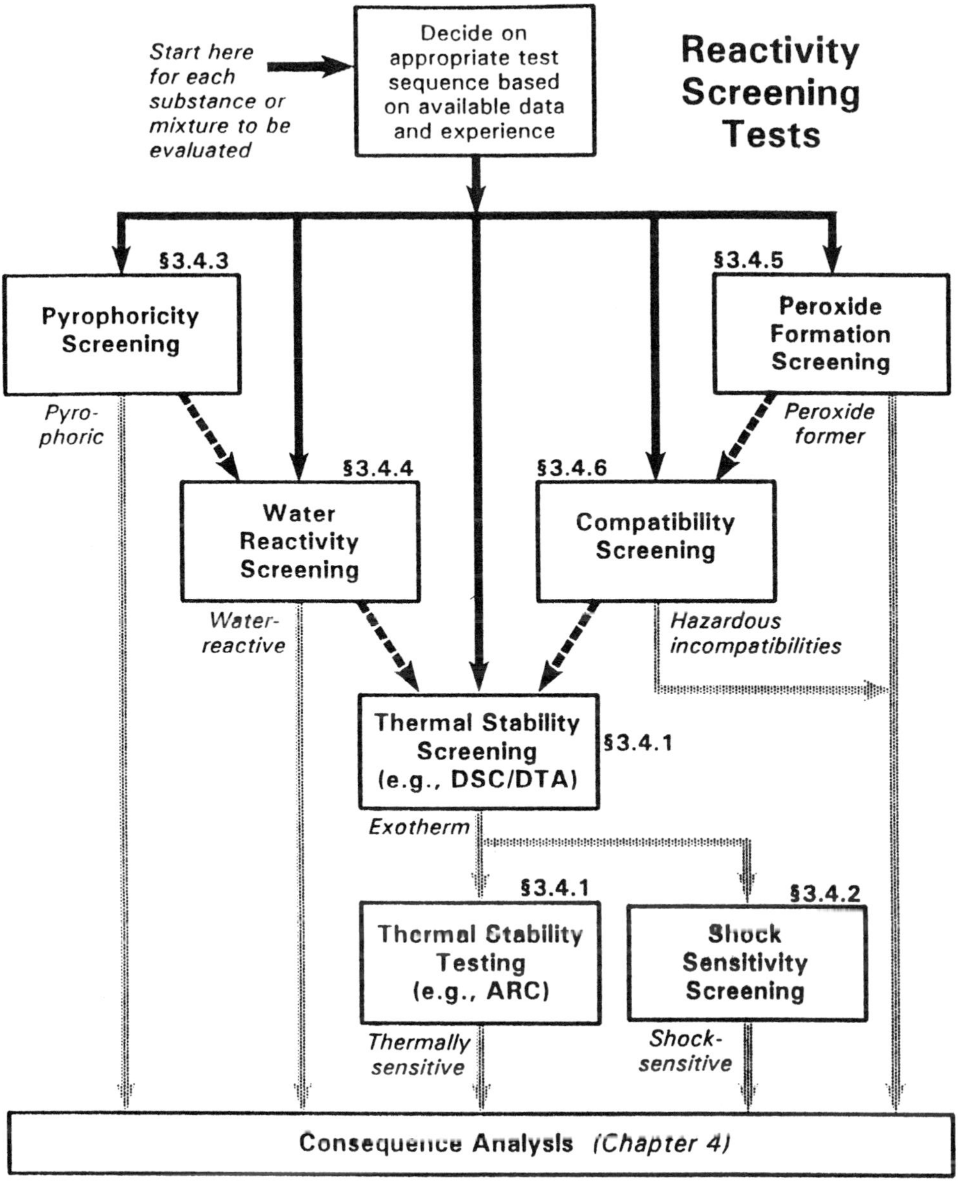

Notes: 1. Tests may be conducted in parallel (solid lines) or sequentially (dashed lines).
2. Expert determination may deem one or more tests to be not necessary.

FIGURE 3.5. **Strategy for Conducting Reactivity Screening Tests.**

Pyrophoricity screening, as discussed in Section 3.4.3, is suggested as the first test to be conducted on new substances if the pyrophoricity tests are deemed to be warranted. This is because pyrophoric materials can, by definition, spontaneously ignite in air, and thus the identification of pyrophoric behavior early in the test sequence may enhance the safety of the testing procedures. Likewise, water reactivity screening (Section 3.4.4) is suggested as the second test, due to the possibility of the material inadvertently coming into contact with water at some time during its testing, storage, and/or handling. Early screening for peroxide formation (Section 3.4.5) and incompatibilities (Section 3.4.6) may also identify hazards that will necessitate precautions while handling even small quantities of the material; however, peroxide formation and compatibility screening can be done at any time during the test sequence.

Thermal stability screening is the cornerstone of reactivity screening. The most common tests employed for initial thermal stability screening are differential scanning calorimetry (DSC) and differential thermal analysis (DTA). If the initial screening tests indicate exothermic activity, then more sophisticated thermal testing, such as by the use of accelerating rate calorimetry (ARC), can provide better quantitative data on onset temperature, reaction enthalpy, instantaneous heat production as a function of temperature, maximum temperature and/or pressure excursions as a consequence of runaway, and additional data useful for facility design and operation. Section 4.2.1 summarizes test methods and results for thermal stability tests—both screening tests and more detailed testing. More details regarding screening and other reactivity hazard tests can be found in CCPS/AIChE (1995).

The results of every test discussed in this Section must be individually interpreted, due to limitations and variations in test conditions. It is essential to recognize that these tests are conducted on small-scale samples under specified conditions that do not include all the aspects of the storage and handling conditions. Therefore, it is recommended that an expert on the type of tests involved be consulted, both before testing is requested and again for interpretation and evaluation of the test results. This latter point is particularly significant.

3.4.1. Thermal Stability Screening Tests

The first aim of a thermal stability screening test is to obtain data on the possibility of exothermic reaction (for mixtures) or self-reaction (for single substances) by measuring the heat of reaction or heat of decomposition. The screening tests can also measure reaction onset temperatures and, to a lesser extent, the kinetic parameters of the reaction. Thermal stability screening tests can be performed both with and without the presence of air, in order to differentiate between self-reaction thermal hazards such as thermal decomposition and oxygen-reaction hazards such as flammability.

Reliable and internationally used techniques for thermal stability screening include differential scanning calorimetry (DSC) and differential thermal analysis

(DTA). These methods are able to determine exothermic heats of reactions and their observed onset temperatures and to generate, in some cases, approximate reaction kinetic data by calculation.

Thermal Stability Screening Criteria

In general, a substance should be considered as energetic when the experimental heat of decomposition $-\Delta H_d$, measured in the absence of air, is greater than 50 to 70 cal/g (0.2 to 0.3 kJ/g). This range may expand to 30 to 100 cal/g (0.13 to 0.4 J/g) depending on the rate of reaction, rate of pressure increase and system design considerations. In the worst case, a heat of decomposition of 50 to 70 cal/g results in a maximum adiabatic temperature rise of approximately 100 to 200°C. As a rule of thumb, temperature rises in this range are not regarded as critical, as long as the substance does not easily produce a significant amount of gas or otherwise lead to hazardous situations (Grewer et al., 1989). However, this must be determined for each individual operation.

DSC and DTA

According to NFPA 704 (1990), DSC is considered to be the primary screening test for assessing reactivity hazards. It is used chiefly as a screening tool to indicate when more detailed testing may be warranted.

In DSC, the difference in the energy inputs into the sample and reference material required to keep their temperatures equal is measured as a function of temperature, while the sample and reference material are subjected to a controlled increase in temperature. The main advantages of DSC as a screening test are:

- It is a fast technique, taking only a few hours for a complete scanning experiment.
- Only a small sample size is required (several milligrams). The small sample size permits the use of DSC even for very energetic systems in a regular laboratory setup.
- DSC is applicable over a wide temperature range. With DSC, the complete temperature range of thermal decomposition or reaction can be scanned, up to 400°C or even higher with some instruments.
- Heats of decomposition or reaction can be established from a single scanning experiment, although multiple scans are generally required to determine kinetics.

Small sample sizes are desirable from a safety and availability standpoint; however, care must be taken that such a small sample is representative of the material to be tested.

Differential thermal analysis (DTA) is a technique in which the temperature difference between a substance and reference material is measured as a function of temperature, while the substance and reference material are subjected to a controlled increase in temperature. Classical DTA has been developed into heat-flux DSC by the application of multiple sensors (e.g., a Calvet-type arrangement) or with a controlled heat leak (Boersma-type arrangement). Another type of DSC instrument is the power-compensation DSC, which means that the energy input to the reference and sample is adjusted to remove the temperature difference between sample and reference. Details of DSC and DTA testing apparatuses and comparisons with other thermal analysis methods are given in CCPS/AIChE (1995).

The purpose of differential thermal systems is to record the difference in the enthalpy change that occurs between reference and tested sample, when both are identically heated. A number of publications are available concerning the theoretical aspects and application of various techniques of thermal analysis, including DSC (e.g., McNaughton and Mortimer, 1975; Wendtlandt, 1974; Mackenzie, 1972).

Most DSC equipment can be used in the temperature range of 25 to 500°C. Most can be cooled as well, a feature required for investigating samples that are unstable at ambient conditions. DSC instrumentation is usually sufficient for indicating thermal hazards of stirred systems and small-scale unstirred systems, provided the reaction is kinetically controlled under normal operating conditions. The resulting data must be carefully used if mixing or mass transport is important.

Performing a DSC Experiment

To perform a DSC test, a representative sample of 10 mg or less of test material is placed in the sample cup. This cup is, if necessary, sealed and inserted in the DSC equipment on the sample sensor side. The reference cup, identical to the sample cup but containing inert reference material (glass beads) in the same amount and with approximately the same heat capacity as the sample, is placed on the reference sensor side. In studying the temperature range of exothermic decomposition, a heating rate of 5 to 10°C/min. (scanning mode) is generally used.

The observed onset temperature, T_0, is the temperature at which the substance or mixture first shows an observable instrumental response due to decomposition or reaction. T_0 depends on the sensitivity of the apparatus, the sample mass, the atmosphere, the confinement and the heating rate. When the experiment is designed to establish T_0 of the exotherm more accurately, a heating rate of 1 to 5°C/min is appropriate (Grewer et al., 1989). However, it should be emphasized that the onset temperature strongly depends on the instrument sensitivity and that application of kinetic data to large-scale calculations may introduce large errors.

A typical DSC curve with an exothermic peak is represented in Figure 3.6. The shape of the DSC curve depends on the reaction order; the occurrence of autocatalytic decomposition; parameters such as heating rate, heat capacity and the heat conductivity of both example and reference material; etc. The peak area is proportional to the total heat of reaction or decomposition. Because of the limited sensitivity of the instrument and potential secondary decompositions at

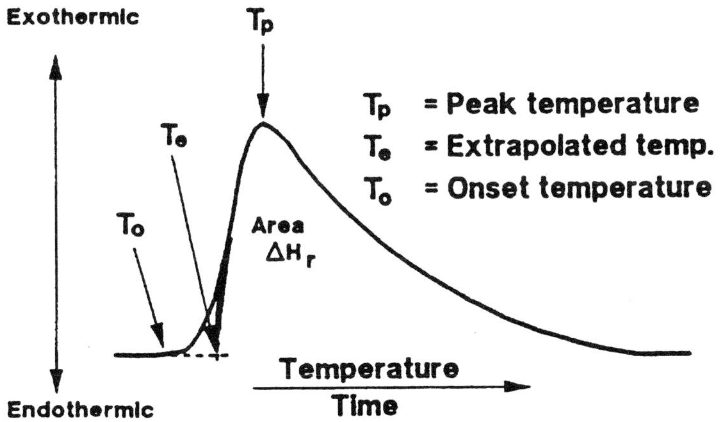

FIGURE 3.6. **Typical Scanning DSC Curve of an Exothermic Decomposition.**

higher temperatures, DSC tests should be run at least 150°C past the maximum operating temperature in the facility.

DSC can be conducted in the isothermal mode as well. In such an experiment, the sample cup containing the substance being studied is inserted into a DSC that has been preheated to the desired test temperature. This type of experiment may be performed to examine systems with induction periods such as autocatalytic reactions or inhibitor depletion times. Furthermore, kinetic parameters can be determined more accurately by isothermal DSC experiments.

For systems without a solvent barrier, i.e., without a temperature maximum imposed by the boiling point of a solvent, the time of testing at the test temperature is important, so that the selected period of time and temperature should be based on the worst-case operating conditions. After conducting an isothermal DSC experiment, it is useful to run a standard scanning DSC experiment on the isothermally aged material to confirm whether or not the material has fully decomposed. Combinations of isothermal and scanning DSC experiments are required to thoroughly examine the kinetics of the test system.

Assessment of DSC Results

The first thing determined from a scanning DSC experiment is the observed exotherm onset temperature T_0. Since the observed onset depends on the scanning rate and on the sensitivity of the instrument, its use is limited when establishing a "safe operating region." Such a determination is usually done with an understanding of the operational and equipment specifications, combined with the kinetics of reaction or decomposition. As a rule of thumb in testing by DSC, a substance is indicated to be sufficiently thermally stable if the T_0 of the exotherm, as determined in the DSC, exceeds the highest operating temperature by at least 100°C, recognizing other factors such as short time periods and the existence of good heat transfer conditions. The highest operating temperature must be defined under normal or upset conditions (such as loss of cooling media), should consider possible external heat input causes, and should have a suitable safety margin. However, exceptions to this rule of thumb have been shown to exist (Hofelich and Thomas, 1989). Impurities and catalysts may decrease T_0 significantly; a decrease of 100°C is not unusual. The material of the sample cup itself may act as a catalyst, resulting in surface-induced decomposition that is even promoted by the sample/surface area ratio in the DSC cup.

In DSC instruments, heat production can be determined directly as a function of temperature. The shape of the heat production curve is also important for hazard identification. A sharp rise in energy release rate (i.e., a steep slope of the exotherm), whether due to a rapid increase of the rate constant with temperature or to a large heat of reaction, indicates that the substance or reaction mixture may be hazardous.

In principle, DSC also offers the capability of calculating the kinetics of an (exothermic) reaction or decomposition from isothermal or scanning experiments. A number of calculation methods for establishing reaction kinetics from the scanning DSC are discussed in the literature (e.g., Duswalt, 1974; Skvara and Satava, 1970). However, the reliability of kinetic data obtained from scanning DSC for scaling calculations is questionable.

Accelerating Rate Calorimeter Testing

If the thermal stability screening tests (DSC/DTA) show that a substance is thermally unstable close to the temperature range of large-scale storage or handling (using criteria provided in the description of the respective screening test), further investigations by specific, more accurate, more sensitive and more dedicated tests are recommended. In particular, an accelerating rate calorimeter (ARC) is often used within the context of thermal stability screening. In many companies, ARC is used instead of DSC/DTA for screening purposes, due to the higher-quality and more useful results. Other thermal stability tests such as isothermal storage tests and gas evolution tests are addressed in the context of a consequence analysis (Section 4.2).

Quantities that can be determined by thermal stability tests include

- precise thermal stability (including autocatalysis and inhibitor depletion, compatibility, etc.)
- maximum operating temperature to avoid hazardous decomposition
- maximum rate of reaction
- runaway behavior
- adiabatic temperature rise
- gas evolution (mass flux) during decomposition and/or runaway reaction
- behavior of the material under external heat load such as an external fire
- induction time to the start of exothermic reaction; e.g., when inhibitor depletion is involved
- compatibility with metals, additives and contaminants.

To investigate the gas evolution during decomposition and/or runaway, the ARC simultaneously records the temperature rise and the pressure rise, which is usually proportional to the gas evolution during reaction or decomposition.

The accelerating rate calorimeter provides the temperature–time and pressure-time data for tested materials under adiabatic conditions. ARC is particularly well-suited for determining a useful onset temperature T_0 for exothermic activity. Due to its greater sensitivity, the T_0 determined by ARC will nearly always be significantly lower than T_0 values measured by DSC/DTA. Fenlon (1984) indicates that thermal stability is acceptable in systems where the maximum operating temperature is 50°C or more lower than the T_0 value measured by an ARC test (as compared to the 100°C rule of thumb for DSC/DTA), as long as the material(s) are held at this maximum operating temperature only for a short time. If the temperature difference is less than 50°C, or if the maximum operating temperature is maintained for an extended time, then an isothermal aging test in the ARC is warranted. Details of the Accelerating Rate Calorimeter and similar devices are given in CCPS/AIChE (1995).

3.4.2. Shock Sensitivity Screening

If a material exhibits a tendency to decompose energetically when heated, then it may also be shock-sensitive, and screening tests for shock sensitivity may be called for. In particular, if the heat of decomposition $-\Delta H_d$ of the tested material is relatively high, then it is also more likely to be shock-sensitive. In this Section, the terms *deflagration* and *detonation* will be defined and related to the thermal stability screening tests previously discussed, and methods for shock-sensitivity screening will be presented.

Deflagrations

A *deflagration* is the energy release from a self-sustaining chemical reaction that propagates through the unreacted material at subsonic velocity, after being initiated either spontaneously or by an ignition source. A deflagration propagates by thermal energy transfer.

A substance or mixture is considered potentially capable of deflagration if it has a $-\Delta H_d$ larger than 250 cal/g (1.05 kJ/g). Deflagration tests (CCPS/AIChE, 1995) should be considered when this is the case, as well as tests for sensitivity to impact and friction. The rule of thumb value of $-\Delta H_d$ greater than 250 cal/g (calculated maximum value) as a criterion for deflagration properties is an approximate value. A number of values, varying from 170 to 300 cal/g, are stated in the literature (Fierz and Zwahlen, 1989; Grewer, 1988; Seaton et al., 1974; Yoshida, 1987). This $-\Delta H_d$ range is valid for most of the known substances that are able to deflagrate.

Propagation rates of deflagrations vary from very slow (1 to 10 mm/min) to very fast (10 to 1000 m/s). For example, some fertilizers show very mild deflagrations, whereas some organic peroxides, di- and tri-nitro compounds (at higher temperatures) and oxidizer/reducer mixtures show very fast deflagrations resulting in violent pressure rises. The severity of the deflagration is not only related to the type of substance, but it is also dependent on the magnitude of confinement or pressure (Merzhanov, 1969) and the bulk temperature of the material.

Detonations

A *detonation* is the energy release from a violent, self-sustaining, propagating reaction where the reaction front (shock wave) advances faster than the speed of sound in the unreacted material. Generally, a substance or mixture is considered capable of detonating if it has a calculated exothermic $-\Delta H_d$, measured without the presence of oxygen, that is larger than 700 cal/g (3.0 kJ/g). A minimum theoretical $-\Delta H_d$ value of 700 cal/g is generally accepted as a criterion for detonation properties. In the CHETAH program (discussed in Section 3.2.6), a value $-\Delta H_d$ greater than 700 cal/g is classified as "high" related to energy hazard potential. However, this is an approximate value, and a number of exceptions are known. For example, initiating explosives like azides that have $-\Delta H_d$ values lower than 475 cal/g (Meyer, 1977) are able to detonate and are very sensitive to mechanical shock and friction. Ammonium nitrate in some formulations and a number of organic peroxides are also able to detonate, notwithstanding a $-\Delta H_d$ lower than 475 cal/g.

Furthermore, the $-\Delta H_d$, if determined by one of the screening tests, will often be significantly less than the heat of explosion. Conditions of the screening tests (temperature and pressure) compared with the explosion conditions are less extreme and, as a result, the decomposition products may not be the same, resulting in the differing $-\Delta H_d$ values. Therefore, the experimental value must be reviewed critically. If there is any doubt about the detonative potential of a substance or a mixture, detonation testing is necessary.

Detonation testing is discussed in CCPS/AIChE (1995). It is emphasized that detonation tests and a number of deflagration tests are specialized tests that must be run by experts in specialized facilities (Kohlbrand, 1987, 4.69; Gygax, 1990, 54).

If a substance is able to deflagrate very fast, it is possible that the reaction front can accelerate to detonation velocities. This is called a *deflagration-to-detonation transition*, or DDT. If necessary, this phenomenon can also be investigated by special tests (CCPS/AIChE, 1995).

Mechanical Sensitivity Testing

It is important to know how sensitive a particular substance is to mechanical stimuli, if the substance has exhibited some tendency towards decomposition. In practice, a substance can be wedged between surfaces such as a lid and packaging or between flanges. The maximum temperature increase due to friction between non-melting surfaces can be approximately 2000°C. This may lead to initiation of a sensitive substance. For substances that are sensitive to impact, dropping may also lead to initiation.

Mechanical sensitivity is divided into sensitivity to mechanical shock, also called sensitivity to impact, and to friction. One large chemical company uses impact testing by a drop weight test as its initial screening test for shock sensitivity, then proceeds to additional testing such as confinement cap sensitivity and adiabatic compression tests if the drop weight test is positive (Kohlbrand, 1985). Equipment that is available to investigate these properties is summarized in CCPS/AIChE (1995).

The determination of mechanical shock (impact) sensitivity is applicable to both solids and liquids. The principle of the test is that a drop weight falls from a certain height onto the confined sample. The load can be varied by changing the drop height and mass of the drop weight. Observations are made concerning sample decomposition or explosion (e.g., report, smoke and/or fire), cup deformation and possible gas generation. Examples of impact apparatus are the Bundes Anstalt für Materialprufung (BAM) fallhammer, the Rotter test, the 30-kg fallhammer test and the Bureau of Mines impact apparatus (Mason and Aiken, 1972). The latter apparatus is illustrated in Figure 3.7.

Determination of friction sensitivity is applicable to solids, pastes and gel-type substances. To determine the friction sensitivity, a thin sample is placed under a load between two roughened surfaces, and the surfaces are rubbed together. The load can be varied. Results such as smoke, cracking or discoloration are observed. Examples of friction apparatus are the BAM friction apparatus, the Rotary friction test and the Allegany Ballistics Laboratory (ABL) friction test.

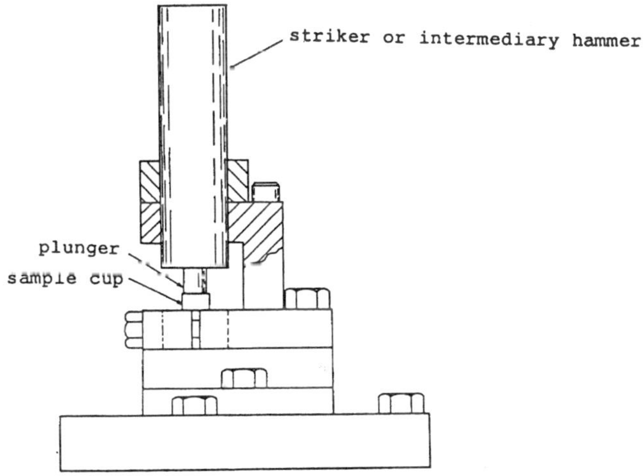

FIGURE 3.7. **The Bureau of Mines Impact Apparatus (lower portion)** (Yoshida, 1987).

Assessment of the Results

From the mechanical sensitivity tests, two types of results are obtained, namely no reaction or decomposition with or without explosion. The magnitude of friction and impact sensitivity is reported as the smallest load at which a positive result was witnessed.

The aim of mechanical sensitivity testing is to establish whether or not the substance is sensitive under normal handling conditions. However, the test results may not reflect the operating conditions because most testing is done at ambient temperatures and pressures. Since results are dependent on the type of test apparatus, the interpretation of the results requires much experience in this field of testing.

3.4.3. Pyrophoricity Screening

Previous experience with the handling of most industrial substances will indicate whether the substance is pyrophoric or not. Nevertheless, rudimentary tests have been described to screen whether a material is pyrophoric (EEC, undated; UN, 1991). The U.S. DOT test for pyrophoricity is given in Appendix E to 49 CFR Part 173 ("Materials Liable to Spontaneous Combustion").

In principle, 1 to 2 cm^3 of powder or 5 cm^3 of liquid is poured from a height of about 1 meter onto a noncombustible surface, while visual observation is made whether the substance ignites during dropping (powders) or within five minutes. The test is repeated six times. Furthermore, with liquids, 0.5 cm^3 is delivered from a syringe to dry filter paper and observations are made to see if ignition or charring occurs on the filter paper within five minutes (UN, 1991). This test is repeated three times unless a positive reaction is observed. If a substance ignites in one of the tests, or if a liquid chars the filter paper, the substance is considered to be pyrophoric.

3.4.4. Water Reactivity Screening

As with pyrophoricity, previous experience will usually indicate whether a substance is water-reactive or not. Nevertheless, water reactivity screening may be desirable in some situations.

Mason and Cooper (1972) proposed an apparatus for water reactivity screening that measures both temperature rise and rate of temperature rise. A schematic of the apparatus is shown in Figure 3.8 (from Stull, 1977, 95). The thermocouple has a small square of copper sheet soldered to it, for measurement of the sample temperature as a function of time. Water is added from a calibrated syringe, and the maximum temperature and time to maximum temperature are recorded. The time to maximum temperature can range from nearly instantaneous to several minutes. Visual observation is made as to whether spontaneous ignition occurs. In addition, since the off-gases from the reaction with water can be flammable and/or toxic, the off-gases can be collected and analyzed.

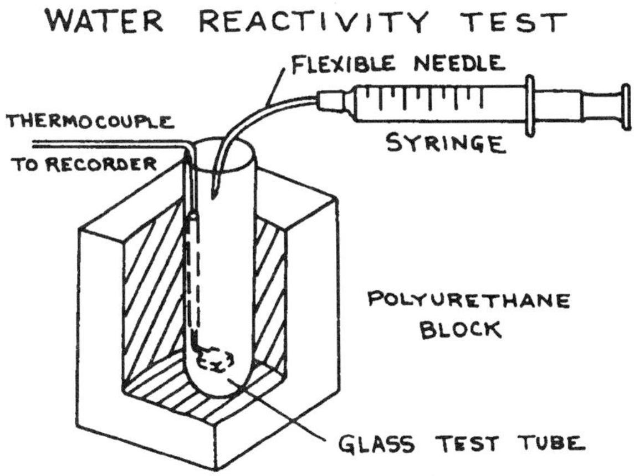

FIGURE 3.8. **Water Reactivity Test Apparatus.**

A more recent arrangement, the "2-drop" mixing calorimeter described by Hofelich et al. (1994), was developed for screening incompatibility hazards. It is expected that this arrangement could also be used for water reactivity screening.

If prior experience and/or the above or similar test indicate no immediate or violent reaction with water, further water reactivity screening can be performed by thermal analysis by mixing the substance with one or more proportions of water and observing whether an exotherm results upon heating of the mixtures. For example, ARC runs might be made at 10:1 and 1:10 ratios of water to the substance being tested.

3.4.5. Peroxide Formation Screening

No tests are known to exist to directly screen for the ability of substances to peroxidize. However, tests can be performed to detect the presence of peroxides that have already been formed by a peroxidizable substance. These tests may be especially warranted if the chemical structure of the substance (if known) contains one or more features known to be susceptible to peroxide formation (Section 1.5.5; see also Section 2.4).

A visual check may detect the presence of crystalline solids or a viscous liquid layer in or around a storage container of material suspected as being a peroxide former. If such a situation exists, extreme caution should be exercised in case the material is a shock- or friction-sensitive peroxide.

At least two kinds of peroxide test strips are commercially available (Wray, 1992) that are easy to use and can be suitable for peroxide screening. These peroxide test strips must be kept dry in order to give valid results.

Jackson et al. (1970) summarize two iodide test methods and one ferrous thiocyanate test method for detecting peroxides in relatively simple chemicals. More complex organic substances may act as oxidizing agents and give false positive test results for these peroxide screening tests.

Specific tests for detecting peroxides in ethyl ether, tetrahydrofuran and dioxane are summarized in Steere (1971, 251). Johnson and Siddiqi (1970) and Mair and Hall (1971) give additional methods for detecting organic peroxides.

Chemical tests may have limitations as screening tests for peroxide formers. For example, Davies (1961) indicates that dialkyl peroxides can only be detected after they have been converted to hydroperoxides with a strong acid. For certain types of peroxides, no suitable chemical detection tests may be available. However, in addition to chemical tests, organic peroxides may be able to be detected by such methods as chromatography, ion exchange, polarography and absorption spectra (Davies, 1961).

3.4.6. Compatibility Screening

Compatibility screening is necessary to determine if the combination of substances that may not be significantly hazardous in themselves can produce a hazardous result. These substances may include not only the material(s) to be stored and handled, but also any or all of the following:

- environmental or ubiquitous substances (air; water/humidity; any environmental contaminants present in significant concentrations in the actual storage/handling location)
- utilities (steam, heat transfer fluids, nitrogen, etc.)
- materials of construction and gasket materials, including any having a reasonable likelihood of being substituted (intentionally or otherwise) for the design materials sometime during the life of the facility
- contaminants (rust, scale, lubricating oil, impurities, etc.)
- any other chemicals that may be introduced into the operation inadvertently, such as other raw materials that are unloaded at the same or adjacent truck unloading station or other materials handled in similar drums.

A chemical interaction matrix, as described in Section 4.1, can be used to identify all binary combinations that may occur in a given operation. Many of the combinations can be addressed by prior experience or by using compatibility databases such as those listed in Section 1.8. Use of a computer program such as CHETAH (Section 3.2.6) may also prove useful in predicting reaction potential. The combinations where the consequences are unknown may require compatibility screening tests.

Screening for reactivity with air and water have already been addressed under pyrophoricity screening (Section 3.4.3) and water reactivity screening

(Section 3.4.4), respectively. Reactivity screening for other combinations can be performed in a three-step procedure:

1. In the laboratory, mix small quantities of the material combinations in question to check for immediate ignition, fast reaction and/or significant gas evolution, taking all necessary safety precautions. Make note of all observed effects, including temperature rise. This can be done, for example, in a "2-drop" calorimeter as described by Hofelich et al. (1994).
2. For those combinations not exhibiting immediate ignition or fast reaction, test samples of the mixtures by thermal analysis such as DTA/DSC or ARC, with the tests covering an appropriate range of proportions or concentrations.
3. The substance in question and the combinations being tested can also have tests run with potential contaminants, such as rust, added to the mixture to see whether any catalytic effects are observed.

Testing for compatibility with solids such as materials of construction or gaskets is complicated by such factors as the form of the solid material (the more finely divided the material, the greater the active surface area exposed for reaction) and the degree of passivation. Compatibility testing may only be able to identify problems associated with gross contamination or obvious reactions that occur in a short time frame.

Bibliography

Bretherick, L., *Bretherick's Handbook of Reactive Chemical Hazards,* 4th ed., Butterworths, London, 1990.

Center for Chemical Process Safety, *Guidelines for Chemical Reactivity Evaluation and Application to Process Design,* American Institute of Chemical Engineers, New York, 1995.

Merck & Co., *The Merck Index,* 11th ed., Merck & Company, Rahway, NJ, 1989.

Sax, N. I. and R. J. Lewis, *Dangerous Properties of Industrial Materials* (3 volumes), 7th ed., Van Nostrand Reinhold, New York, 1989.

U.S. Department of Transportation, "CHRIS Hazardous Chemical Data," U.S. Coast Guard, MI6465-12, U.S. Government Printing Office, Washington, D.C., 1978.

See also Appendix A *for Reactive Chemicals Literature Sources.*

References

CCPS/AIChE (1995). *Guidelines for Chemical Reactivity Evaluation and Application to Process Design.* American Institute of Chemical Engineers, New York.

Benson, S. W. (1969). *Chem. Review* 69, 279.

Benson, S. W. (1982). *The Foundation of Chemical Kinetics.* Robert E. Krieger Publishing Co., Malabar, Florida.

Bretherick, L. (1986). *Reactive Chemical Hazards.* Butterworths, London.

Bretherick, L. (1987). "Reactive Chemical Hazards: An Overview." *Proced. Intl. Symp. on Preventing Major Chem. Accidents.* Center for Chemical Process Safety/American Institute of Chemical Engineers, New York.

Bretherick, L. (1990). *Bretherick's Handbook of Reactive Chemical Hazards,* 4th ed. Butterworths, London.

Craven, A. D. (1987). "A Simple Method of Estimating Exothermicity by Average Bond Energy Summation," in "Hazards from Pressure: Exothermic Reactions, Unstable Substances, Pressure Relief and Accidental Discharge." *Inst. Chem. Eng. Symp. Ser. 102,* 97.

Cruise, D. R. (1964). *J. Phys. Chem. 68,* 3794-3802.

Davies, A. G. (1961). *Organic Peroxides.* Butterworths, London.

Davies, C. A., Freedman, E., Frurip, D. J., Hertel, G. R., Seaton, W. H., and Treweek, D. N. (1990). "CHETAH Version 4.4: The ASTM Chemical Thermodynamic and Energy Release Evaluation Program (2nd Edition)." ASTM Data Series Publication DS 51A. American Society of Testing and Materials, Philadelphia, Pennsylvania.

Davies, C. A., Kipnis, I. M., Chase, M. W., and Treweek, D. N. (1985). "The Thermochemical and Hazard Data of Chemicals, Estimation Using the ASTM CHETAH Program." ACS Symp. Ser. No. 274, Chemical Process Hazard Review, J. M. Hoffman and D. C. Maser, eds. American Chemical Society, Washington, D.C.

DIPPR (1993). Design Institute for Physical Property Data chemical database. Available from American Institute of Chemical Engineers, New York.

Domalski, E. S. (1972). *J. Phys. Chem. Ref. Data 1,* 221.

DOT (1978). "CHRIS Hazardous Chemical Data." U.S. Department of Transportation, U.S. Coast Guard, MI6465-12. U.S. Government Printing Office, Washington, D.C.

Duswalt, A. A. (1974). "The Practice of Obtaining Kinetic Data by Differential Scanning Calorimetry." *Thermochim. Acta 8,* 56.

EEC (undated). E.E.C. Directive 79-831, Annex V (84/449/EEC), Sixth Amendment to Dangerous Substances Directive. European Economic Community, Brussels.

Fenlon, W. J. (1984). "A Comparison of ARC and Other Thermal Stability Test Methods." *Plant/Oper. Prog. 3*(4, October), 197.

Fierz, H., Zwahlen, G. (1989). "Deflagration in Solids-A Special Hazard in Chemical Production." *6th Loss Prev. and Safety Prom. in Process Ind.,* Oslo.

Frurip, D. J. (1992). "Using the ASTM CHETAH Program in Chemical Process Hazard Evaluation." *Plant/Oper. Prog. 11*(4, October), 224.

Grewer, T. (1988). *Exothermic Decomposition: Investigation of the Characteristic Properties* (in German). VDI Verlag, Dusseldorf, Germany.

Grewer, T., Klusacek, H., Loffler, U., Rogers, R. L., Steinbach, J. (1989). "Determinations of the Characteristic Values for Evaluation of the Thermal Safety of Chemical Processes." *J. Loss Prev. Proc. Ind. 2.*

Gygax, R. W. (1990). "Scaleup Principles for Assessing Thermal Runaway Risks." *Chem. Eng. Prog. 86*(2, February), 53-60.

Hofelich, T. C., Frurip, D. J., and Powers, J. B. (1994). "The Determination of Compatibility via Thermal Analysis and Mathematical Modeling." *Proc. Safety Prog. 13*(4, October), 227-233.

Hofelich, T. C. and Thomas, R. C. (1989). "The Use/Misuse of 100-Degree Rule in the Interpretation of Thermal Hazard Tests." *Proced. Intl. Symp. on Runaway Reactions.* American Institute of Chemical Engineers, New York.

Houghan, D. A., Watson, K. M, Ragatz, R. A. (1959). *Chemical Process Principles,* 2nd Ed., Part II: Thermodynamics. John Wiley, New York.

Jackson, H. L., McCormack, W. B., Rondesvedt, C. S., Smeltz, K. C., and Viele, I. E. (1970). "Control of Peroxidizable Compounds." *J. Chem. Ed.* 47(3), March.

Johnson, R. M., and Siddiqi, I. W. (1970). *The Determination of Organic Peroxides.* Pergamon Press, New York.

Kohlbrand, H. T. (1987). "The Relationship Between Theory and Testing in the Evaluation of Reactive Chemical Hazards." *Proced. Intl. Symp. on Prevention of Major Chemical Accidents.* American Institute of Chemical Engineers, New York.

Mackenzie, R. C. (1972). *Differential Thermal Analysis,* Part I and II. Academic Press, New York.

Mair, R. D., and Hall, R. T. (1971). *Organic Peroxides,* Vol. II. Wiley, New York.

Mason, C.M. and Aiken, E. G. (1972). "Methods of Evaluating Explosives and Hazardous Materials." U. S. Bureau of Mines IC 8541.

Mason, C.M. and Cooper, J. C. (1972). Report No. 4160 (March). Pittsburgh Mining and Safety Research Center, Pittsburgh, Pennsylvania.

McCloskey, C. M. (1989). "Safe Handling of Organic Peroxides: An Overview." *Plant/Oper. Prog. 8,* 185.

McNaughton, J. L. and Mortimer, C. T. (1975). IRS: Physical Chemistry Series 2, Vol. 10. Butterworths, London.

Merzhanov, A. G. (1969). *Combustion and Flame 13.*

Meyer, R. (1977). *Explosives.* Verlag Chemie, New York.

NFPA 325M (1994). *Fire Hazard Properties of Flammable Liquids, Gases, and Volatile Solids.* National Fire Protection Association, Quincy, Massachusetts.

NFPA 43B (1993). *Storage of Organic Peroxide Formulations.* National Fire Protection Association, Quincy, Massachusetts.

NFPA 49 (1994). *Hazardous Chemical Data.* National Fire Protection Association, Quincy, Massachusetts.

NFPA 491M (1991). *Manual of Hazardous Chemical Reactions.* National Fire Protection Association, Quincy, Massachusetts.

NFPA 704 (1990). *Recommended System for the Identification of the Fire Hazards of Materials.* National Fire Protection Association, Quincy, Massachusetts.

Pedley, J. B., Naylor, R. D., and Kirby, S. P. (1986). *Thermochemical Data of Organic Compounds,* 2nd ed. Chapman and Hall, London, 128.

Perry and Green, eds. (1984). *Perry's Chemical Engineers' Handbook,* 6th ed. McGraw-Hill, New York.

Sax, N. I. and Lewis, R. J. (1989) *Dangerous Properties of Industrial Materials* (3 volumes), 7th ed. Van Nostrand Reinhold, New York.

Seaton, W. H. (1989). "Group Contribution Method for Predicting the Potential of a Chemical Composition To Cause an Explosion." *J. Chem. Ed.* 66(5), A137-A140.

Seaton, W. H., Freedman, E., Treweek, D. N. (1974). "CHETAH: The ASTM Chemical Thermodynamic and Energy Release Potential Evaluation Program." American Society of Testing and Materials, Philadelphia, Pennsylvania.

Shanley, E. S., and Melhem, G. A. (1995) "The Oxygen Balance Criterion for Thermal Hazards Assessment." *Proc. Safety Prog.* 14(1, January), 29-31.

Skvara, F. and Satava, V. (1970). "Kinetic Data from DTA Measurements." *J. Thermal Analysis 2,* 325.

Steere, N. V., ed. (1971). *CRC Handbook of Laboratory Safety,* 2nd Ed. CRC Press, Boca Raton, Fla.

Stull, D. R. (1970). "Identifying Chemical Reaction Hazards." *Loss Prev. 2.* Chemical Engineering Progress Technical Manual, AIChE, New York.

Stull, D. R. (1977). *Fundamentals of Fire and Explosion.* AIChE Monograph Series No. 10, Vol. 73. AIChE, New York.

Stull, D. R., Westrum, E. F., Jr., and Sinke, G. C. (1969). *The Chemical Thermodynamics of Organic Compounds.* John Wiley, New York.

UN (1991). "Recommendations on the Transport of Dangerous Goods" ST/SG,AC.10/1 Rev. 7. United Nations, New York.

Weast, R. C., ed. (1985). *Handbook of Chemistry and Physics,* 66th Ed. CRC Press, Boca Raton, Fla.

Wendtlandt, W. (1974). *Thermal Methods of Analysis,* 2nd Ed. John Wiley, New York.

Wray, T. K. (1992). "Danger: Peroxidizable Chemicals." *Hazmat World* (November), 48.

Yoshida, T. (1987). *Safety of Reactive Chemicals.* Industrial Safety Series, Vol. 1. Elsevier, Amsterdam.

4

Consequence Analysis

Consequence analysis is the estimation of the injury and loss that could result if an accidental event occurred, such as a vessel rupture explosion due to overpressurization. Consequence analysis seeks to answer the question

If a reactive chemical incident occurs, how severe are the consequences likely to be?

The answer to this question has a direct bearing on what safeguards may be required to safely store and handle a given reactive material. Thus, consequence analysis is the bridge between the identification of reactivity hazards (Chapter 3) and the design and operation of safe storage and handling facilities (Chapters 5 through 7).

In the context of this book, consequence analysis generally consists of the following three activities in sequence:

1. Development of reactive chemical accident **scenarios**, including an estimation of the quantity or rate of hazardous material release or the magnitude or rate of energy release. **Severity testing** of the reactive materials of concern may be necessary in order to assess the magnitude or rate of energy release.
2. Estimation of the **extent** of toxic/corrosive effects, heat effects, explosion blast effects, and/or any other significant effects with the potential for injury or loss, as consequences of the identified accident scenarios.
3. Estimation of the potential **impact** of each accident scenario; i.e., the **severity** of consequences in terms of on-site and off-site effects such as injuries, environmental damage, property damage, and business interruption losses.

Consequence analyses can range in resolution from order-of-magnitude estimates to detailed, quantitative studies, depending on the information available and the decisions to be made. These guidelines emphasize methods that can be used to make an initial quantitative assessment of the severity of consequences.

4.1. Identifying Potential Accident Scenarios

The first step in consequence analysis is identifying and developing potential accident scenarios that are related to the storage and handling of the reactive chemical(s) of concern. These accident scenarios then form the basis for the remainder of the consequence analysis.

Included in this section are a few means by which reactive chemical accident scenarios can be identified: process hazard analysis, accidental event checklist, chemical interaction matrix, industry experience review, and local site experience review. These methods are not mutually exclusive; in some cases it may be decided to use two or more approaches, or an even simpler approach such as assuming that the maximum on-site inventory of a reactive chemical is suddenly discharged to the environment.

4.1.1. Process Hazard Analysis

The most comprehensive means of identifying chemical accident scenarios is by conducting a thorough process hazard analysis of the storage and handling operation, using techniques such as hazard and operability (HAZOP) studies. The full range of process hazard analysis methodologies is given in *Guidelines for Hazard Evaluation Procedures, Second Edition with Worked Examples* (CCPS/AIChE, 1992). However, it is generally not necessary to conduct a complete process hazard analysis to identify significant reactive chemical accident scenarios that can serve as the basis of design efforts and safety reviews. Indeed, the detailed process hazard analysis for a new design is generally conducted only after a final design is available, whereas essentially all of the design safety features for handling reactive chemicals should be included in the final design, thus necessitating the identifying of accident scenarios well before the process hazard analysis is conducted.

4.1.2. Checklist of Potentially Hazardous Events

Included in Figure 4.1 is a checklist of accidental events that can occur upon release of reactive material or energy to the surroundings of a storage and handling operation. A scenario can be developed by (1) identifying which accidental events on the checklist could occur in a given facility if something goes wrong and all safeguards fail; (2) identifying one or more credible causes that could lead to each accidental event, such as equipment failures, operator errors, or external events such as a moving vehicle impacting a pipeline; and (3) documenting the scenarios as the basis for the remainder of the consequence analysis.

Reactive Chemicals Accidental Events Checklist	
Accidental Event	*Examples*
EXPLOSIONS	
☐ **Vessel rupture explosion**	Unrelieved runaway polymerization reaction Unrelieved runaway decomposition reaction Unrelieved gas buildup Unrelieved internal deflagration
☐ **Confined-area explosion**	Reaction of pyrophoric or water-reactive material inside an enclosure
☐ **Condensed-phase explosion**	Detonation of shock-sensitive high explosive Decomposition of organic peroxide initiated by shock, friction, or heat
FIRES	
☐ **Pool or pile fire**	White phosphorus or chlorosulfonic acid spill
☐ **Fireball**	Rupture of volatile pyrophoric liquid tank with immediate ignition
☐ **Flash fire**	Release, and delayed ignition of pyrophoric liquid or solid
☐ **Jet fire**	Pressurized leak of pyrophoric gas to the atmosphere
UNCONTROLLED REACTIONS	
☐ **Violent reaction**	Inadvertent mixing of incompatible materials
☐ **Oxygen depletion**	Reaction with oxygen in confined space
ACCIDENTAL RELEASES	
☐ **Toxic vapor cloud generation**	[not included in the scope of this book]
☐ **Toxic/corrosive/flammable liquid spill**	[not included in the scope of this book]
☐ **Obscuring cloud generation**	Titanium tetrachloride release

FIGURE 4.1. **Checklist of possible Reactive Chemical Accident Events.**

Explosion Types

The types of explosions that may occur depend on the confinement of the reactive material, its energy content, its kinetic parameters, the mode of ignition (self-heating or induced by external energy input), etc. (CCPS/AIChE, 1995). Explosions are characterized as physical or chemical explosions. Physical explosions such as pressure vessel failures due to physical overpressurization and boiling-liquid–expanding-vapor explosions (BLEVEs) are not caused by chemical reaction.

Chemical explosions are characterized as deflagrations, detonations and thermal explosions. Deflagrations and detonations are discussed in Section 3.4.2. Both of these types of explosions are often called heterogeneous explosions because of the existence of a reaction front that completely separates reacted and unreacted material. By contrast, in a thermal explosion, no reaction front is present and it is, therefore, called a homogeneous explosion. Initially, in a thermal explosion, no reaction front is present and the material has a uniform temperature distribution. Then a runaway reaction occurs throughout the material as a result of heat accumulation because of insufficient transfer of the heat generated by the reaction. The runaway reaction can lead to overpressurization and possible explosive rupture of the vessel.

Explosion phenomena have occurred in all types of confined and unconfined units—separation and storage units, filter systems, pipelines, etc. Typical reactions that may cause explosions are oxidations, decompositions, nitrations and polymerizations. Examples of chemicals and processing system characteristics that increase the potential for an explosion are the following (CCPS/AIChE, 1995):

- high decomposition or reaction energies
- high rates of energy generation
- insufficient heat removal, such as with excess quantities of substances
- the presence of an initiation source, such as a "hot spot" on the vessel or pipe surface
- substances with an oxygen balance close to zero (see Section 3.2.2)
- confinement
- production of gases.

4.1.3. Chemical Interaction Matrix

Chemical hazards can be identified by examining the characteristics of each chemical in a process, one at a time. Information on the hazardous, or reactive, properties of chemicals can usually be found on material safety data sheets (MSDSs) or other common hazardous chemical data references. However, examination of the individual chemicals handled in a process may not identify all important process hazards, since many hazards are related to interactions of process chemicals, either inadvertent or intentional, with each other and with their surroundings. Thus, a complete process hazards analysis will often need to supplement a review of the chemical safety data and the process parameters with a means of systematically examining possible chemical interactions.

Developing a *chemical interaction matrix* is one effective means of finding the potential interactions in an operation that may lead to a fire, explosion, or hazardous material release. Dangerous interactions can then be assessed as part of process hazard analyses, included in emergency response plans, and incorporated into employee awareness programs.

This section describes an approach to generating chemical interaction matrices. This approach is similar to the method used by one large U.S. chemical company, as summarized by Gay and Leggett (1993).

What to Consider in a Chemical Interaction Matrix

Chemical interactions and their consequences can be systematically studied for a storage and handling operation by considering the following in a matrix format:

- all stored/handled chemicals, which may include raw materials, intermediates, products, by-products, and catalysts
- any other chemicals that may be introduced into the operation inadvertently, such as other raw materials that are unloaded at the same truck unloading station serving a storage operation, or piping containing other materials that is tied into the same transfer system
- all utilities (steam, compressed air, nitrogen, natural gas, heat transfer media, refrigerants, service water, etc.) that could potentially interact with the operation
- common environmental substances: air, water/humidity, and any environmental contaminants present in significant concentrations in the actual storage/handling location
- likely process contaminants such as dirt, rust, scale, and lubricating oil
- all materials of construction and gasket materials used in the process, including those having a reasonable likelihood of being substituted (intentionally or otherwise) some time during the life of the facility
- other materials that may contact process chemicals, such as absorbents and insulation
- all operating conditions that pertain to the given facility, such as elevated temperature or reduced pressure
- other sources of energy in or around the process, such as static electricity or kinetic energy (e.g., associated with pneumatic transfer)
- in some situations, conditions such as "confinement" and "adiabatic compression" may be pertinent.

A matrix that would include all of the above items for a given process can be quite large. If it is necessary to restrict the effort involved in developing the interaction matrix, judgment can be exercised in limiting the scope of the study or including only those substances and conditions that have a reasonable likelihood of being present and causing reactivity concerns.

Setting Up a Chemical Interaction Matrix

To conduct a chemical interaction study, a matrix is set up that has each of the above items listed along both the horizontal and vertical axes. The cells are then filled in, either above or below the diagonal running from top left to bottom right, with the consequences expected if each interaction occurs. An example is given in Figure 4.2.

The cells along the diagonal are the intersections of each chemical with itself. These cells can be used to document any self-reactivity potential, such as a propensity to decompose at elevated temperatures or to polymerize.

C_4H_6 1,3-butadiene	flammable; peroxidizes; polymerizes; decomposes						
Cl_2 chlorine	fire, toxic gas generation	oxidizer; toxic vapor; cryogenic liquid spill					
HF anhydrous hydrogen fluoride	heat generation, violent polymerization	heat generation, liberating toxic vapors	strong acid; corrosive; toxic vapor and liquid				
NH_3 anhydrous ammonia	heat generation, violent polymerization	explosive NCl_3 formed with excess chlorine or heat	heat generation, liberating toxic vapors	combustible; toxic vapor; cryogenic liquid spill			
Fe, etc. carbon steel	none predicted	iron/chlorine fire if above 250°C (or 100°C with impurities)	hydrogen blistering between steel laminations	none predicted	material of construction		
H_2O 150 psig steam	antioxidant consumed, leading to polymerization	none predicted	heat generation, liberating toxic vapors	heat generation, liberating toxic vapors	none predicted	elevated pressure, temperature	
combined with...	**C_4H_6 1,3-butadiene**	**Cl_2 chlorine**	**HF anhydrous hydrogen fluoride**	**NH_3 anhydrous ammonia**	**Fe, etc. carbon steel**	**H_2O 150 psig steam**	

Descriptions along diagonal are properties of materials by themselves.

FIGURE 4.2. **Example Interaction Matrix for Identifying Scenarios.**

Using the Chemical Interaction Matrix to Identify Scenarios

Once the interaction matrix is complete, it should be examined for severe consequences such as violent reactions, generation of toxic gases, or significant fire hazards, and particularly for interactions that were previously not recognized as having hazardous consequences. These interactions should then be studied, by a team of knowledgeable persons, to develop accident scenarios by determining what could cause each hazardous chemical interaction, where and when each interaction might occur, and what safeguards exist to prevent the occurrence of the interaction and/or deal with the consequences of the interaction. This can be accomplished as part of a process hazard analysis. Any matrix cells with missing data or unknown consequences will indicate where research or testing may be required.

Managing the Chemical Interaction Data

A chemical interaction matrix for relatively simple storage/handling operations may fit in a one-page table or on several pages that fit together. For larger or more complex operations with many chemicals and other matrix items, three formats that have been used to capture and present the chemical interactions are (a) database programs, (b) spreadsheets, and (c) word processing programs having table-generating features. Each has its advantages:

- Database programs are most useful for very large matrices, where the power of the database program to search and retrieve a given combination is needed.
- Spreadsheets have the advantage of being able to put the entire matrix in a true matrix form, thus making it easier to see which intersections still need to be filled.
- The tables format allows the matrix to be directly incorporated into a report, and enlarges cells automatically to accommodate multiple lines of text.

Some chemical interaction studies have found it worthwhile to have a dedicated program written to access the data, such as the REACT program referenced below. Gay and Leggett (1993) describe a computerized approach to storing interaction information and printing compatibility charts; this approach includes a mixing hazard rating from 0 to 4 that parallels the NFPA 704 ratings for health, flammability, and reactivity (NFPA 704, 1990). This computer shell program, known as CHEMPAT, is available from AIChE and serves as an aid to organizations for establishing compatibility charts

Chemical Interaction Data Sources

Many chemical interactions are obvious, such as acid–base reactions resulting in heat and gas generation. Many other interactions will be known to have no significant consequences. However, there are usually many potential interactions for which the results are not immediately known. In addition to materials testing in a properly equipped laboratory, data can often be obtained from chemical suppliers or from many literature sources such as the following.

- *Bretherick's Handbook of Reactive Chemical Hazards* (Bretherick, 1990)—a compilation of reactivity and incompatibility hazards of 4600 different elements and compounds; an electronic version is also available from the publisher
- *Manual of Hazardous Chemical Reactions*, NFPA 491M (1991)—a listing of 3550 mixtures of two or more chemicals reported to have the potential to cause fires or explosions at ordinary or moderately elevated temperatures
- REACT Hazardous Chemical Reactivity code (Wertz, 1989)—a program for predicting the likely consequences of interactions for nearly 1300 chemicals using chemical reactivity groups, based on a U.S. Environmental Protection Agency report pertaining to the compatibility of hazardous waste streams (Hatayama et al., 1980)
- *Chemical Hazard Response Information System (CHRIS) Hazardous Chemical Data* (DOT, 1984)—reactivity group, water reactivity, and reactivity with common materials are given, as well as a bulk cargo compatibility matrix
- *Dangerous Properties of Industrial Materials* (Sax and Lewis, 1989)—3-volume publication with incompatibility information on numerous hazardous materials
- "Hazardous Materials Car Placement in a Train Consist" (Thompson et al., 1992)—2-volume report examining all binary combinations of the top 101 hazardous commodities by rail volume movement in the U.S., plus fuming nitric acid
- Organic and inorganic chemistry textbooks
- Experience and resources of company organic and/or inorganic chemists
- Literature search on the particular chemicals.

When the consequences of a given interaction are unknown, that fact should immediately raise a red flag, since unintentional chemical interactions must be identified and controlled for continued safe operation of chemical processes.

4.1.4. Industry Experience

Industry-wide experience is another good potential source of accident scenarios that may be used in a severity analysis. Since industry-wide experience for a given reactive material may be based on a large collection of operations, it has the prospect of pointing to realistic incident scenarios that may only have a small probability of occurrence at an individual location.

Industry experience can take a number of different forms. Loss-of-containment incidents can be reviewed by inquiring available databases such as the U.S. Environmental Protection Agency's ARIP (Accidental Release Information Program) database. Numerous case histories have been reported in the *Loss Prevention Bulletin* (Institution of Chemical Engineers, Rugby, England), *Plant/*

Operations Progress and *Process Safety Progress* periodicals, and the *Loss Prevention Symposium* series (American Institute of Chemical Engineers, New York). Other articles related to a particular reactive material might be found by a literature search. Bretherick (1990) summarizes incidents related to the handling or inadvertent mixing of a large number of reactive materials. Other possible sources of industry experience include manufacturers associations such as The Chlorine Institute (Washington, D.C.) and individual manufacturers or suppliers.

4.1.5. Local Site Experience

Experience with storing and handling a reactive material at any of a company's facilities can be used to further identify potential accident scenarios. The primary documentation to be reviewed in this regard is the facility's **incident reports**. Any incident reports relating to the specific material being handled should be reviewed, not only to identify scenarios that actually occurred in the past, but also to ask what could have happened if conditions were somewhat different or if a protection system would have failed. This implies that near misses should be reviewed, as well as incidents that actually caused injury and/or loss.

Where incident reports are lacking, or if a facility just recently began a documented incident investigation program, operations and maintenance personnel can be interviewed to glean from their experiences with the handling of the reactive material. The interviews should be held in a nonthreatening manner so that personnel will feel free to recall incidents and near misses without fear of recrimination or blame. The aim of the interviews and review of incident reports is to explore what could happen based on what has happened in the past, in order to develop realistic accident scenarios.

For example, an interview with personnel responsible for testing and inspection of relief valves on a series of reactive material storage tanks might uncover the fact that, on a couple of occasions, the inlet line to the relief valve was found to be partly blocked with polymer when the relief valve was removed for annual testing. The inlet line could have been cleaned and the relief valve returned to service each time without incident, and thus such an occurrence may not be reflected in the facility's incident investigation reports. Many companies document and investigate such occurrences as safety incidents, even if an actual accidental event did not occur. However, it could point to a scenario where a storage tank was involved in an external fire or other cause of pressure buildup, with blockage of the relief valve inlet resulting in an unrelieved overpressurization of the storage tank and a vessel rupture explosion.

4.2. Severity Testing

In order to adequately define possible accidental event scenarios associated with the storage and handling of reactive chemicals, it may be necessary to conduct severity testing on the hazardous materials or combinations of materials involved in the scenarios. Typical data that may be needed to define the accident scenarios are:

- onset temperature of runaway reaction
- time to onset, such as if an inhibitor is present
- reaction kinetics; temperature increase as a function of time once onset temperature is reached
- heat of decomposition or heat of reaction
- maximum temperature rise
- pressure data; gas generation rate.

These data may be needed for one or more of several purposes in the consequence analysis:

- to see whether a given accident scenario is feasible in the actual process arrangement
- to determine the expected progression of the potential accidental event scenario
- to determine what safeguards are expected to be effective towards reducing the likelihood of the accidental event (e.g., whether an operator has time to respond to a given alarm and bring the system back under control before a runaway reaction starts)
- to be used in determining the extent of the accidental event consequences; i.e., the area expected to be affected if the accidental event were to occur.

Test methods that can be used to obtain the needed severity testing data are reviewed in this section. Specialized tests may also need to be conducted to answer specific questions raised during the development of the reactive chemical accident scenarios.

4.2.1. Calorimetric Testing for Consequence Analysis

Most of the needed data for the consequence analysis that is not already known from the literature, theoretical evaluations, or prior testing can be obtained from one or more calorimetric tests performed in commercially available apparatus. It should be noted that many companies determine the needed severity testing data at the same time that screening tests (as described in Chapter 3) are performed; this is particularly true when ARC testing is done.

Table 4.1 gives an overview of several typical calorimetric techniques that can be used to obtain consequence analysis data. The "Principal Application" column shows only the major applications of the respective calorimetric techniques. Each of the test methods in Table 4.1 is described in detail in *Guidelines for Chemical Reactivity Evaluation and Application to Process Design* (CCPS/AIChE, 1995).

TABLE 4.1
Overview of Calorimetric Tests for Accident Scenario Data (CCPS/AIChE, 1995)

Test Method	*Typical Sample Mass (g)*	*Typical Sensitivity (W/kg)*	*Thermal Inertia Phi-Factor*	*Principal Applications*	*Data Acquired*	*Advantages/ Disadvantages*
DSC/ DTA	0.005 to 0.010	1 to 20	—	Screening	T_o; ΔH_d; reaction kinetics (isothermal mode only)	*Advantages:* quick; small sample quantity required; wide temperature range covered. *Disadvantages:* insensitivity (to T_o); hard to obtain representative sample.
ARC	2 to 10	0.5	1.5 to 4.5	Thermal stability; study autocatalysis, contaminations, inhibitor depletion; complete runaway behavior and simultaneous pressure measurement; time to maximum rate of reaction	T_o; ΔH_d; reaction kinetics; pressure data	*Advantages:* quick; small sample quantity required; wide temperature range covered; sensitivity (to T_o). *Disadvantage:* medium to high Phi-factor.
Isothermal storage tests, TAM	5 to 20	0.0001 to 0.005		Very sensitive thermal stability; study autocatalysis, contaminations, inhibitor depletion; time to maximum rate of reaction	T_o; ΔH_d; reaction kinetics; pressure data when performed in closed dewars or autoclaves	*Advantages*: sensitivity (to T_o); accurate overall kinetics. *Disadvantages:* time-consuming; restricted temperature range; more test runs required.
Isoperibolic calorimetry	5 to 30	0.5	3.5	Thermal stability; study autocatalysis, contaminations, inhibitor depletion	T_o; ΔH_d; reaction kinetics; pressure data when performed in closed dewars or autoclaves	*Advantage:* accurate overall kinetics. *Disadvantages:* time-consuming; restricted temperature range; medium to high Phi-factor.
Dewar tests, adiabatic storage tests	200 to 700	0.015	1.05 to 3.5	Sensitive thermal stability; study autocatalysis, contaminations, inhibitor depletion; reaction due to oxidation; runaway behavior (initial phase); time to maximum rate of reaction	T_o; reaction kinetics; pressure data when performed in closed dewars or autoclaves	*Advantages:* sensitivity (to T_o); low Phi-factor; accurate overall kinetics. *Disadvantages:* time-consuming; large quantities required; restricted temperature range.

Caution: Small-scale tests are useful for screening purposes. However, the application of small-scale test results to actual process-scale and equipment arrangements can lead to significant error. Engineering principles must be used to scale up the laboratory data to plant equipment. For example, the exotherm onset temperature T_O determined from DSC testing will generally be significantly higher than the actual T_O for a full-scale batch reactor. Induction time may also be quite inaccurate. More fundamental thermodynamic values such as the heat of reaction or heat of decomposition should be minimally affected by scale-up. Screening test limitations were discussed in Section 3.4.1.

4.2.2. Self-Accelerating Decomposition Temperature

An important measurement of the storage stability for potentially unstable materials is the temperature at which an uncontrolled decomposition reaction can be initiated under actual storage configurations. The following basic description of the self-accelerating decomposition temperature (SADT) is taken from NFPA 49 (1994):

> Certain compounds, such as organic peroxides and [some] swimming pool chemicals, when held at moderate ambient temperatures for an extended period of time, may undergo an exothermic reaction that accelerates with increase in temperature. If the heat liberated by this reaction is not lost to the environment, the bulk material increases in temperature, which leads to an increase in the rate of decomposition. Unchecked, the temperature grows exponentially to a point at which the decomposition cannot be stopped or slowed. The minimum temperature at which this exponential growth occurs in a material packed in its largest standard shipping container is defined as the self-accelerating decomposition temperature. Self-accelerating decomposition temperature is a measure of the ease in which decomposition occurs under normal storage conditions. It is not an indicator of the violence of any decomposition reaction under conditions of fire exposure or contact with incompatible materials.

Test apparatus for measuring SADT, as well as a method for determining SADT from accelerating rate calorimetry results, are presented in Fisher and Goetz (1991). Whitmore and Wilberforce (1993) discuss the use of a Swedish thermal activity monitor (TAM) device in conjunction with ARC data to test the ARC data extrapolations that are necessary to obtain SADT values from ARC testing. The measurement of SADT for self-reactive materials is further discussed in an article by Fisher and Goetz (1993), with the conclusion that pressure effects, material viscosity, and reactant depletion need to be considered when determining the SADT for a reactive material.

4.2.3. Isoperibolic Calorimetry

Isoperibolic (constant jacket temperature) measuring techniques are designed to investigate the thermal behavior of substances and reaction mixtures under process conditions. The detected onset temperatures obtained from isoperibolic calorimetry are generally similar to those found in the ARC and significantly lower than DSC.

An isoperibolic apparatus consists of a sample container that is placed in a fan-assisted oven or heater. Typically, the sample container is a small tube or beaker that contains a 5-g to 30-g sample. The temperature range of a commercial instrument is about 0 to 300°C. Some instruments include miniature autoclaves or small stirred vessels; the pressure range of a test-tube autoclave is about 0 to 1500 psig (0 to 10 MPa). Some test setups will provide gas flow through the sample.

In an isoperibolic experiment, the jacket temperature of the sample container (or the surroundings of the container, i.e., the oven temperature) is held constant. On attaining a steady state, the temperature difference between the sample or reaction mixture and the jacket is measured. The temperature difference will be zero (within the detection limit of the equipment) if no energy is released from the sample and positive if energy is released due to chemical reaction. If no temperature difference is recorded after a fixed time interval, the oven temperature is increased (typically an increase of 5198C) and held constant once again. This procedure is repeated until an exothermic event is observed.

For the identification of the onset temperature of the exotherm, the steady state temperature difference may be plotted against the sample temperature. A typical plot of isoperibolic temperature T_I and temperature difference ΔT against time is illustrated in Figure 4.3. After calibration, the evolved heat can be estimated.

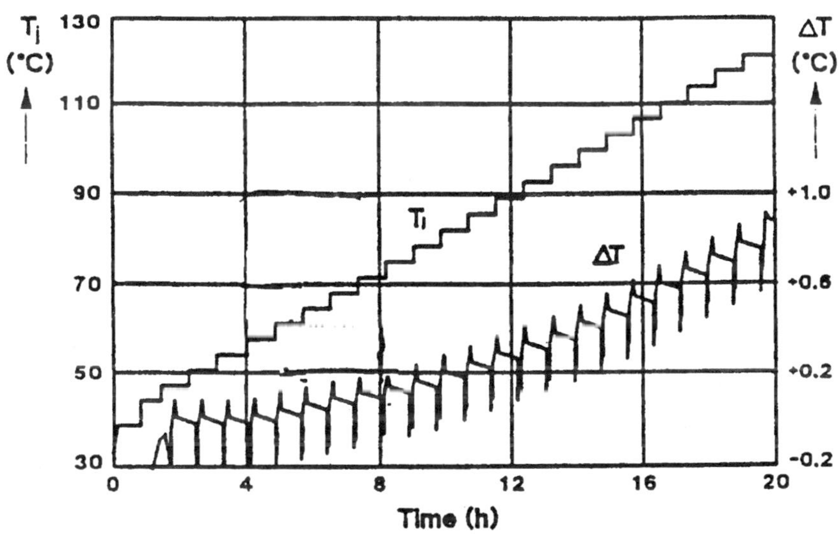

FIGURE 4.3. **Isoperibolic Measurement.**

A number of isoperibolic instruments have been built in-house, basically consisting of a glass tube or Carius tube placed in an oven. SEDEX (Hakl, 1984) is an example of a commercially available instrument, which will also run in isothermal and scanning modes.

4.2.4. Assessment of Maximum Pressure and Temperature

Where the potential exists for an uncontrolled reaction, the maximum possible pressure and temperature that the reaction can generate is one good measure of the magnitude of the hazard associated with the reaction. Even though emergency relief systems are employed, or heat may be removed from the system under normal storage and handling conditions, the maximum pressure and temperature rise attainable by a complete reaction is useful for developing scenarios where emergency relief or external heat transfer may be inadequate. The maximum pressure can be obtained by nearly all of the test methods listed in Table 4.1 if the apparatus is properly configured and equipped to measure pressure. Adiabatic devices can also give maximum temperature data.

Calculational methods can be used either with, or instead of, testing to estimate maximum pressure and temperature. If adiabaticity (no significant heat transfer out of the system) is assumed, the maximum pressure and temperature can be calculated if the reactants and products are known or can be predicted from a minimum Gibbs free energy approach (Stull, 1977, 16–17). Since heat capacity is temperature-dependent, a trial-and-error approach is employed to determine the maximum temperature. First, the heat of reaction is calculated (if not already known experimentally) from the difference between the heats of formation of the products and of the reactants. This is compared to the enthalpy of the products from 250°C to the estimated final temperature; the final temperature is correct when the heat of reaction equals the enthalpy of the products to this final temperature. The maximum pressure can then be calculated by multiplying the initial absolute pressure by the ratio of final to initial moles of gases in the closed system and by the ratio of the final to the initial absolute temperature (assuming ideal gas behavior). Several illustrations of these calculations are given in Stull (1977).

4.3. Where to Find Methods for Estimating Immediate Consequences

The next step in a full consequence analysis for the storage and handling of reactive chemicals is to determine how far the dangerous effects of a reactive chemical accident could extend. Detailed calculational methods are beyond the scope of this book; however, the following sections indicate how to get started and where to find descriptions of the more detailed approaches, if needed.

Several different kinds of accidental event consequences might be recognized as part of the identification of hazards and event scenarios described earlier in this chapter. The most commonly encountered accidental events in the Figure 4.1 checklist are discussed in turn.

4.3.1. Reactive Chemical Explosions

As Figure 4.1 indicates, several kinds of explosions may involve reactive chemicals. If an uncontrolled reaction generates sufficient overpressure inside a storage tank or other vessel, piping, or containment to reach the burst pressure of the containment, then a vessel rupture explosion will be the result. Similarly, an accidental heat- and gas-generating reaction inside a building, room, gearbox, or other structural enclosure can result in a confined-area explosion, such as the blowing out of the walls of a room due to a rapid, unrelieved pressure buildup inside the room. If the rapid decomposition of a reactive material such as a high explosive or an organic peroxide is initiated, a condensed-phase explosion can result.

Pressure Vessel Explosion

The energy released when a overpressurized vessel ruptures is due to the sudden expansion of the compressed gases and vapors inside the vessel. This energy is converted primarily to a shock wave, as discussed below. Bodurtha (1980, 61) considers the fraction of this energy imparted to missiles to be insignificant for rupture of a gas-filled vessel; however, as discussed in CCPS/AIChE (1994, 223–239), between 20% and 50% of the available explosion energy may be transformed into kinetic energy of fragments and liquid or solid contents. If major vessel sections become projectiles, values as high as 60% of the energy release have been used for the fraction imparted to missiles (Saville, 1977).

The most important parameters in determining the possible extent of vessel rupture explosion effects are the compressed gas pressure P and volume V at the time of rupture. (Note that, for an overpressurized vessel, some bulging and expansion will occur prior to rupture at low to moderate rates of pressure rise, and the volume at the time of rupture can be significantly greater than the normal internal vessel volume.) The ratio of specific heats for the expanding gases also affects the energy release E, using the following equation that calculates E assuming isentropic expansion of the gas to atmospheric pressure (Bodurtha, 1980):

$$E = \frac{PV}{\gamma - 1}\left[1 - \left(\frac{P_S}{P}\right)^{(\gamma - 1)/\gamma}\right] \qquad (4.1)$$

This equation is in metric units as presented, with E in MJ and V in m^3. Both P and the pressure of the surrounding gas P_s are absolute pressures in MPa. (One MPa is equal to 145 psi.) Specific heats ratios can be found in references such as

Perry and Green (1984, 3–144). Note that the volume of liquid in the vessel is not included in this equation, due to the incompressibility of liquids relative to gases.

Dividing E by the heat of decomposition of TNT of about 4.19 MJ/kg gives a TNT equivalence in kg TNT that is useful in determining the extent of the expected blast effects using standard approaches documented by the U.S. Army (1990). CCPS/AIChE (1989, 103ff and 163ff) discusses the use of the TNT equivalence approach in determining the blast overpressure as a function of distance from the ruptured vessel.

Simplified consequence analysis programs include the above approach for determining the energy release from a vessel rupture explosion. An alternative approach, based on the assumption of isothermal expansion of ideal gases, is given by CCPS/AIChE (1989, 114). Crowl (1992) shows that an isothermal assumption results in a significant overprediction of the blast energy, and that the isentropic assumption of Equation (4-1) is not accurate at high pressures. He gives instead an approach based on the "batch availability," or the maximum amount of stored mechanical energy that can be converted into blast overpressure. Assuming an ideal gas stored at pressure P at the same temperature as its surroundings and converting into the same form as Equation (4-1):

$$E = PV\left[\ln\left(\frac{P}{P_s}\right)\right] \tag{4.2}$$

Both theory and experimental work related to defining the explosion energy are discussed in detail in CCPS/AIChE (1994, 185–199).

It should be noted that the follow-on effects of an explosion can be more devastating than the blast itself. For example, rupture of a storage tank containing a reactive material that is also flammable can result in a massive fire. The impacts (severity of consequences) from explosions and other consequences are discussed in Section 4.4.

Confined-Area Explosion

The energy release from a confined-area explosion can be evaluated essentially the same as for a vessel rupture explosion. The explosion cause, such as an uncontrolled reaction or deflagration, is likely to affect any persons inside the enclosure at the time of the incident. In addition, the initial structural damage and flying debris can pose a severe threat to people in the immediate area of the explosion. A strategy for evaluating confined-area explosions is given in CCPS/AIChE (1989, 136).

Condensed-Phase Explosion

For a liquid or solid material capable of explosive decomposition, such as many organic peroxides, the mass of the material involved in the decomposition reaction combined with the heat of decomposition of the material determines the energy release. This energy release is usually reported in equivalent pounds or kilograms of TNT. Determination of the heat of decomposition for a material was discussed in Chapter 3.

4.3.2. Reactive Chemical Fires

The most important parameters in determining the extent of fire consequences are the quantity of flammable or combustible material (or pool size for a pool fire), the heat of combustion of the material, and some measure of how fast the material will burn. For pool fires, the latter is the burning rate; for flash fires where a flammable gas or vapor is premixed with air before ignition, an empirical fireball duration is generally used. Other parameters that also enter into consequence analysis of fire hazards are flammable limits, flame luminosity, wind speed and direction, and atmospheric transmissivity. These parameters are used to determine the heat flux and duration around the fire location.

Calculation of the heat flux and duration will be dependent on the type of fire—pool or pile fire, jet fire, flash fire, or fireball (see Glossary for definitions). Description of techniques used for fire consequence calculations are given in sources such as CCPS/AIChE (1989, 139ff), Lees (1980), and Mudan (1984). A conservative upper bound on the received thermal radiation Q_x in kW/m^2 at a distance of x meters from a point source of thermal radiation Q_R in kW is

$$Q_x = \frac{Q_R}{4\pi x^2} \tag{4.3}$$

assuming an atmospheric transmissivity of 100%. For a pool fire, QR can be determined as a point-source quantity by multiplying the burning rate (kg/m^{2}s) by the pool area (m^2) and the heat of combustion of the material (kJ/kg), noting that 1 kW is equal to 1 kJ/s. For a jet fire, QR can be estimated assuming complete combustion by combining the leak rate in kg/s with the heat of combustion in kJ/kg. These calculations also conservatively assume that all of the heat of combustion is converted to thermal radiation.

Note that this point-source simplification is only valid at distances far from the flame. At closer distances, more rigorous approaches should be used that consider such parameters as flame tilt and view factors (e.g., CCPS/AIChE, 1989, 139–147; CCPS/AIChE, 1994, 176–180).

4.3.3. Toxic Releases

The area likely to be affected by a toxic gas or vapor release is most often determined by dispersion modeling. Before the dispersion modeling can be performed, the *source term* (release characteristics such as release rate, duration, and phase) must be calculated. The CCPS book *Guidelines for Use of Vapor Cloud Dispersion Models* (CCPS/AIChE, 1987) presents the use and limitations of some current source and dispersion models.

Both chemical properties and physical parameters are required to perform dispersion calculations. The minimal data required for a continuous, neutral-buoyancy, ground-level plume using a Gaussian dispersion model are the release rate, atmospheric stability, wind speed, and concentration of concern. Parameters such as surface roughness and mixing layer height may also be needed, depending on the specific model. More detailed models that take into account such phenomena as buoyancy, jet mixing, vapor density, vapor reactivity, vapor hygroscopicity, aerosol entrainment, multicomponent releases, and time-dependent source terms will, of course, have far greater input requirements and calculational demands.

The concentration or inhalation dose boundaries for the areas of concern will be dependent on the toxicity of the material that could be released. Industry-consensus toxicity values for several industrially important toxic vapors have been developed as Emergency Response Planning Guidelines (ERPGs), available from the American Industrial Hygiene Association (AIHA, 1992). The ERPGs are for one-hour exposures, and are given at three levels:

- The **ERPG-1** is the maximum airborne concentration below which it is believed that nearly all individuals could be exposed for up to 1 h without experiencing other than mild transient adverse health effects or perceiving a clearly defined objectionable odor.
- The **ERPG-2** is the maximum airborne concentration below which it is believed that nearly all individuals could be exposed for up to 1 h without experiencing or developing irreversible or other serious health effects or symptoms that could impair their abilities to take protective action.
- The **ERPG-3** is the maximum airborne concentration below which it is believed that nearly all individuals could be exposed for up to 1 h without experiencing or developing life-threatening health effects.

Thirty-minute worker "immediately dangerous to life and health" (IDLH) values as published by NIOSH (1990), or a fraction of the inhalation dose predicted to be fatal to 50% of a given population (LD_{50}) are often used where ERPGs are not available.

Another more detailed approach that gives the probability of fatality or other health effect as a function of both concentration and exposure time is the probit method. A description of the probit method, along with probit constants for 20 common industrial chemicals with inhalation toxicity, is given in CCPS/AIChE (1989, 148ff).

4.4. Where to Find Methods for Estimating Immediate Impact

Once an assessment, as outlined in Section 4.3, has determined the area likely to be affected by a reactive chemical accident and the characteristics of the energy and/or material release and dissipation, the consequence analysis is completed by using this information to assess the degree to which people and property may be affected by the accident. Models that can be used to evaluate the severity of consequences are fairly well-developed, and depend on the mechanism; i.e., explosion, fire, hazardous material release.

4.4.1. Explosion Effect Models

The primary mechanisms by which explosions cause injuries and damage are direct blast overpressure, missiles (projectiles/flying debris), structural collapse, bodily translation, startle reaction, and follow-on effects. Which mechanism is most important will vary from scenario to scenario, so more than one mechanism may need to be evaluated.

Comprehensive reviews of the effects of blast overpressure and missiles can be found in Lees (1980), Baker et al. (1983), and CCPS/AIChE (1994). A compilation summarizing various levels of damage produced by blast overpressures is shown in Table 4.2. It should be noted that very short-duration blast overpressures, such as by detonation of condensed-phase explosives or rupture of a pressure vessel, will cause less damage at the same overpressure level than longer-duration overpressures such as may be caused by vapor cloud explosions. The potential for bodily translation (i.e., the blast wave throwing a person against another object and thus causing impact injuries) is likewise a function of both blast overpressure and duration. A startle reaction from the loud noise of the explosion may be of concern for persons working at elevated locations, but is more subjective to evaluate. Follow-on effects, particularly fires and the rupture of other lines and vessels that may contribute additional flammable or combustible material to such fires, are often the most devastating of the explosion consequences. Follow-on effects can also include toxic gas and toxic combustion product releases.

4.4.2. Thermal Effect Models

Thermal effects on people and structures are relatively well-established. API RP 521 (1990) reviews the effects of thermal radiation on people and gives recommended design flare radiation levels. For example, a permissible design level for incident radiation of 3000 Btu/h·ft^2 (9.5 kW/m^2) excluding solar radiation (which is about 320 Btu/h·ft^2 or 1 kW/m^2 on a clear, hot summer day) must be limited to a few seconds, sufficient for escape only. Second-degree burns can result at this radiation intensity after only 20 s. Nearly 8,000 Btu/h·ft^2 (25 kW/m^2) is required to ignite wood without an additional ignition source, but plastic tubing can be melted at half this intensity (World Bank, 1985).

Suggested thresholds for significant injury and fatalities are charted by Mudan (1984), following Eisenberg et al. (1975), as a function of incident thermal flux Q_x in kW/m^2 and exposure time t in s. The resulting correlations in the 0.1-s to 100-s time frame are:

TABLE 4.2
Damage Produced by Blast (Clancey, 1972; Glasstone and Dolan, 1977)

Side-on Overpressure		*Description of Damage*
(psig)	*(kPa)*	
0.02	0.15	Annoying noise (137 dB if of low frequency 10–15 Hz)
0.03	0.2	Occasional breaking of large glass windows already under strain
0.04	0.3	Loud noise (143 dB), sonic boom glass failure
0.1	0.7	Breakage of small windows under strain
0.15	1	Typical threshold for glass breakage
0.3	2	"Safe distance" (probability of 0.95 of no serious damage beyond this value); projectile limit; some damage to house ceilings; 10% window glass broken
0.4	3	Limited minor structural damage
0.5–1	3.5–7	Large and small windows usually shattered; occasional damage to window frames
0.7	5	Minor damage to house structures
1.0	8	Partial demolition of houses, made uninhabitable
1–2	7–15	Corrugated asbestos shattered; corrugated steel or aluminum panels fastenings fail, followed by buckling; wood panels (standard housing) fastenings fail, panels blown in
1.3	10	Steel frame of clad building slightly distorted
2	15	Partial collapse of walls and roofs of houses
2–3	15–20	Concrete or cinderblock walls, not reinforced, shattered
2.5	18	Lower limit of serious structural damage; 50% destruction of brickwork of houses
3	20	Heavy machines (3000 lb, 1400 kg) in industrial building suffered little damage; steel frame building distorted and pulled away from foundations
3–4	20—28	Frameless, self-framing steel panel building demolished; rupture of oil storage tanks
4	30	Cladding of light industrial buildings ruptured
5	35	Wooden utility poles snapped; tall hydraulic press (40,000 lb or 18 t) in building slightly damaged
5–7	35–50	Nearly complete destruction of houses
7	50	Loaded tank cars overturned
7–8	50—55	Unreinforced brick panels, 8–12 in (200–300 mm) thick, fail by shearing or flexure
9	60	Loaded train boxcars completely demolished
10	70	Probable total destruction of buildings; heavy machine tools (7000 lb, 3200 kg) moved and badly damaged
300	2070	Limit of crater lip

Significant injury threshold	$\log t = -1.35 \log Q_x + 2.80$
1% fatalities	$\log t = -1.35 \log Q_x + 3.14$
Mean 50% fatalities	$\log t = -1.35 \log Q_x + 3.62$
Near 100% fatalities	$\log t = -1.35 \log Q_x + 3.87$

All logarithmic values are base ten. These results are quite conservative, in that no credit is given for protection afforded by clothing or for medical treatment. A thorough discussion of burn injury as a consequence of various scenarios is given by Lees (1994), taking into account the human response, the effective incident radiation, the protection afforded by clothing, the ignition of clothing, the severity of burns, and the medical treatment of burns.

4.4.3. Toxic Gas Effect Models

Toxic solids and liquids will only have a very limited area of effect, unless the material is finely divided and will disperse before settling out or unless the material is dissolved or otherwise enters a drain or other system where environmental damage can result. Thus, the effects of toxic gases and vapors are generally the most important effect to be considered in a consequence analysis, and the inhalation toxicity of the material is likewise the most often-used parameter. It should always be remembered that inhalation toxicity for acute (short-duration, high-concentration) exposures is a function of both the concentration of the material inhaled and the duration of the exposure, thus giving an inhalation "dose" that can be compared to available toxicity data.

The plume isopleths generated by a toxic gas dispersion analysis can be overlaid on a map of the surrounding area to determine the populations that could be affected by a toxic gas release above an inhalation dose of concern. The number of potentially affected persons can be estimated based on census data or population densities, for determining the potential severity of consequences in terms of the likely number of people affected. A more detailed analysis can look at site-specific meteorological data to include wind speed, wind direction, and atmospheric condition frequencies in the modeling effort. Other factors that can be taken into consideration are day/night population and meteorology variations, the ability of persons to escape cross-wind from the cloud and, thus, reducing the duration of exposure, and sheltering in place to take into account the reduction in vapor concentration inside a building or enclosure compared to the outside atmosphere, particularly for short-duration releases.

4.4.4. Modeling Systems

All of the above-described effects models have been included in modeling systems, mostly computerized, that are available from various organizations. CCPS/AIChE (1989, 172–173) gives a summary of selected modeling systems that can be used for consequence analysis. The ARCHIE code (EPA, 1989) contains simplified calculational models for each of the accident types described above. Other simplified approaches include The Dow Chemical Company's *Chemical Exposure Index Guide* (CEI) for screening toxic release potentials (AIChE, 1994a) and Dow's *Fire and Explosion Index Hazard Classification Guide* (F&EI) for empirically predicting effect areas and loss estimates for fire/explosion hazards (AIChE, 1994b).

4.4.5. Caveats

The primary purpose for a first-cut consequence analysis is to estimate what accidental event consequences could occur when storing or handling a reactive chemical, and how severe the ensuing accident consequences could be. As such, most results should be considered of order-of-magnitude accuracy, and this should be taken into account when designing systems to contain and control reactive chemicals.

The traditional approach to dealing with the uncertainty and inaccuracies in the analysis of hazards is to err on the conservative side and build safety factors into the process design. In many cases it would be worth the effort to refine the consequence analysis by using more rigorous models, with the potential benefit of needing less conservatism (wall thicknesses, relief capacity, barricade strength, redundancy of safeguard systems, etc.) in the process design. However, caution should be used not to extend any of the available models beyond their original applicability. For new chemicals or process arrangements, larger-scale hazards testing may need to be conducted to achieve a high degree of confidence in a particular design or sensitive application.

4.5. Applications of Consequence Analysis

Consequence analysis generally requires a significant amount of knowledge and effort. This effort pays off in a number of ways, as indicated by the examples that follow.

4.5.1. Selection of Size, Quantity, and Location of Facilities

Consequence analysis can aid in the proper siting and layout of facilities, by indicating how large an area can be impacted by a reactive chemical incident. Conversely, for an already-operating facility, it may be necessary to limit the quantity of reactive materials or size of reaction vessels used in the facility, based on the results of consequence calculations.

One application of this principle is in the use of standard quantity–distance tables for separation of high explosives (U.S. Army, 1990). These tables are based on blast effect principles, as well as the historical record of incidents involving inadvertent ignition of high explosives. Similar quantity–distance relationships are used for minimum spacing of organic peroxide storage facilities (see Chapter 7).

For reactive chemicals that are not condensed-phase explosives, the primary effects of concern are hazardous material releases, fires, and vessel rupture explosions. Siting and layout considerations for protecting against hazardous material releases are considered in *Guidelines for Safe Storage and Handling of High Toxic Hazard Materials* (CCPS/AIChE, 1988). For fires and bursting vessel explosions, additional considerations are required, such as ignition source control, shielding, access for fire fighting equipment, blast shielding, and control room construction.

A checklist of facility siting issues is presented in Figure 4.4. This checklist shows the breadth of issues that must be addressed when reactive chemicals are to be stored and handled. This checklist can also be used to address facility siting issues as part of process hazard analyses.

4.5.2. Selection of Dedicated Safeguard Systems

If the consequence analysis indicates that the potential effects of a major reactive chemical incident are severe but alternatives are not available for significantly reducing the hazard, then importance must be placed on developing highly reliable **prevention, protection,** and **mitigation** systems to reduce the likelihood of the major incident and/or reduce the severity of consequences if the incident occurs, as shown in Figure 4.5. Johnson (1990) describes how this reliability can be gained by multiple "safety layers" in reactive chemical systems. The consequence analysis can also point to what particular consequence needs to be guarded against the most; i.e., which scenarios pose the greatest risk of injury and loss.

Design of safeguard systems must be based on sound design information. The consequence analysis can provide the information needed for the design of prevention,

protection, and mitigation systems such as those described in Chapters 5 and 7. For example, if it is decided that a blast barricade is warranted due to an explosion hazard, then the explosion severity information from the consequence analysis (such as TNT equivalence) can be used to design a barricade to withstand an accidental explosion involving the reactive material. Other parameters that are useful from the consequence analysis include maximum operating limits, sizing basis of emergency relief systems, design basis of fire protection systems, and design criteria for mitigation systems such as water curtains and personal protective equipment.

4.5.3. Basis for Emergency Response Systems and Planning

The accident event scenarios developed as part of the consequence analysis are an excellent resource for emergency response planning. The scenarios indicate

PROCESS HAZARD ANALYSIS—FACILITY SITING REVIEW

- *As a team, discuss each of the following siting considerations as they relate to the process being reviewed.*
- *Identify any calculations, charts, and other documents that can show that facility siting has been considered.*
- *Document all recommended action items that are generated during the facility siting checklist discussions.*

General Considerations	1 ☐	Location of on-site populations relative to the unit
	2 ☐	Location of critical systems
	3 ☐	Dominant wind direction
	4 ☐	Climate and weather extremes; earthquake, flooding, windstorms
	5 ☐	Site topography
	6 ☐	External hazards or threats (fire/explosion/toxic release from nearby process or facility; aircraft; subsidence; sabotage)
	7 ☐	On-site traffic flow patterns and clearances
	8 ☐	Security and reliability of all critical feeds and utilities
	9 ☐	Command center and alternate command center locations
	10 ☐	Evacuation routes, emergency exits, safe rally spots
Control Room	11 ☐	Minimum occupancy; only essential functions during emergencies
	12 ☐	Control room construction
	13 ☐	Fresh air intakes location/isolation; temporary safe havens
	14 ☐	Control room location relative to unit, columns, and pipe bridges
Process Facilities	15 ☐	Area electrical classification
	16 ☐	Accessibility for mechanical integrity (sampling, maintenance, repairs)
	17 ☐	Protection of piping and vessels from vehicles and forklifts
	18 ☐	Protection of small-bore lines, fittings from external impact, personnel
	19 ☐	Routing of process piping, critical controls cable trays, critical utilities
	20 ☐	Vent, drain, and relief valve discharge locations
Loading/ Unloading and Storage Facilities	21 ☐	Incompatible materials segregated; storage, dikes, sumps, drains, waste
	22 ☐	Siting, labeling of unloading spots for incompatible materials
	23 ☐	Storage tank separation distances (to process, between tanks)
	24 ☐	Spill control, drainage direction, destination, treatment capacity
Fire Protection	25 ☐	Access for fire fighting and any other emergency services
	26 ☐	Ignition sources (continuous, occasional/intermittent, uncontrolled)
	27 ☐	Access to hydrant, monitor, and deluge valves
Accident Mitigation	28 ☐	Detection of leaks/ruptures
	29 ☐	Emergency shutdown switch locations
	30 ☐	Accessibility of isolation valves
	31 ☐	Potential for fire/explosion in unit affecting other equipment
	32 ☐	Critical controls, mitigation, communication, and fire protection systems functional and accessible after initial explosion or release
Personnel Protection	33 ☐	Passageways, pedestrian traffic patterns vs. hazardous locations
	34 ☐	SCBA/respirator locations; accessibility on all shifts

FIGURE 4.4 **Facility Siting Checklist** (courtesy of Battelle).

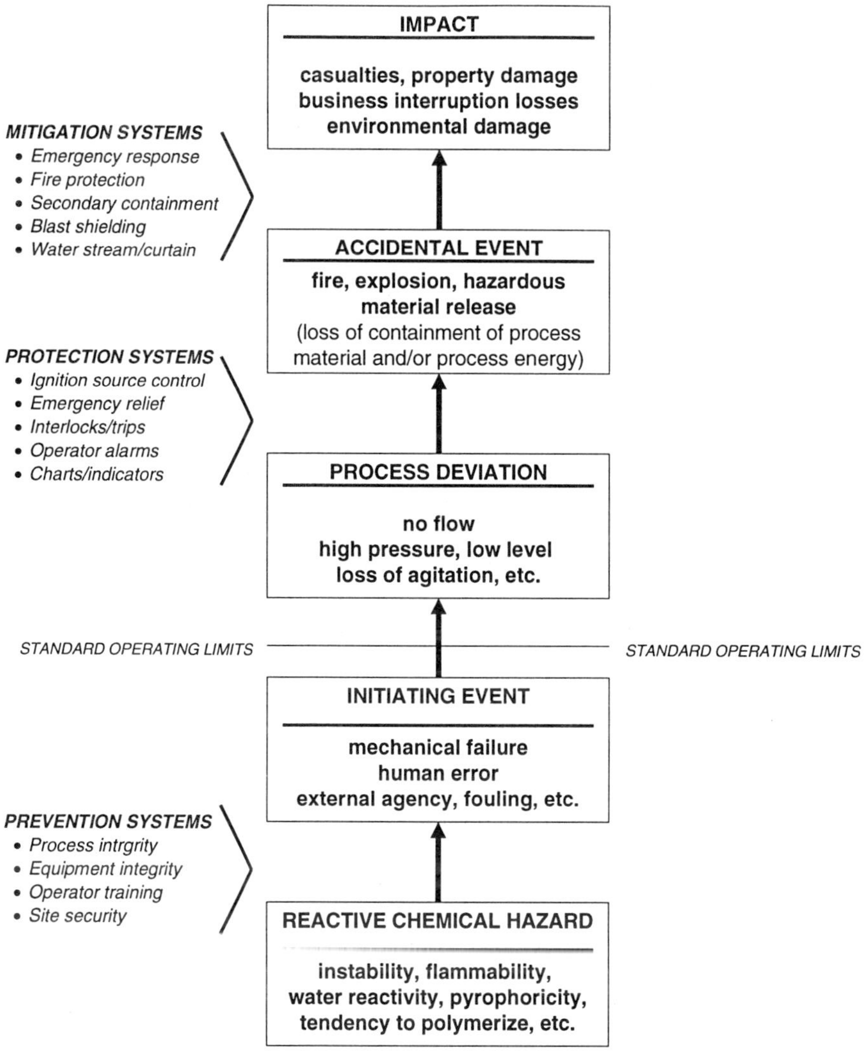

FIGURE 4.5. **Prevention, Protection, and Mitigation Meassures** (Johnson, 1990).

the range of possible consequences that may have to be handled by emergency response systems. The consequence analysis points to scenarios that could have off-site or severe on-site impacts; these scenarios can also be used in drills and tabletop exercises. Judgment must be used in selecting scenarios that have the most severe potential consequences but also have a nonnegligible likelihood of occurrence some time during the life of the facility.

4.5.4. Better Understanding of the Hazard and the Consequences

The identification of reactive chemical hazards and their potential consequences will point out where gaps exist in an understanding of what can go wrong and how. The use of hazard analysis techniques such as HAZOP studies, where the accident scenarios are developed in a rigorous, systematic way, is particularly valuable in looking at the operation from a different perspective; i.e., from the point of view of "What can go wrong?" "How likely is it to occur?" "How severe are the consequences?" and "What do we need to do about it?" If a facility overlooks a potential consequence, then the design of the facility, the training of operating personnel, and the facility's emergency response systems could all be inadequate to guard against that consequence.

4.5.5. Significant Step toward a Well-Managed Operating Facility

All facilities operate under the constraint of limited resources. These resources must be allocated in such a way as to do the most good. Consequence analysis can be a key element of the management of process risks

- (a) by ensuring that all significant risks have been identified;
- (b) by prompting an evaluation of what risks are really necessary to be managed and what hazards may be able to be reduced or eliminated, such as by using a less hazardous material where feasible;
- (c) by providing a sound basis for good design and operating practices when reactive chemicals do need to be stored and handled; and
- (d) by helping to sort out which risks are most important and warrant significant risk reduction efforts, versus the risks that are comparatively minor and warrant less attention on a relative basis.

Bibliography

Center for Chemical Process Safety, *Guidelines for Chemical Process Quantitative Risk Analysis,* American Institute of Chemical Engineers, New York, 1989.

Center for Chemical Process Safety, *Guidelines for Chemical Reactivity Evaluation and Application to Process Design.* American Institute of Chemical Engineers, New York, 1995.

Center for Chemical Process Safety, *Guidelines for Evaluating the Characteristics of Vapor Cloud Explosions, Flash Fires, and BLEVEs.* American Institute of Chemical Engineers, New York, 1994.

References

AIChE (1994a). *Dow's Chemical Exposure Index Guide.* American Institute of Chemical Engineers, New York.

AIChE (1994b). *Dow's Fire & Explosion Index Hazard Classification Guide,* 7th ed. American Institute of Chemical Engineers, New York.

AIHA (1992). Emergency Response Planning Guidelines. American Industrial Hygiene Association, Fairfax, Virginia, (703) 849-8888.

API RP 521 (1990). *Guide for Pressure-Relieving and Depressuring Systems.* Recommended Practice 521. American Petroleum Institute, Washington, D.C.

Baker, W. E., Cox, P. A., Westine, P. S., Kulesz, J. J., and Strehlow, R. A. (1983). *Explosion Hazards and Evaluation.* Elsevier, New York.

Bodurtha, F. T. (1980). *Industrial Explosion Prevention and Protection.* McGraw-Hill, New York.

Bretherick, L. (1990). *Bretherick's Handbook of Reactive Chemical Hazards,* 4th ed. Butterworths, London.

CCPS/AIChE (1987). *Guidelines for Use of Vapor Cloud Dispersion Models.* American Institute of Chemical Engineers, New York.

CCPS/AIChE (1988). *Guidelines for Storage and Handling of High Toxic Hazard Materials.* American Institute of Chemical Engineers, New York.

CCPS/AIChE (1989). *Guidelines for Chemical Process Quantitative Risk Analysis.* American Institute of Chemical Engineers, New York.

CCPS/AIChE (1992). *Guidelines for Hazard Evaluation Procedures, Second Edition with Worked Examples.* American Institute of Chemical Engineers, New York.

CCPS/AIChE (1994). *Guidelines for Evaluating the Characteristics of Vapor Cloud Explosions, Flash Fires, and BLEVEs.* American Institute of Chemical Engineers, New York.

CCPS/AIChE (1995). *Guidelines for Chemical Reactivity Evaluation and Application to Process Design.* American Institute of Chemical Engineers, New York.

Clancey, V. J. (1972). "Diagnostic Features of Explosion Damage." *6th International Meeting on Forensic Sciences.* Edinburgh, Scotland.

Crowl, D. A. (1992). "Using Thermodynamic Availability to Determine the Energy of Explosion for Compressed Gases." *Plant/Oper. Prog. 11*(2, April), 47–49.

DOT (1984). *Chemical Hazard Response Information System (CHRIS) Hazardous Chemical Data.* Commandant Instruction MI6465-12A. U.S. Department of Transportation, U.S. Coast Guard, 14 November. U.S. Government Printing Office, Washington, D.C.

Eisenberg, N. A., Lynch, C. J., and Breeding, R. J. (1975). *Vulnerability Model: A Simulation System for Assessing Damage Resulting from Marine Spills.* Report CG D-136-75, Enviro Control, Rockville, Maryland.

EPA (1989). Automated Resource for Chemical Hazard Incident Evaluation (ARCHIE), Version 1.00. Federal Emergency Management Agency, U.S. Department of Transportation, and U.S. Environmental Protection Agency, Washington, D.C.

Fisher, H. G. and Goetz, D. D. (1991). "Determination of Self-Accelerating Decomposition Temperatures Using the Accelerating Rate Calorimeter." *J. Loss Prev. Proc. Ind.* 4(5, October), 305–316.

Fisher, H. G. and Goetz, D. D. (1993). "Determination of Self-Accelerating Decomposition Temperatures for Self-Reactive Substances." *J. Loss Prev. Proc. Ind.* 6(3, June), 183–194.

Gay, D. M., and Leggett, D. J. (1993). "Enhancing Thermal Hazard Awareness with Compatibility Charts." *J. Testing and Evaluation* 21(6, November), 477–480.

Glasstone, S., and Dolan, P. J. (1977). *The Effects of Nuclear Weapons,* 3rd Edition. U.S. Department of Defense, Washington, D.C.

Hakl, J. (1984). "Advanced SEDEX (Sensitive Detector of Exothermic Processes)." *Thermochim. Acta 80,* 109.

Hatayama, H. K., Chen, J. J., DeVera, E. R., Stephens, R. D., and Storm, D. L. (1980). "A Method for Determining the Compatibility of Hazardous Wastes." EPA-600/2-80-076, U.S. Environmental Protection Agency, Washington, D.C., April. Available through National Technical Information Service, Springfield, Virginia.

Johnson, R. W. (1990). "Building Safety Layers Around Reactive Chemical Processes." Paper 58d. AIChE Spring National Meeting, 21 March.

Lees, F. P. (1980). *Loss Prevention in the Process Industries,* 2 volumes. Butterworths, London.

Lees, F. P. (1994). "The Assessment of Major Hazards: A Model for Fatal Injury from Burns." *Trans. IChemE* 72(Part B, August), 127–134.

Mixter, G. (1959). "Thermal Radiation Burns Beneath Fabric Systems." *Ann. N. Y. Acad. Sci. 82,* 701–713.

Mudan, K. S. (1984). "Thermal Radiation Hazards from Hydrocarbon Pool Fires." *Proc. Energy Combust. Sci. 10*(1), 59–80.

NFPA 49 (1994). *Hazardous Chemical Data.* National Fire Protection Association, Quincy, Massachusetts.

NFPA 491M (1991). *Manual of Hazardous Chemical Reactions.* National Fire Protection Association, Quincy, Massachusetts.

NFPA 704 (1990). *Recommended System for the Identification of the Fire Hazards of Materials.* National Fire Protection Association, Quincy, Massachusetts.

NIOSH (1990). *NIOSH Pocket Guide to Chemical Hazards.* DHHS (NIOSH) Publication No. 90–117. U.S. Department of Health and Human Services, National Institute for Occupational Safety and Health, Cincinnati, Ohio, June.

Perry, R. H. and Green, D. W. (eds.) (1984). *Perry's Chemical Engineers' Handbook,* 6th ed. McGraw-Hill, New York.

Saville, G. (1977). *High Pressure Safety Code.* High Pressure Technology Association, Imperial College, London.

Sax, N. I. and Lewis, R. J. (1989). *Dangerous Properties of Industrial Materials* (3 volumes), 7th ed. Van Nostrand Reinhold, New York.

Stull, D. R. (1977). *Fundamentals of Fire and Explosion.* AIChE Monograph Series No. 10, Vol. 73. AIChE, New York.

Thompson, R. E., Zamejc, E. R., and Ahlbeck, D. R. (1992). "Hazardous Materials Car Placement in a Train Consist" (2 volumes). DOT/FRA/ORD-92/18-1 and -2, U.S. Department of Transportation/Federal Railroad Administration, Washington, D.C. Available through National Technical Information Service, Springfield, Virginia.

U.S. Army (1990). *Structures to Resist the Effects of Accidental Explosions (Rev. 1).* Technical Manual TM5-1300. U.S. Department of Defense Explosives Safety Board, Defense Technical Information Center, Alexandria, Virginia.

Wertz, D. (1989). REACT Hazardous Chemical Reactivity code. Uniformed Services University of the Health Sciences, 4301 Jones Bridge Road, Bethesda, Maryland 20814-4799, (202) 295-3323.

Whitmore, M. W., and Wilberforce, J. K. (1993). "Use of the Accelerating Rate Calorimeter and the Thermal Activity Monitor to Estimate Stability Temperatures." *J. Loss Prev. Proc. Ind.* 6 (2, March), 95–101.

World Bank (1985). *Manual of Industrial Hazard Assessment Techniques.* Technica Ltd. for Office of Environmental and Scientific Affairs, World Bank, Washington, D.C.

5

General Design Considerations

This chapter gives a summary of current good industry practices related to the design of reactive materials storage and handling facilities. The information in this chapter was developed by surveying the nearly 100 CCPS sponsor companies using a survey form that paralleled the sections in this chapter. Twenty-eight survey forms were completed and returned; these responses are summarized in the tables of Appendix B and form the basis for the information presented in this chapter. Survey responses were obtained according to category of reactive chemical hazard (polymerizing, shock-sensitive, etc.), so that design considerations specific to a hazard of concern could be identified. Since the survey responses tend to be brief and without much explanation, explanations have been added in an attempt to make the practices and reasons clearer.

It must be kept in mind that the responses to a survey of this type report the practices that have been adopted for specific situations; also, that there may be other, equally important practices that were not reported. It may also be that the practices reported may not be appropriate for other situations. In addition, there is no representation that the practices reported do not violate any law, regulation or other legal requirement. Neither AIChE, CCPS, the Subcommittee, the writers/editors/reviewers, nor any other party participating in this project can be held responsible for the application of the reported practices or for any omissions or errors in this report.

After summarizing some general strategies for designing safe facilities, this chapter covers design considerations related to the following topics covered in the industry survey:

- compatibility
- storage time and shelf life
- storage quantity and configuration
- air and moisture exclusion
- storage monitoring and temperature control
- manual handling and on-site transfer piping

- last-resort safety features
- passive mitigation systems
- detection, warning, and isolation
- fire prevention and protection
- postrelease mitigation
- hazard reviews
- codes and standards

Design practices relating to specific reactive materials are covered in Chapter 7.

5.1 Summary of General Design Strategies

Listed below are widely used strategies for designing safe reactive chemical facilities. The many topics covered in the remainder of this chapter generally follow the same order as the approaches described in this section.

5.1.1. Reduce the Inherent Hazards

The necessity for handling significant quantities of reactive materials to achieve various operational goals is often taken for granted. However, opportunities for using less reactive materials and otherwise reducing the underlying hazards of an operation (i.e., increasing the *inherent safety* of the operation) should be examined at the earliest concept and planning stages. The designer of a storage and handling facility should always consider what the minimum inventory is, consistent with smooth and efficient operation and without excessive conservatism. Another concept in the inherent safety of a chemical processing plant is to keep the plant as simple as possible. Complexity in equipment, piping, instrumentation, and process control can lead to more opportunities for reactive chemicals incidents to occur. For example, as compared to a single valve, having several valves in a chemical transfer system poses a greater chance of leakage, mechanical failure, and valving errors that could lead to a reactive chemical incident.

Other inherent safety approaches should also be considered early on, such as storing and handling a reactive material in a diluted form and providing safe separation distances to worker and off-site populations. *Guidelines for Engineering Design for Process Safety* (CCPS/AIChE, 1993, Chapter 2) presents an inherent safety checklist and describes several approaches.

5.1.2. Build Reliable Safety Layers

Reactivity hazards that cannot be eliminated must be contained and controlled for ongoing safe operation of a facility. Principles of defense in depth, i.e., the use of multiple safety layers (barriers), should be applied when designing a storage and handling facility (Johnson, 1990).

- Storage and handling incidents can be **prevented** by designing and maintaining a high-integrity primary containment system; by installing, testing, and conducting preventive maintenance on reliable equipment and instrumentation; by quality control of all incoming materials and by labeling and segregating incompatible materials; by operator training to minimize operational errors; and by considering possible external influences in the design and layout of the facility.
- Potential storage and handling deviations can be **controlled** with reliable manual or automatic control systems, adequate redundancy, and operator training to compensate for abnormal conditions.
- Facilities can be **protected** by alarms, interlocks, and last-resort safety systems such as emergency relief systems. Perhaps, even designing the primary containment to contain all excess energy should be considered as the first safety layer.
- A wide array of strategies is used to ensure the consequences of a reactive chemical incident are **mitigated**; i.e., the severity of consequences is lessened if loss of containment of reactive material/energy does occur. Secondary containment such as a double-walled vessel or piping is sometimes used to contain all material/energy that might escape from the primary containment. Excess energy can be directed to a safe location using emergency relief systems and physical barriers such as blast walls. Water sprays may be able to knock down a cloud of vapor. Fire protection systems can reduce the damage and spread of a fire. Personal protective equipment can reduce personnel injury, and emergency response actions such as sheltering and evacuation can likewise reduce the number of people affected by an accidental event. Physical separation (distance) is very effective as a passive mitigation strategy.

Passive engineered features such as concrete dikes and weak-seam roofs are generally more reliable than active engineered features such as purge systems and interlocks. Procedural safety systems that require the active intervention of an operator to detect, diagnose, and promptly respond to a process deviation are generally the least reliable, although often a first defense against upset conditions (CCPS/AIChE, 1993, 7).

Some storage and handling facilities with relatively minor reactivity hazards will require only a minimum of defenses. The storage and handling of a large quantity of a highly reactive material, such as a large storage tank of styrene monomer, requires multiple safety layers and more careful attention to ensure their continued reliability.

5.1.3. Conduct In-Depth Reviews

Periodic in depth reviews of reactive chemical facilities, both at the design stage and during ongoing operation, can identify previously unrecognized hazards, design weaknesses, operational problems, and lapses in administrative controls. Many methodologies are used to conduct in-depth reviews; systematic team reviews using methods such as hazard and operability (HAZOP) studies are often warranted for complex systems involving multiple safety layers.

5.1.4. Use Previous Experience

Codes and standards embody a wealth of technical know-how and lessons learned from past experience and previous incidents. Although few codes and standards relate to particular reactive materials, many deal with hazardous characteristics such as flammability and management systems such as emergency planning and response. A designer should consider the use of existing codes and standards as the starting point from which to design a safe facility.

5.2 Compatibility

As described in Chapter 1, the reactive properties of chemicals can be generally categorized as *self-reactivity* (polymerizing, decomposing, isomerizing, disproportionating) or *reactivity with other materials* (pyrophoric, peroxide-forming, and water-reactive, as well as other more general reactivities such as flammability, acidity, and toxicity that are outside the scope of these guidelines). Incompatibility concerns for self-reactive materials focus mostly on impurities and other materials that can act as initiators or catalysts for an uncontrolled self-reaction. Incompatibility concerns for other-reactive materials focus more on bulk or contact mixing with specific process, surrounding, or inadvertently introduced materials.

Safe storage and handling of all reactive chemicals requires the **identification** and **exclusion** of any materials that can initiate an uncontrolled reaction or self-reaction. After highlighting a number of methods used to identify potential incompatibility problems in Section 5.2.1, a summary is given in Section 5.2.2 of design and operating practices aimed at excluding materials that may result in an incident if inadvertent or uncontrolled contact occurs.

5.2.1. Identifying Potential Incompatibility Problems

As the survey results shown in **Summary 1** (page 284) indicate, a wide variety of means are used to identify potential incompatibility hazards. It is apparent from the survey responses that chemical interaction matrices are used quite broadly. The use of material safety data and other chemical literature, hazard reviews and other team-based reviews, corporate experience, development work and experimental results, identification of key chemical groups, and known or suspected chemistry and kinetics also apply to the identification of incompatibility hazards.

Chemical Interaction Matrix and Process Hazard Analysis

Process hazard analysis methodologies such as the What-If Analysis have been commonly employed to rigorously examine an operation in order to identify potential incompatibility scenarios. More systematic methodologies such as hazard and operability (HAZOP) studies are often reserved for operations where a higher level of risk is perceived, such as where a potential risk to off-site populations exists or where a complex control system is employed.

Use of a *chemical interaction matrix* is generally supplemental to standard process hazard analysis methodologies. A chemical interaction matrix examines all possible interactions among process chemicals, materials of construction, and any other materials that may be in or around a storage and handling operation and that the process material may reasonably contact. Use of a chemical interaction matrix is described in detail in Section 4.1.3.

Incompatibility Data

Existing data sources that list common incompatibilities are widely used to identify potential chemical incompatibilities. These data sources include material safety data sheets (MSDSs), hazardous material references (Section 4.1.3), and individual articles in the chemical literature that can be found by any one of several electronic search services. Such data sources are most useful for identifying incompatibilities associated with bulk mixing with process materials or reagents; however, the most common or most dangerous incompatible impurities and contaminants are often indicated as well. Impurities are also identified through management-of-change protocols, in which any changes to be made to a process must be reviewed for potential impact on the safety of the operation, including the possible introduction of incompatible impurities such as rust, oils, or metals.

Experience versus Testing

The expertise of chemists and corporate safety consultants, particularly within large companies, can prove invaluable in identifying potential chemical incompatibilities. However, for companies not having expertise available (e.g., in-house or from a supplier or insurance company) or for new chemical formulations or combinations, it may be necessary to have small-scale experiments conducted to determine the behavior of a given combination of materials. Depending on the nature of the materials, such as may be indicated by earlier calorimetry testing, fairly extreme safety precautions may need to be taken when conducting such trials. A typical test involves the mixing of small quantities of the materials in question in a calorimeter that will measure the heat generation and gas evolution (or pressure buildup) rates.

Service Streams and Materials of Construction

Incompatibilities with heat transfer fluids are sometimes less obvious than with other materials that reactive chemicals may come into contact with under leak or break-through conditions. For this reason, process hazard analyses such as HAZOP studies are sometimes employed to identify incompatibilities with heat transfer fluids.

Incompatibilities with materials of construction are identified using compatibility charts or, where such information is insufficient, through materials evaluation testing. Compatibility with corrosion products is assessed through materials evaluation testing also. Identification of incompatible materials of construction and corrosion products are explicitly addressed in HAZOP studies as performed by some companies.

It may be necessary to also look at potential incompatibilities with insulation, gasketing, or other materials that reactive chemicals may come into contact with. Polymerizable materials in particular may be affected by autocatalysts; the introduction of autocatalysts into such operations is also explicitly addressed in HAZOP studies as performed by some companies.

5.2.2. Compatibility with Process Materials/Reagents

Design and operating practices currently used in industry to prevent incompatibility incidents related to process materials and reagents are listed in **Summary 2** (page 285). These survey results indicate that physical isolation/segregation is more commonly achieved by facility design than by operating practices, although both are employed.

Dedicated Equipment and Isolation/Segregation

The use of dedicated equipment for storage and handling of reactive materials is the most common measure used for preventing the mixing of incompatible process materials and/or reagents. Dedicated equipment can be used throughout a storage and handling operation, from receipt of raw materials at a dedicated unloading spot with unique fittings or connectors, to transfer via dedicated piping runs to dedicated storage tanks, to the loading of products using dedicated lines and loading spots. This use of dedicated equipment significantly reduces, but does not totally eliminate, the likelihood of a shipping/receiving or operational error or a mechanical failure resulting in the unloading or transfer of an incompatible material.

The avoidance of cross-connections is an important aspect of physical isolation. For example, a cross-connection could possibly result in the indirect introduction of water or steam into a process stream containing a water-reactive material. Physical disconnects can be used in many places to further reduce the potential for cross-contamination of process and/or service streams. Where physical disconnects are not feasible, a positive means of isolation such as a double block and bleed should be provided.

The use of dedicated equipment can most effectively be integrated into the design of a continuous operation. For many batch operations, such as specialty chemical manufacture, avoidance of all possible cross-connections or multiple usage of storage and piping facilities is much more difficult, and generally requires stronger administrative controls.

Labeling and Operating Practices

Labeling of lines and connections is a key strategy for preventing incompatibility incidents. It can supplement, but not replace, the need for maximum physical isolation or segregation. In addition to all process materials and reagents being labeled, such as tank trucks of raw materials and drum quantities of inhibitors, all connections, piping runs, and valves should be clearly labeled to further reduce the likelihood of a wrong hook-up or transfer being made by operating personnel.

As the above indicates, the avoidance of mixing of incompatible process materials and/or reagents is accomplished by both administrative controls and design features. If the consequences of an incompatible mixing scenario are severe (such as the generation and release of a toxic gas), then redundancy of controls and design features is often required to further reduce the likelihood of an undesired event. For example, in some instances where an operator error could lead to an incompatible mixing scenario, both the operator and an area supervisor may need to sign off on the procedure indicating that it was performed correctly. If unloading of a reagent such as ammonium hydroxide to the wrong storage tank could result in a severe incident, then escorting the supplier, checking the supplier's manifest, and dual sign-off of the log before transfer is started may be warranted. Well-defined standard operating procedures that are followed without exception on all shifts, along with a high level of operator training, will further reduce the likelihood of operational errors that could result in incompatibility incidents.

One of the most certain methods of avoiding the mixing of reactive materials with incompatible materials is to not have the incompatible materials on site at any time. This may not be feasible for chemicals such as water-reactive materials; however, it can apply to materials of construction and other potentially incompatible materials as well as process chemicals. For example, if the process material being used is incompatible with copper, then it may be feasible to exclude the use of all copper and copper-containing materials, such as tubing and conduits, from the entire site. This exclusion will require on-going administrative controls, such as inventory checks, to be maintained.

5.2.3. Compatibility with Impurities

By far the most common approach to avoid undesired reactions due to impurities and contaminants, as seen in **Summary** 3 (page 286), is quality control of incoming materials through sampling and analysis procedures. A broad range of approaches are used to accomplish this quality control. The most common strategy appears to be by lab analysis at all critical points in the handling of the materials, such as before transfer to storage.

Typically, the possibility of incompatible impurities requires analysis of incoming materials before unloading. Many times, however, analysis is not performed on each incoming load. Some companies will accept a certificate of analysis after review with the supplier. The acceptability of this practice will depend on the severity of consequences of receiving a contaminated shipment of material and the degree of confidence in the supplier's analysis and quality practices.

Impurities can also be introduced via services such as compressed air or nitrogen. In some instances involving polymerizing materials, dedicated nitrogen blanketing systems from dedicated nitrogen vaporizer systems have been used.

A reactive material such as a peroxide former may be a product that will need to be stored before packaging or shipment. Such products must be analyzed for impurities before transfer to storage. It is also good practice to only store fresh product. For example, if a quantity of product is returned by a distributor or customer, it may have become contaminated at the distributor's or customer's site or in transit. Acceptance and storage of such material without thorough analysis and prompt reprocessing or disposal has caused numerous storage and handling incidents.

5.2.4. Compatibility with Heat Transfer Fluids

Some heat transfer fluids are obviously not the appropriate choice for a given application, such as the use of water or steam in an operation involving a water-reactive material. The survey results in **Summary** 4 (page 287) indicate that identification of incompatible heat transfer fluids (e.g., by HAZOP studies) is the most important aspect of dealing with them. A compatible heat transfer fluid should then be specified or an alternate design proposed. For example, nonaqueous heat transfer arrangements such as air coolers are used with water-reactive materials. Other fluids used in specific situations include refrigerants and molten salt.

If it is indeed not feasible to avoid using an incompatible heat transfer fluid, then the system design should have special features to substantially reduce the likelihood of the heat transfer fluid coming into contact with the process material. For example, double-tubesheet heat exchangers are used where incompatibility (either the heat transfer fluid leaking into the process or the process fluid escaping to the transfer-fluid side) is a significant concern.

5.2.5. Compatibility with Materials of Construction and Corrosion Products

Similar to the results for Summary 4 for heat transfer fluids, the survey results in **Summary** 5 (page 288) indicate that identification of incompatibilities and corrosion problems with specific materials of construction is key to avoiding material-of-construction incompatibility problems. The actual corrosion of equipment may not be as important in reactive material systems as the potential for corrosion products to cause contamination that could initiate or catalyze an undesired self-reaction such as thermal decomposition.

Specific methods of identifying such incompatibilities include corrosion testing such as by the use of "corrosion coupons"—small plates of the material of construction in question that are exposed to the actual reactive material composition under the actual process conditions (but not installed in the actual process), to check the corrosion effects over time. This testing is conducted in the laboratory, where the process conditions can be varied to check the possible effects of process deviations on the material's corrosion resistance. Further testing can be performed using such methods as differential thermal analysis or accelerating rate calorimetry, whereby suspected corrosion products are added to a reactive material and the effects are measured and compared to the stability of the pure material.

Engineering design and material specification is most often guided by past experience and known incompatibilities. Standard practices exist within some companies and for a few reactive materials. (Some external codes and standards organizations are listed in Summary 45 of Appendix B.) "Standard practice" can include maintaining a compilation of past test results and operating experience, such as maintaining detailed listings of materials of construction that should and should not be used with a given reactive material. The best source of this information is often the supplier of the reactive material, if it is purchased rather than manufactured on-site. A company's internal design practices will specify the actual materials of construction (metal, alloy, plastic, lined piping, bolt material, hardness, etc.) and schedule/rating/thickness to be used when handling a specific reactive material within specified operating conditions of pressure and temperature.

As can be seen, incompatibilities with materials of construction are primarily avoided by proper attention to facility design. However, operating and maintenance practices may also be important: operating practices to avoid deviating from the design envelope, and maintenance practices both to avoid the later introduction of incompatible materials of construction during plant modifications and to detect incipient corrosion problems by such means as internal inspections.

5.2.6. Compatibility with Insulation

Only a few of the survey respondents indicated the use of a specific design or operating practice to prevent incompatibility incidents related to insulation. The practices that were reported are indicated in **Summary 6** (page 289). As with materials of construction, some internal design practices specify the type of insulation to be used for specific reactive materials, particularly where significant incompatibilities could exist if a leak occurred in the piping system. Insulation can also be included in the chemical incompatibility matrix for the process, and laboratory testing can be conducted where definitive information is not available on an insulation material's compatibility.

5.2.7. Compatibility with Fire-Extinguishing Agents

Consideration must be given to potential incompatibilities during unusual, upset, or emergency conditions as well as during normal operation. The most obvious incompatibility with fire-extinguishing agents is the use of water and water-containing substances when dealing with fires involving water-reactive or pyrophoric materials, as indicated in **Summary 7** (page 290). Less-obvious sources of water (such as aqueous foams and building sprinkler systems) need to be identified and eliminated or controlled when dealing with water-reactive and pyrophoric materials. It is possible to use deluge quantities of water to fight minor fires involving some water-reactive materials that react slowly, rather than violently, with water.

Information on fire-extinguishing-agent compatibilities often can be found in hazardous material response references, NFPA references such as NFPA 49 (1994) and NFPA 325M (1994) for water-reactive materials, material safety data sheets, or information from suppliers. Otherwise, it may be necessary to test the compatibility of reactive materials of concern with fire-extinguishing agents being considered for use in storage and handling areas.

5.2.8. Compatibility with Other Materials

In addition to obvious potential incompatibilities, such as with materials of construction, compatibility needs to be established with all fluids and other materials that a reactive chemical might come into contact with. A few examples are indicated by the **Summary 8** (page 291) survey results: decontamination and cleaning fluids; gasketing, sealing, and packing materials; piping and valve linings; sealing fluids; lubricating oils; and instrument fluids.

5.2.9. Other Compatibility-Related Practices

A few additional considerations for avoiding incompatibility problems are listed in **Summary 9** (page 291). Environmental effects may need to be taken into account when designing a reactive chemicals facility; for example, some products must be kept out of direct sunlight to avoid initiating a self-reaction by either the light itself or by the increased temperature. Such materials may need to be handled inside buildings or enclosures during all storage and handling operational phases. Likewise, rain water must be excluded from water-reactive materials. Reported practices include covering or enclosing all storage tanks and unloading areas, keeping containment dikes drained, and preventing infiltration of rain water by leaks, drainage, or other pathways.

5.3. Storage Time and Shelf Life

Polymerizing, thermally sensitive, and other self-reactive materials often have a finite shelf life (maximum storage time) due to slow self-reaction or decomposition over time. Likewise, peroxide formers have finite shelf lives due to the slow generation of potentially unstable peroxides over time. The addition of reaction inhibitors, control of the storage and handling temperature, and exclusion of contaminants and other materials that may catalyze a self-reaction or otherwise degrade the reactive material are important design and operating considerations when handling such materials. For some materials, degradation of the material may only be a quality-related concern. For many others, such as peroxide formers, significant safety issues may be involved.

5.3.1. Storage Time Limitations

Summary 10 (page 292) lists many different considerations associated with storage time/shelf life limitations for reactive materials. If the reactive chemicals are purchased from a supplier, then the supplier's recommendations for storage time should be followed. Additional considerations are generally required when storing and handling reactive materials in larger quantities at producing sites.

Polymerizing materials such as butadiene and styrene must have an inhibitor added prior to storage to prevent slow polymerization; such slow polymerization can, if not adequately controlled by removal of the free radicals and/or heat of reaction, lead to a violent, uncontrolled polymerization with the potential for rupture of the storage vessel. Oxygen must be present to activate some polymerization inhibitors.

With respect to storage time, most shock-sensitive materials have fairly long shelf lives unless contaminated or exposed to elevated temperatures. Thermally decomposing materials generally have lower activation energies, and the rate of decomposition is dependent on the temperature (as well as the purity) of the material. Degradation occurs at all temperatures, but can be negligibly slow if the temperature is kept low enough for a given material (as established by material testing such as accelerating rate calorimetry).

Isomerizing and disproportionating chemicals will be dependent on the specific chemical and operating conditions, and no generalizations can be made. Because the reaction of pyrophoric and water-reactive materials is so rapid with air and/or water, storage time is much less important a consideration for these materials as air/water exclusion.

Shelf life is a key parameter for peroxide-forming chemicals, due to their propensity to generate unstable peroxides slowly over time. Management of small quantities of peroxide-forming chemicals (e.g., in laboratories) is straightforward by following shelf-life specifications as in Table 2.6 (page 54).

5.3.2. Practices for Increasing Shelf Life

Several practices for increasing shelf life are used in industry. The most common practices, as indicated by **Summary 11** (pages 293–294), are adding inhibitors, sampling and monitoring, managing and minimizing inventories, refrigerating or otherwise controlling material temperature, and increasing material purity.

Inhibitors

Inhibitors are used mostly for polymerizable chemicals; they can also be used effectively with some peroxidizable compounds to prevent or slow peroxide formation. Inhibitors are added to directly interfere with the polymerization reaction or to scavenge free radicals or other polymerization-inducing species.

Most inhibitors are consumed over time as they perform their inhibiting function. Hence, either sufficient inhibitor must be added to ensure it will last for the maximum expected storage time, or the inhibitor concentration must be monitored and replenished as necessary. The latter is more likely to be necessary for bulk storage arrangements. Following are other considerations for inhibitors:

- It is possible in some systems or storage containers to add extra inhibitor beforehand if extended storage or higher-than-normal storage temperatures are anticipated.
- Adequate mixing of the inhibitor throughout the material to be protected must be ensured. For production facilities, the inhibitor can be added to the reactive material and thoroughly dispersed before the reactive material is sent to storage.
- Care must be taken to prevent freezing in liquid systems, since the inhibitor may be separated from the material upon freezing and a dangerous condition would exist upon subsequent temperature rise.
- Some inhibitors require the presence of oxygen to "activate" them. Hence, although inert-gas purging in such systems might be proposed as a means of reducing a flammability risk, the total exclusion of oxygen will actually increase the risk of a runaway reaction. An intermediate oxygen level is used by some companies in such situations; the oxygen concentration for reactive materials such as styrene monomer can be reduced enough to have insufficient oxygen to sustain vapor combustion, yet still be high enough to maintain inhibitor activation. Generating and maintaining an inerting source with a controlled oxygen concentration may be needed in some cases to minimize purging errors.

Sampling Systems

The sampling of reactive materials for parameters critical to shelf life, such as inhibitor concentration, varies considerably depending on the nature of the material to be sampled. For example, for polymerizing chemicals, industry practice as indicated by the Appendix B survey results ranges from continuous analysis for inhibitor to a daily hydrogen cyanide color check by visual observation of the reactive material to weekly monitoring of storage tank contents. Some sampling may even be done on a less frequent basis. The three primary points at which sampling and analysis take place are (1) upon receipt of reactive materials into the storage and handling facility, (2) as an ongoing part of on-site storage when the storage time can be sufficiently long to potentially allow the inhibitor to be consumed or self-reaction to otherwise be initiated, and (3) upon transfer of the reactive material out of the storage and handling facility.

Inventory Records and Inventory Control

The maintaining of current inventory records for all types of reactive chemicals is the most common practice indicated by the survey as being related to storage time and shelf life. Depending on the quantity and diversity of chemicals to be monitored, it may be necessary to use a computerized inventory control system or data base to keep track of inventories, sampling/analysis scheduling and results, and duration of storage. Such a system can facilitate "first-in, first-out" practice and can flag any stored material that approaches its preassigned shelf life. The inventory control system can also be employed to keep inventories to a minimum.

Ongoing attempts should be made to minimize both the quantity and the time of all reactive materials in storage. Some practices along these lines include maintaining a low enough storage quantity that the turnover rate of material is relatively high in order to keep material fresh, avoiding arrangements that will allow some material to remain in storage without being rotated (for example, by the use of dished-bottom tanks and bottom unloading to reduce the likelihood of material settling in the bottom of a storage tank for an extended time), working with suppliers to arrange for less in-transit time, and decreasing handling time such as by using quick-unloading rail cars.

Temperature Control

Since the shelf life of most reactive materials is highly dependent on the storage temperature, it is common to extend the maximum possible storage time by using refrigerated storage or at least maintaining the stored material in a temperature-controlled building or enclosure. Some means of temperature control can be started even before receipt of the reactive material at the storage and handling facility, such as transport from the supplier by refrigerated truck, cooling of tank cars before leaving the supplier's site, or precooling of material manufactured on-site before transfer to storage.

Several methods extst for refrigeration for bulk storage, such as internal or external cooling coils, recycle of material through a heat exchanger, and/or by use of a vapor condenser. Refrigeration of materials such as organic peroxide products stored in small containers can be achieved by the use of limited-capacity warehouses that are maintained at a controlled temperature. Materials with low self-accelerating decomposition temperatures (SADTs) may even need to be stored in freezers, with backup refrigeration capability. Recommended practices for low-temperature storage of organic peroxides are given in NFPA 43B (1993).

Material Purity

The stability and shelf life of many reactive chemicals is dependent on the level of impurities or contaminants. It may be possible to obtain raw materials with greater purity from the same or a different supplier. Improving the cleanliness of storage and handling equipment, such as by pretreatment or periodic cleaning, may also be feasible. Impurities may be able to be removed within the storage and handling facility by one of many purification techniques such as by continuous filtration.

5.3.3. Handling and Disposal of Too-Old Material

Labeling, inventory records, sampling/analysis, or unusual characteristics such as off-color or gas generation may occasionally indicate that a reactive material has been stored too long or has exceeded acceptable limits for undesirable chemical changes such as polymerization, decomposition, or peroxide formation. In such cases, it is necessary to safely deal with the material before a dangerous situation exists.

Since it is difficult to predict all conditions under which reactive materials may need to be disposed of, such situations are often handled on a case-by-case basis with no significant pre-planning. However, it is better to anticipate the most likely situations that might be encountered, and have a preestablished plan of action for each situation. Action plans for minor events, such as discovery of a sealed container of peroxide-forming material slightly older than the specified maximum storage time, can usually be incorporated into the operating procedures for the storage and handling operation. More dangerous situations, such as an overpressurized container or a thermal decomposition, may also need to be included in the facility's emergency response plan.

Many different means are used for disposing of outdated or off-spec reactive materials. Several approaches are indicated in Summary 12 (page 295). These include recycling or reworking; destroying by controlled polymerization, decomposition, reaction, or thermal oxidation; disposing of as a hazardous waste in an approved manner; quenching; diluting; and adding extra inhibitor. Returning outdated material to the supplier is also practiced; however, this has caused incidents in the past both in transit and at the supplier's site and should be avoided. Any return of material should require prior approval from the supplier, proper precautions in case of a problem developing in transit, and the material being labeled such that it is immediately dealt with once it is received by the supplier.

5.4. Storage Quantity and Configuration

Closely associated with shelf life is how reactive materials are stored and in what quantities.

5.4.1. Determining Maximum Inventory

As indicated in **Summary 13** (page 296), production requirements and usage rates are the most common determinants of maximum reactive material inventories. Closely associated with production requirements are supplier capabilities; if a supplier must have a significant lead time before delivering a shipment of a given reactive material, or if supplies are unreliable for some reason, then usage rates may necessitate more on-site storage. Likewise, if a material is only economically supplied in rail car quantities, then on-site storage will necessitate the ability to off-load at least one full rail car.

Many companies have means of ensuring inventories of reactive materials are kept to minimum essential levels, both by design and by administrative controls. Inventories can be limited by design by imposing physical limitations on storage capacity. In some cases these limitations have been established by careful review and analysis; in many other cases, the physical limitations have been reached by means such as debottlenecking other parts of the production facility. Engineering controls such as high-level interlocks can be incorporated into an existing facility design; however, administrative controls are also necessary in such cases to ensure the interlocks are not bypassed and the settings are not changed. The general principal to be followed is to design and operate the storage facility with the minimum inventory of reactive materials that is consistent with a safe and stable production operation.

Trade-offs between minimizing inventories and maximizing production up-time have been studied using various risk analysis methods. Likewise, consequence analysis has been used to determine a maximum safe inventory for a given population of concern. For example, it is possible to evaluate the maximum inventory that, if suddenly and accidentally released under worst-case or reasonable-worst-case conditions, will not result in a significant health threat beyond the fenceline of the facility. For some reactive materials, an empirical method such as the Dow Fire/Explosion Index (AIChE, 1994a) can indicate an expected effect radius for a major fire/explosion event.

Regulatory, insurance, or corporate guidelines sometimes dictate maximum inventories for on-site storage of reactive materials. For example, fire regulations can limit the maximum inventories of a thermally decomposing material based on maximum deluge or sprinkler system rate of discharge.

5.4.2. Storage Configurations

Summary 14 (page 297) indicates that the dominant container types for storing reactive materials are tanks of various kinds. This may reflect the fact that many of the surveyed companies are producers of reactive materials; end-users of such materials may more commonly store reactive materials in smaller shipping containers. Depending on the nature of the reactive material, tanks may have special requirements such as being constructed with no rough internal surfaces or having refrigeration or nitrogen inerting. Other bulk-quantity storage configurations include rail cars, tank trucks, large storage spheres, and barges.

Drum storage is used very commonly for smaller quantities of reactive materials, due to the portability and uniform size of drums. However, many companies are moving away from metal drums (such as the typical 55-gallon drum) to either disposable or more easily reusable containers. Typical disposable containers are cartons, paper sacks and bags, and smaller pails, drums, or container cans. Establishing a safe and allowable disposal method for such containers after use is always a key consideration when selecting the type and size of container to ship or receive nonvolatile reactive chemicals. Because of such considerations, several reusable kinds of containers have gained popularity in recent years, including tote tanks, tote bins, and super sacks.

Volatile reactive chemicals in either the liquid or gaseous state may require pressurized containers for storage and transfer. The most common type of container for volatile reactive chemicals is the pressurized cylinder.

5.4.3. Top versus Bottom Discharge

In determining the design of a tank to store reactive chemicals, a decision must be made as to whether outlet nozzles will be installed on the top or bottom of the tank. For some tank designs, company engineering standards or external recommended practices will dictate the nozzle configuration. Some transportation or local regulatory requirements may actually forbid a given discharge configuration, such as bottom outlets.

The industry practice presented in **Summary 15** (page 298) indicates that both top and bottom discharge are used with storage tanks for nearly all types of reactive chemicals, with bottom discharge somewhat more common. The decision as to whether bottom or top discharge is better for a given chemical has been hotly debated in many companies. On the one hand, bottom discharge prevents the aging of any material in the bottom of the tank, and the tank can be emptied by gravity instead of by pressure. However, if a leak develops at the bottom nozzle or flange (the chances of which can be diminished by careful design and more robust construction), then the leak is very difficult to isolate and control before the tank is completely emptied. For this reason, some companies prefer to use top discharge for many materials, including highly toxic materials. Leakage can be rapidly stopped by depressuring the tank, thus halting transfer of the material if the material is not a liquefied gas at the storage conditions. If the tank must be deinventoried, instead of pressurizing the tank vapor space to provide the driving force for liquid flow, it may be possible to generate a vacuum in the receiving vessel to achieve the same differential pressure driving force.

5.4.4. Facility Siting

As **Summary 16** (page 299) indicates, industry practice varies widely as to siting practices for reactive materials storage. Only a few standards, company guidelines, building codes, and local ordinances exist that pertain to siting of reactive materials storage, and most pertain to fire protection considerations. Widely used quantity-distance tables are also available where separation of shock-sensitive storage locations is warranted. The most common siting practices that are employed for reactive materials storage are discussed below.

NFPA Guidelines
The NFPA recommended practices that pertain to facility siting provide guidelines primarily from a fire protection standpoint. NFPA 30 (1993), for example, addresses the diking of storage tanks and remote impoundments handling flammable liquids. NFPA 43B (1993) addresses separation distances for various classes of organic peroxides, based on quantity and reactivity.

Insurance Guidelines
The industrial insurance industry has also provided guidance for separation distances between storage vessels and from storage to other on-site facilities. Industrial Risk Insurers (IRI) provides minimum separation distances for various kinds of storage tanks used in the oil and chemical industries, which are generally based on multipliers of the largest tank diameter (IRI, 1992). These minimum separation distances are also given in *Guidelines for Engineering Design for Process Safety* (CCPS/AIChE, 1993, 74). Factory Mutual also provides minimum siting/spacing requirements (FM, 1993). These guidelines appear to be established on more of an experiential than a scientific basis, and care must be exercised in using any of these guidelines to apply to reactive chemical storage facilities that may have special considerations due to the nature of the specific reactive material.

Consequence/Risk Analysis
The use of consequence and risk analysis methods as aids for making siting decisions has become more widespread in recent years, both by recognition of their benefits and due to regulatory requirements. *Consequence analysis* models worst-case or worst-reasonable-case scenarios involving fires, explosions, and/or hazardous material releases, and determines a radius or area of dangerous effects. Layout and siting can then be arranged, particularly at the early design stages, to eliminate or minimize potential effects on nearby populations. Likewise, the consequence analysis can help avoid the potential for domino effects by assessing the potential impact of a reactive chemical accident on surrounding facilities. *Risk analysis* goes one step further in assessing both the likelihood of occurrence and the severity of consequences of potential accidental events. Where quantitative risk guidelines have been established, either on a corporate, community,

statewide or national basis, risk analysis can indicate where siting or storage configuration may need to be modified in order to reduce the accident risk to meet the guidelines. Both consequence analysis and risk analysis are fully addressed in *Guidelines for Chemical Process Quantitative Risk Analysis* (CCPS/AIChE, 1989). The Dow Fire/Explosion Index (AIChE, 1994a) gives an empirical method for determining the radius of effects for reactive chemical fire/explosion consequences.

Other Siting Practices

When dealing with the storage and handling of reactive chemicals, distance is one of the best "inherent safety" approaches. Many companies emphasize the separation and segregation of reactive materials from sensitive populations and equipment, especially where large quantities are involved. Where separation is not feasible, engineered approaches such as the use of barricades or blast walls for explosion protection can give a substantial amount of risk reduction. Separation/segregation of incompatible materials, as discussed in Section 5.2.2, should also be prominently addressed when considering facility siting issues.

Some materials in unreacted form or as products generated by a reactive chemical incident may form dangerous vapor clouds. This potential should also be considered during facility siting. The book *Guidelines for Use of Vapor Cloud Dispersion Models* (CCPS/AIChE, 1987) gives methods for calculating downwind vapor cloud concentrations. Fire/explosion consequences of vapor clouds are addressed in *Guidelines for Evaluating the Characteristics of Vapor Cloud Explosions, Flash Fire, and BLEVEs* (CCPS/AIChE, 1994a). Design considerations such as plant layout and control room location are treated in *Guidelines for Engineering Design for Process Safety* (CCPS/AIChE, 1993, 63–75) and an article by Englund (1991).

5.4.5. Restrictions on Container Shape or Configuration

As can be seen in **Summary 17** (page 300), the only significant restriction on the shape or configuration of containers used for reactive chemicals storage is *free space*, or the vapor space above the stored liquid (or solid). Sufficient free space must be maintained to avoid container overflow or overpressurization due to thermal expansion of stored liquids under both normal and anticipated abnormal conditions. Chemicals with large coefficients of thermal expansion have larger free space requirements than materials that expand only slightly with an increase in temperature. For example, portable tanks for anhydrous ammonia are only allowed to be filled to 82 liquid volume percent maximum (CGA, 1990, 245).

Various means are employed to prevent the overfilling of tanks or containers, such as the use of high-level interlocks and dip tubes. Hydrostatic relief valves may be required wherever thermal expansion of an overfilled container or a blocked-in liquid can pose a containment rupture hazard.

In addition to protecting against thermal expansion, free space must be maintained for some polymerizing materials in order to provide oxygen for activation of the polymerization inhibitor. One survey response indicated that the free space in drums of such materials must be adequate for the oxygen supply to last at least one year.

For some materials, such as phosphorus and other pyrophoric chemicals, air contact must be avoided, and so the material is stored in a manner which avoids any free space above the material. One way of accomplishing this is to store the material immersed in a compatible liquid that is less dense than the pyrophoric material. Such an arrangement may even be feasible in a liquid–liquid system; for example, carbon disulfide, which is an extremely flammable liquid that is also significantly denser than water, is sometimes stored under water or stored in a tank with a pool of water beneath the tank such that any liquid carbon disulfide leaking from the tank will immediately sink below the surface of the pool of water and be isolated from potential ignition sources.

5.4.6. Mixing and Recirculation

Summary 18 (page 301) shows that circulation of material in storage is practiced mainly with polymerizing materials. This recirculation is accomplished generally by withdrawing liquid from the discharge line and pumping it through a heat exchange system before returning it to the storage vessel. A few respondents use mixing eductors (in connection with pumped recirculation) or agitators to facilitate mixing in the tank. The use of a mixing eductor in the tank gives quicker/better homogenization of the contents, so the temperature is more uniform and inhibitor added to the recirculation stream is more uniformly distributed. The use of an eductor can also facilitate the injection of short stop to an incipient runaway reaction.

Not all recirculation is maintained on a continuous basis. One company recirculates polymerizable material on a four-week intermittent basis instead. When the vapors from a tank are condensed for return, the condensate may lack inhibitor and may produce undesirable materials that can be removed by filtration.

5.5. Air and Moisture Exclusion

For many kinds of reactive materials, most notably pyrophoric and water-reactive substances, it is essential to have a highly reliable system for excluding air and/or moisture from the storage and handling operation. Various design and operating practices used by industry for air/moisture exclusion are discussed in this section.

5.5.1. Air Exclusion Practices

It must be noted that air should not be excluded from the storage containment of all reactive substances. In particular, polymerization inhibitors such as tert-butylcatechol (TBC) for styrenic monomers and monomethyl ether of hydroquinone (MEHQ) for acrylic monomers require some oxygen for activation of the inhibitors. If the material is flammable, it may be possible to use a reduced-oxygen air space above the stored liquid, at a high enough concentration to activate the inhibitor but not sufficient to exceed the minimum oxygen concentration for a flammable vapor–air mixture. Eight percent or less may be sufficient for inhibitor activation, whereas organic vapor–air mixtures will generally not propagate flame if the oxygen concentration is below about 10.5% by volume (at room temperatures and one atmosphere pressure). Bodurtha (1980, 14) shows how to estimate the minimum oxygen concentration from the combustion stoichiometry if the lower flammable limit is known.

The most common situations where air exclusion is necessary are for pyrophoric materials and for flammable/combustible materials that may be stored or handled above their flash points. **Summary 19** (pages 302–303) gives several industry approaches to excluding air from reactive materials storage and handling operations. Inert gas padding is the most widely used approach, with various other containers filling and vapor control procedures also employed.

Inert Gas Purging/Padding

The displacement of air in the containment free space with an inert gas such as nitrogen is used selectively with all types of reactive chemicals where air exclusion is warranted. The inert gas atmosphere can have the double benefit of both air and moisture exclusion, such as by using nitrogen with a low dew point from cylinders or from a dedicated air plant. It is generally necessary both to purge lines with the inert gas before introduction of the reactive material and to maintain a slight pad pressure of the inert gas during continuous operation. Vent streams and relief headers also must be considered for inerting. NFPA 69, *Deflagration Prevention Systems* (1992), gives detailed information on inerting gases, methods for designing and installing inerting systems, and recommended purge gas concentrations for various flammable liquids and combustible dusts. This information is also summarized in the National Fire Protection Association *Fire Protection Handbook* (NFPA, 1994).

Container Filling Procedures

Several survey respondents indicated that container filling procedures are used as a means to exclude air from reactive materials containment. The primary consideration is for shipping containers or vessels that have been opened for cleaning, maintenance, etc. If it is necessary to remove the air from a vessel (generally, this means to reduce the oxygen partial pressure to a predetermined level), this is most commonly done by purging with an inert gas and/or "pulling" a vacuum before introducing material into the vessel. If a vacuum is created before transfer but not maintained during the transfer, the pressure will increase during the transfer. In this case, the vacuum must be specified so that the partial pressure of oxygen at the end of the transfer is below the permissible level.

5.5.2. Moisture Exclusion Practices

Moisture exclusion from reactive chemical storage and handling may be necessary from both a quality and a safety standpoint. Moisture exclusion is of primary importance when dealing with water-reactive materials. **Summary 20** (pages 304–305) gives some industry practices for moisture exclusion.

Start-Up Procedures

As for air exclusion, the removal of all residual moisture from containers, tanks, or other equipment that will be contacted by the reactive material is crucial to the safe storage and handling of water-reactive materials as well as many other types of reactive materials. The use of dry nitrogen, either heated or at ambient temperature, appears to be the most common means of removing residual moisture. The nitrogen or other purge gas can be checked for moisture content before introducing the reactive material. Final flushing of equipment with a compatible hydrophilic liquid is an alternate means of removing residual moisture. As for air exclusion, a systematic analysis of start-up or container filling procedures can be effectively used to identify possible deviations from the procedure that could result in sufficient moisture being left inside, or being introduced to, the containment to create the potential for a reactive chemical incident.

Air or Pad Gas Drying Systems

If air contact with a given reactive material is normal, such as the storage of a nonvolatile liquid that is not reactive with air in an atmospheric storage tank, then moisture exclusion may be able to be sufficiently achieved by installing a desiccant dryer on the tank breather vent, which would also necessitate having a means to periodically check the activity of the desiccant. Some installations may require drying of the nitrogen or other pad gas.

Waterproofing, Drainage, and Flood Control

Waterproof packaging and sealed containers are especially necessary for shipping or receiving container-size quantities of water-reactive materials; such packaging is employed for other reactive materials also. For water-reactive materials, the container size and process demands should be matched, if possible such that the entire container contents are used all at one time.

Buildings intended for water-reactive material storage must be very carefully waterproofed. This includes not only the building roofing but also necessitates water exclusion from sumps, drains, or other places to which an accidental spill of the material could drain. Consideration should be given to storing containers of material above floor level, such as on pallets or even higher, where the possibility of building or area flooding exists.

Other Practices for Moisture Exclusion

Operation of a storage and handling facility above atmospheric pressure is another approach to moisture exclusion, particularly for a continuous operation, since in-leakage of air or moisture is avoided. Consideration should also be given as to what other process materials or utilities will be added to, connected to, or mixed with the reactive material. For situations where moisture exclusion is important, measures must be taken to ensure moisture is not introduced by way of the added material or connected stream.

5.6 Monitoring and Control

The most common operating parameters that are monitored and controlled in reactive materials storage facilities are oxygen, humidity/moisture, pressure, and especially temperature. The only other storage parameter that was specifically mentioned in the industry-survey responses was tank level; level indicators, alarms, and sometimes interlocks are used with some bulk storage facilities. Other operating parameters, such as purity, are checked more by sampling and analysis than by monitoring and control.

5.6.1. Oxygen Concentration Monitoring

The oxygen concentration in the vapor space of a reactive chemicals storage/handling operation may need to be monitored for one of two reasons. For certain monomers such as acrylic monomers, the oxygen concentration needs to be maintained **above** a minimum level in order to ensure the polymerization inhibitor remains activated. For liquids stored or handled above their flash points, the oxygen concentration may need to be maintained **below** a certain level to avoid a flammable vapor–air mixture inside the containment. For some flammable and polymerizing liquids, both of the above may apply.

Oxygen concentration may need to be monitored prior to the introduction of reactive materials into a tank or container, as well as during subsequent storage. As **Summary 21** (page 306) indicates, direct measurement and monitoring of oxygen concentration is the exception rather than the rule. Where a minimum oxygen concentration needs to be maintained for polymerization inhibitor activation, normal tank breathing is usually considered sufficient atmospheric exchange to supply the necessary oxygen. However, if a reduced-oxygen atmosphere is maintained (at 8% oxygen, for example), then ongoing monitoring of the oxygen concentration, either directly or indirectly, will be needed in order to maintain the oxygen concentration within the allowable range.

For most flammable liquids, inert gas purging and padding according to predetermined procedures is generally considered to be adequate assurance of oxygen reduction when required. A low pressure alarm is often added to the inert gas supply line to provide additional assurance of adequate purging.

In the relatively few situations where direct monitoring of oxygen concentration is considered to be warranted, many instruments are available to measure oxygen concentration of a gas stream. This measurement can either be on-line, by means of a permanently installed oxygen analyzer, or in the field, by means of a hand-held gauge or meter such as is used for checking for a flammable atmosphere before hot work is conducted or vessel entry is permitted.

5.6.2. Humidity/Moisture Content Monitoring

The industry survey results in **Summary 22** (page 306) indicate that humidity/moisture is monitored even less commonly than oxygen concentration. Other design and operating practices are relied upon to exclude humidity/moisture when necessary, such as dry inert-gas purges and container sealing. One response indicated that the humidity/moisture content of a polymerizing material was monitored periodically, such as on a monthly basis, but only to ensure the material remained within its specified range for moisture content.

5.6.3. Pressure Monitoring

The **Summary 23** survey results (page 307) indicate two primary situations when the pressure is monitored in a reactive chemical storage facility: first, to detect loss of inert gas supply to a purge connection or loss of inert gas padding on the storage vessel; second, to detect overpressurization of the storage vessel, particularly during a filling operation. However, there are many storage facilities that have no pressure monitoring, such as for smaller containers or when a storage tank can be maintained at atmospheric pressure and have multiple venting systems to ensure adequate and reliable vapor displacement upon filling or discharging of the vessel.

A variety of instruments are used to monitor inert gas or internal tank pressure. Pressure gages and differential pressure transmitters are employed to measure the pressure; chemical seals are commonly used to isolate the pressure sensor from the reactive material. Depending on the criticality of the pressure monitoring, redundant devices may be employed. Pressure readings may be taken either manually, such as during routine operator patrols, or recorded by a chart recorder. Depending on the reason for the pressure monitoring, either a low-pressure or a high-pressure alarm and possibly a process interlock may also be included to give immediate warning or action upon detection of a high or low pressure deviation.

5.6.4. Temperature Monitoring

Temperature monitoring is especially important for thermally decomposing materials, and also for polymerizable materials where elevated temperature can initiate self-polymerization. Reduced temperature may also initiate polymerization in some cases, by causing the inhibitor to freeze out of solution. As indicated in **Summary 24** (page 308), a variety of temperature monitoring configurations are used in industry, both for monitoring the temperature of reactive materials in tank storage and for monitoring the temperature inside refrigerated or otherwise temperature-controlled storage buildings. For storage of reactive materials in small containers, the latter obviously is an indirect monitoring of the material temperature, and may be insufficient under some situations to detect a dangerous temperature rise in thermally sensitive materials. For example, such a material may be stored below its self-accelerating decomposition temperature (SADT) determined for the container size used, but may inadvertently be contaminated by a substance that will cause a gradual but self-accelerating temperature rise during extended storage. In addition to sampling and analysis, this situation shows the importance of both maintaining a safety margin between the actual storage temperature and the SADT and maintaining minimum inventories and rapid turnover of stored material whenever possible.

Temperature monitoring in temperature-controlled storage buildings generally includes both temperature indication (local or control-room) and high temperature alarm. Redundancy in both temperature measurement and high-temperature alarms is common. The alarms may be located either at the storage building itself or in the control room, or both. Proper procedures for dealing with a high-temperature situation must be established ahead of time, such as by starting up a backup cooling system or by moving the stored material to another temperature-controlled building where feasible.

Temperature monitoring in tank storage ranges from no monitoring to multiple redundant sensors and alarms. The general approach is to design the temperature monitoring system, whether field or control room, such that a temperature deviation outside safe limits will be detected and the operator will be alerted in time to regain control before a reactive chemical incident occurs. Depending on the nature of the reactive material and the storage arrangement, "regaining control" might necessitate anything from adjusting a controller setpoint to starting a backup cooling system, adding inhibitor to the tank contents, dumping or quenching the tank contents, or transferring the material to an alternate location where temperature control can be re-established.

5.6.5. Temperature Control

The temperature of stored reactive materials may need to not only be monitored but actively controlled to maintain the material within safe operating limits. As for temperature monitoring, temperature control may either be accomplished as part of the configuration of storage tanks or cylinders or by controlling the internal temperature of the building enclosure where the reactive material is stored. Industry practice for temperature control is given in **Summary 25** (pages 309–310); the responses indicate that tank/cylinder heating is almost as common as tank/cylinder cooling.

Tank/Cylinder Cooling

Heat transfer systems are the most common means employed for providing necessary cooling of storage vessels. Using a white or reflective paint on storage tanks can also reduce heat input from solar radiation. External cooling, such as by fire protection water spray, deluge, or fire water monitors, can be provided as needed for emergency tank cooling.

As discussed in Section 5.2.4, any heat transfer fluid used with a reactive chemical storage arrangement should be compatible with the material being stored, unless special precautions such as double tube sheets are taken to avoid contact between the material and the heat transfer fluid. Chilled water or brine is commonly used for tank/cylinder cooling for non–water-reactive materials; other fluids are used where compatibility problems exist.

Tank/Cylinder Heating

Heating of the contents of a tank or cylinder may be necessary to prevent freezing of the reactive material. Freezing can make certain temperature- or shock-sensitive materials even more easily initiated, and can cause phase separation of polymerization inhibitors. Some heating of surfaces in contact with the vapor space may also be necessary to avoid condensation in some polymerizing materials such as styrene; one method used (that may not always be sufficient, depending on the location of the facility) is to use a black paint for the top of the storage tank to increase the tank surface temperature.

As for cooling, heat transfer systems are also the most common means of providing necessary heating of storage vessels. The same heat transfer system is often used for both heating and cooling, with tempered water used most frequently as the heat transfer fluid. The tempered water can be both cooled by circulating through a cooling tower or chiller and heated by passing through a steam heat exchanger. Tempered water systems have the advantage of more reliably controlling and limiting the temperature of the heat transfer medium. Other heat transfer fluids used, depending on the application, include steam, hot oil, brine, or other compatible material.

Heat tracing of piping systems may be necessary in a few situations such as for high-freezing-point monomers or low-ambient-temperature locations. The heat input can be provided by either steam or electric heat tracing. Note that self limiting electric tracing is **not** generally suitable for thermally sensitive materials, since most such tracing limits the temperature to 200 to 240°F (93 to 116°C), which is well above the initiation temperature for many reactive chemicals. One company reports using a special heat tracing system for thermally decomposing material, with shutdown of the tracing initiated by either the shorting of a thermocouple or the sensing of overtemperature by a separate high-temperature switch.

Building Temperature Control

Refrigerated storage is used in industry for some uninhibited monomers and other shock-sensitive, thermally decomposing, pyrophoric, peroxide-forming, and water-reactive materials. The motivation for having refrigerated storage varies with the type of reactive material; for example, for organic peroxides and other thermally sensitive materials, refrigeration or cold storage (sometimes as low as −15°C or even colder) is required to maintain the material temperature below the self-accelerating decomposition temperature. Peroxide-formers might be refrigerated in order to increase shelf life by slowing the formation of easily initiated peroxides.

Depending on container size and inventory requirements, other reactive materials might be stored in enclosures ranging from properly equipped and labeled refrigerator/freezers to refrigerated warehouse-type storage buildings. Cold rooms, small refrigerated buildings, and enclosures with standard heating/ventilation/air conditioning systems are most commonly used when storing process quantities of reactive materials.

As described in Section 5.6.4, most storage enclosures where temperature control is an important safety-related parameter are also equipped with temperature monitoring and/or deviation alarms. NFPA 43B (1993) gives design requirements, including temperature monitoring, for cold or controlled-temperature storage of organic peroxides. Although the danger of insufficient cooling is obvious when storing and handling thermally sensitive materials, it should be noted that excess cooling can also be dangerous when handling some reactive materials such as inhibited solutions.

Building heating and ventilation are less frequently required than building cooling. Building heating may be required for the same reasons as for tank/cylinder heating as discussed above. The most common heating and ventilation systems used are standard heating/ventilation/air conditioning systems that meet fire code requirements for the stored reactive material.

5.7 Handling and Transfer

Both manual handling and piping system transfers are used for moving reactive materials on-site. The kind of material handling system used for reactive materials is dictated by the process and inventory requirements, the type of container(s) in which the reactive material is received or shipped, and by safety considerations.

5.7.1. Manual Handling

Due to the hazardous nature of reactive materials, they are very often handled in smaller quantities and container sizes than bulk commodity chemicals. For this reason, manual handling of containers ranging from small cartons to drums and tote bins is commonly required. However, manual handling inherently has different incident potentials than piping system transfers. In particular, the close proximity of the operator to the reactive material and the potential for dropping or collision during manual handling are significantly different considerations than potentials such as dead-head pumping or blocked-in material in piping systems. Therefore, some companies avoid all manual handling of certain highly reactive materials. Industry practice with respect to manual handling of reactive materials is represented in **Summary 26** (page 311) and discussed in the following paragraphs.

Pallets with Hand Trucks

Moving pallet-loads of cartons, sacks, or drums is fairly common practice for most types of reactive materials. For short distances where stacking is not required, or in areas in which forklift trucks do not have sufficient room to maneuver, hand trucks or self-propelled but operator-guided pallet movers can be used. They are less expensive and are less hazardous with respect to collision and flammable vapor ignition; generally, they are only practical for single-building use over relatively short distances.

Pallets with Forklift Trucks

Forklifts are used more commonly than hand trucks, such as for transferring drums or cartons from an unloading dock to warehouse storage, from storage to process area, from a packout area to product inventory storage, and/or from storage to a loading dock. It must be ensured that each forklift truck that might be used for handling a given material has the proper safeguards for use with that material, such as its electrical classification for minimizing the likelihood of ignition of leaking or spilled flammable materials.

Forklift incidents are notoriously common in industrial operations. Proper training and qualification of forklift drivers, particularly in situations where reactive materials are being handled, is essential to safe facility operation.

It is good practice to move only sealed drums or cartons of reactive materials by forklift. Some companies have a requirement to open and then re-close the bung of each drum (slowly, while using proper personal protective equipment) before moving in order to detect any unusual pressure buildup inside the drum. However, this may be undesirable for some types of reactive materials such as peroxide formers and for those having significant toxicity and volatility in addition to reactivity.

Other Forklift Transfers

Forklift trucks are also used to handle other types of reactive material shipping or transfer containers. In particular, cylinders, tote bins, super sacks, specially constructed skid-mounted containers, and containers that are moved directly on the forks of a forklift truck are all in use with reactive materials. One company uses even larger portable tanks with certain pyrophoric materials; they are transferred by attaching the tank lifting lugs to a crane. All transfer containers will, of course, need to meet regulatory requirements for the container design if used for off-site transport of reactive materials.

Roller Conveyors

Roller conveyors are used by some facilities for cartons and/or drum transfers over relatively short distances, such as at a drum filling station. Conveyor systems must be carefully designed and the operating procedures thoroughly reviewed to minimize the potential for incidents such as overturning of a full drum or generating sufficient static electricity to create an electrostatic-discharge ignition source for flammable vapors or combustible dusts.

Transfer to Smaller Containers

It may be desirable to transfer a quantity of highly reactive material into a smaller container for on-site transport when a full container is not needed, in order to reduce the handling of the larger quantity. However, transfer to the smaller container while inside a storage enclosure should be avoided, in case spillage of the reactive material could lead to an incident involving the total stored quantity.

5.7.2. Piping Specifications and Layout

Many more design details are required for an operation involving transfer of reactive materials with a piping system than for a manual transfer operation. Codes, industry standards, company standards, regulations, and manufacturers' recommendations may all be available and pertinent for a given reactive material piping system. The exact design details will depend in large part on the physical characteristics of the material being transferred (e.g., the piping schedule will depend on pressure, corrosion allowance, and overdesign considerations, and materials of construction range from carbon steel to high-performance alloys). For this reason, the following sections will only address those generalizations that are common to most reactive materials, based on **Summary 27** (pages 312–314). Details on piping specifications and layout details for several specific reactive materials are given in Chapter 7.

Several layout details are common to nearly all reactive chemical liquid piping systems. Both piping lines and headers are nearly always sloped to allow complete drainage of lines, with no dead legs being allowed. Note that drainage back towards a storage tank creates a risk of contaminating the bulk of the stored material. During operation, reverse flow into equipment containing reactive materials may lead to uncontrolled reactions and severe consequences. Backflow may need to be protected against by installing failsafe controls such as check valves with double block and bleed valve systems or duplication of instrumentation.

Process layout and piping runs should be configured to minimize the length of piping between units. Piping must be routed to prevent impact by external forces such as trucks or cranes on-site, and barriers should be installed where piping might be vulnerable to such forces. For critical piping installations, a suggested practice is not only to review the piping design as diagrammed schematically on piping and instrumentation diagrams (P&IDs), but also to review, as part of the design process and/or the process hazard analysis, the actual piping layout configuration as shown on a piping arrangement drawing or a three-dimensional model (computer-generated or scale model).

5.7.3. Fittings and Connections

Summary 27 indicates a wide variety of practices with respect to fittings and connections in reactive materials service. Some generalizations are discussed in this section.

Piping Connections

Piping system failures nearly always occur at connections, flexible hoses, or small-diameter piping. Screwed connections should be avoided whenever possible, along with potentially vulnerable points such as sight glasses. All-welded connections should be used as much as possible, both to reduce the likelihood of leaks and line failures and to provide better electrical continuity in piping systems where static electricity is of concern. Where flange connections are necessary, higher-strength flanges can be used to improve the integrity of the piping system (such as going from class 150 flanges to class 300 flanges). A ground-strap clamp and/or an electrical continuity check across each flange may be needed to avoid accumulation of static electricity. Connections on lines handling flammable reactive materials inside buildings should meet the fire-resistance requirements of NFPA 30 (1993).

Splash Protection

Where the reactive material being handled is a liquefied gas or corrosive liquid under pressure, splash collars may need to be installed at flanges and valve stems to minimize the likelihood of operator injury if a flange leak occurs or a gasket, seal, or packing fails. Splash protection around pump seals may be warranted to reduce the likelihood of operator exposure if a seal failure occurs; aerosol formation will also be reduced, and some condensation of the liquefied gas is possible.

Unloading Lines and Flexible Hoses

Flexible hoses provide an obvious convenience for unloading connections and other frequently disconnected lines. However, for some safety-critical installations, flexible connectors are avoided and hard-piped flanged connections are used instead. Either hard-piped systems or flexible connections may employ specially designed connectors such as dripless couplings or breakaway seals. For less hazardous materials, features such as quick-disconnect connectors can be used.

Where flexible hoses are used, they are generally reinforced with steel or stainless steel braiding. Flexible unloading hoses in hazardous chemical service are typically visually inspected prior to each usage and replaced in their entirety on a regular basis, such as once a quarter or once a year. Ensuring electrical continuity across flexible connections is a significant concern in systems with static electricity problems.

Unloading connections also provide opportunity for cross-contamination and misdirecting of materials. Dedicated and unique connections and hoses are being used for loading and unloading in some installations where contamination or incompatibility is of concern; for example, the decomposition of some organic peroxides can be initiated by acids, unsaturated organics, and other chemicals. Clear and distinct labeling of lines and connections is also very important where multiple unloading connections exist.

5.7.4. Pumps and Pump Seals

Centrifugal pumps are reported as being in use for all types of reactive materials. When any kind of pump seal fluid is used, even for double mechanical seals or tandem seals with a barrier fluid, it is essential for the seal fluid to be fully compatible with the reactive material being transferred, as well as not contaminating the material with respect to process use or product quality. Depending on the application, nonaqueous liquids such as light oil, ethylene glycol, and methanol are in use as seal fluids.

Since all pumps with seals will have some degree of leakage by virtue of the seal design, sealless pumps have become more widely used in industry to reduce fugitive emissions and to avoid the necessity of using a separate seal fluid. Magnetic-drive and canned pumps are totally enclosed and no seal fluid is required; consequently, incompatibility concerns are avoided and leak potential is reduced. Although sealless pumps have become a preferred option for some companies handling reactive materials, one company specifically reports not using canned pumps for such materials. With canned pumps, the reactive material must be compatible with all pump internals, and not degrade in use to the point of causing contamination or motor failure. If the motor overheats for any reason, it may cause serious overheating of the reactive material in the can.

Positive-displacement pumps such as piston and diaphragm pumps are also in use with some reactive materials. It is essential that the motive fluid for diaphragm pumps be compatible with the process material, in case of leakage or rupture of the diaphragm. Diaphragm pumps have the advantage of the diaphragm being the only internal part of the pump in contact with the reactive chemical.

One of the primary considerations when designing a pumping system for reactive chemicals, especially thermally decomposing materials, is to minimize the potential for pump dead-heading. Various design strategies are used to achieve this objective, such as one or more of the following:

- low-flow pump shutoff with timing delay
- high-temperature pump shutoff
- high-pressure pump shutoff
- valve-position sensors interlocked with the pump
- minimum-flow recirculation line
- emergency relief device(s) for overpressure protection.

For materials that are not self-reactive, it is generally necessary to block in both the suction and discharge lines of the pump to generate sufficient heat and internal pressure to result in a pump or line rupture, unless a positive-displacement pump is used. In addition to dead-heading pumps, overheating of seals or bearings may also be protected against.

The automatic shutoff of pumps can be effectively used as an additional safety feature with many storage and handling systems where pump shutoff will not in itself lead to a hazardous situation. For example, tank filling pumps are often interlocked to shut off upon detection of a high storage tank level. The shutoff of a transfer or unloading pump can also slow the spill of material if a downstream leak or line failure occurs.

5.7.5. Valves

Some considerations regarding valves used in reactive material piping systems are as follows:

- Valves for highly reactive chemical service should be selected that do not have dead spaces and will not cause overheating from friction. Valve types such as diaphragm valves can be selected that have minimal internal parts in contact with the reactive material, or features such as Teflon® lining can be specified where corrosivity or reactivity will preclude exposure to internal metal parts.
- Fire-rated valves may need to be specified when handling a flammable reactive material in or near the process, particularly at locations such as storage tank bottom outlet valves and remote isolation valves.
- Ram-type sample valves can be used to prevent material from collecting at the sample point.

- A double-block-and-bleed valve arrangement can be effective where positive isolation is required without disconnecting a line.
- Drain valves and other terminal valves should have a plug or blind flange on the open side of the valve to avoid an accidental release if the valve leaks through or is inadvertently opened or left open. Sample valves should also have provision to prevent and minimize leakage.
- Control valves and other remotely actuated valves are generally specified as fail-close; however, the specific design followed by a process hazard analysis must determine and verify the proper fail-safe state of each automatic or remotely actuated valve in reactive chemical service.
- Likewise, the influence of each valve on the entire process must be reviewed from a safety perspective, to determine the necessity for such configurations as interlocking a valve with a pump in the same line so that the pump will not operate if the valve is closed. This is especially important for self-reactive materials that may polymerize or decompose, for example, under dead-head pumping conditions.
- Another safety consideration of the valving and piping layout is protection, such as by the use of liquid expansion relief valves, against the total blocking in of a fluid that can thermally expand upon warming and lead to a leak or rupture.

5.7.6. Drain Systems

Consideration must be given ahead of time as to what will be done with material drained from the system for maintenance, shutdown, etc. Because most reactive materials also have some degree of toxicity, draining directly to even a process sewer before treatment may not be desirable or allowable. Industry practice varies from area runoff discharged to stormwater drains for relatively nonhazardous materials (no longer recommended) to avoiding all system drainage for some pyrophoric and water-reactive materials.

5.7.7. Cleaning Equipment

In addition to drainage of material from the storage and handling system, special equipment or piping system design may be required for internal cleaning, such as prior to line breaking or where the potential for line plugging exists. For example, a system handling a highly water-reactive material may require initial flushing with a nonaqueous solution. One company reports designing long lines in such a way that a pigging device can be used when line clearing is required.

5.7.8. Transfer Systems Operating and Maintenance Practices

Industry operating practices related to reactive materials piping systems are represented in **Summary 28** (page 315). Written operating procedures are used by most companies for process operations; however, they are somewhat less commonly used for routine transfers within storage and handling operations. Operations such as start-up of continuous facilities, batch transfer of material, and flushing or blowing out lines after completion of transfers can benefit from procedures that give a consistent, established, and thoroughly reviewed sequence of steps to follow to safely conduct such operations.

Maintenance procedures also need to be established and used as the basis of both training and routine maintenance activities. Maintenance activities in reactive material storage and handling facilities that can benefit from standardized procedures and practices include the following:

- replacing piping, gaskets, seals, or hoses using the proper materials of construction
- replacement, inspection, or testing of transfer hoses, seals, emergency relief systems, alarms, interlocks, and other critical safety devices
- draining and/or flushing lines to remove reactive materials before work is done on the system
- testing the system for adequate decontamination before opening the system
- preparing new equipment for service after installation, such as the pickling of steel before use in certain applications.

Facilities where reactive materials are handled above specified threshold quantities may come under government regulations that mandate operating procedures be developed and used. Process hazard analysis methods such as What-If studies and Hazard and Operability (HAZOP) studies can be effectively used to review procedures and determine where inadequate safeguards exist.

5.8. Last-Resort Safety Features

Last-resort safety systems are intended to be used in many reactive chemical storage and handling operations as "last-ditch" efforts to prevent an accidental event such as an explosion or a hazardous material release from occurring, if the operation exceeds safe operating limits and it is not possible to regain control using the operation's normal control mechanisms. In addition to features such as inhibitor injection and quench systems designed to stop an out-of-control reaction or dump and vent systems designed to transfer reacting material to a safer location, last-resort safety features also include emergency relief systems including relief devices and relief discharge treatment. Industry practices related to each of these approaches are represented in **Summary 29** (page 316) and are discussed in the following sections. They are also discussed as "emergency process abort systems" in *Guidelines for Vapor Release Mitigation* (CCPS/AIChE, 1988, 47–49).

5.8.1. Inhibitor Injection

Inhibitor injection systems are primarily used with polymerizing materials such as vinyl acetate. If the material begins to self-react in an uncontrolled manner, then injection of a polymerization inhibitor can interfere with the reaction before sufficient pressure and temperature are built up to cause a release from the storage/handling containment. The type of inhibitor needed will depend on the nature of the polymerization reaction; for example, a free radical scavenger may be used as an inhibitor for a material that reacts by free-radical polymerization. The inhibitor is often the same inhibitor used for normal storage stability requirements, but injected in a much larger quantity. If a different inhibitor is used that is designed to quickly kill the reaction, it is generally called a "short-stop" system.

An inhibitor injection system consists basically of a supply of inhibitor, a means of delivery of the inhibitor to the reactive material, and a means of sensing the need for the inhibitor injection system and actuating the inhibitor delivery. The supply of inhibitor can be in one or more dedicated tanks or bottles or, if a solid, in a bin or container. The delivery system can range from manual addition, such as through a tank manway or into the top of an open-top vessel, to a highly automated, computer-controlled injection system. Depending on the physical properties and quantity of inhibitor required and the operating pressure of the storage/handling facility, the means of transferring the inhibitor from the supply location to the reactive material location may be:

- by gravity (the most reliable motive force, if in a nonplugging service and if the operating pressure under upset conditions will still allow gravity transfer),
- by pressure (such as by a compressed-gas padding in the vapor space of a liquid inhibitor hold tank), or
- by pump (requiring the reliable starting and operation of a standby pump).

Provision must also be made for ensuring the inhibitor is adequately mixed. The potential for common-mode failures must be carefully evaluated in this regard; for example, in a polymerization process, loss of agitation might both initiate an out-of-control situation and prevent adequate mixing of added inhibitor. A backup mixing system such as a gas sparge may be required in such cases. Other means of inhibitor mixing include injection into a recirculating loop (requiring the proper functioning of the recirculation pump), using a mixing eductor, directly sparging into the storage tank, or by manual agitation in smaller, atmospheric-pressure operations with adequate safety precautions.

Inhibitor injection systems need to be carefully designed and maintained to provide a highly reliable last-resort safety system. Since the inhibitor injection system is on standby and may not be used for months, attention must be paid to how the system components can be functionally and effectively tested on a periodic basis, such as once a month, without excessive disruption of normal operations. *Guidelines for Engineering Design for Process Safety* (CCPS/AIChE, 1993, 273–275) discusses testing of continuous-process safety systems. This functional testing is important not only for the checking of adequate inhibitor supply and properly functioning delivery system, but also for the means of detecting an out-of-control situation and actuating the inhibitor injection system. If the system is actuated manually, the system should be part of periodic operator drills or simulation training. Most inhibitor injection systems are designed so that they can be actuated either automatically, such as upon sensing of high-high temperature or pressure, or manually, by the control room or field operator.

5.8.2. Quench System

Quench systems are used for essentially all types of reactive chemicals. A quench system involves the addition of flooding quantities of water or other quenching medium to the reactive material; the quenching medium might be a subcooled material such as liquid nitrogen or dry ice in special applications.

The means by which a quench system works depends on the nature of the reactive material; for example, for water-reactive materials, a quench system will destroy the material in a last-resort situation and generally form less-hazardous products, and will at the same time absorb some of the heat of reaction. Most quench systems are designed to both cool down and dilute a material that may be reacting uncontrollably; the quenching medium may also actually interfere with the chemical reaction or deactivate a catalyst.

Assuming water is the quenching medium, the water for the quench system can be supplied in a number of ways. Water may be gravity-fed if process conditions allow; the supply of water in this case can be a dedicated quench tank or a plant water tower. The water can also be supplied by a reliable process water system or fire water system. As for inhibitor injection systems, care must be taken to design and maintain a highly reliable means of both detecting the need for the quench system and actuating the system manually and/or automatically.

In a quench system, the water or other quenching medium is generally added to the storage/handling facility in significantly greater quantities than the inhibitor in an inhibitor injection system. For this reason, the system design must take the extra volume of material into account. Also, as for inhibitor injection systems, a means of mixing the quenching medium with the reactive material may be needed for effective quenching action.

5.8.3. Dump System

For an inhibitor injection or quench system, the inhibitor or quenching medium is transferred from an external supply to the reactive material; in a dump system, the reactive material is transferred from the storage/handling facility to a safer location that is the same size or, more commonly, larger than the normal capacity of the facility. This allows depressuring and deinventory of the reactive material from the facility in an out-of-control situation, such as an incipient thermal decomposition.

Detection and actuation of a dump system is similar to an inhibitor injection or quench system, with the dumping typically being started by actuating one or more quick-opening dump valves. Containment of the dumped material may be in an above-ground, below-grade, or "piggyback" tank; a sump; or a larger basin. Design considerations include:

- the time required to dump the system, which should be carefully evaluated and take full consideration of backpressure and two-phase flow
- reliability of the detection and actuation system and the dump valves, including protection against inadvertent dumping
- ensuring adequate volume in the dump tank if one is used
- possible leakage of water or introduction of an incompatible material into the dump tank when not in use
- overpressure protection on the dump tank as well as the reactive material storage tank.

It is common to combine the effects of both dumping and quenching by dumping the reactive material to a location where a quenching medium is already located, such as a pond, basin, or large tank. To maintain an adequate level of quenching medium, periodic checks should be part of the facility's operating instructions and routine patrols. An alternate configuration is an eduction system with a high-rate water supply, such as a fire water system, with the educted material being diverted to an empty tank, sump, or basin.

5.8.4. Depressuring System

Although more common for process operations than for storage and handling facilities, a last-resort depressuring system can be added to a reactive material storage tank to vent off excessive pressure buildup in the tank in a controlled manner before reaching the relief valve or rupture disk set pressure. Such a depressuring system typically consists of a remotely actuated vent valve connected to the vapor space of the storage tank, with the venting discharge directed to a scrubber or other treatment system of adequate capacity (as discussed in Sections 5.8.7 and 5.8.8 below). The system can be designed to be actuated either manually, by a control room or field operator, or by detection of high pressure and/or high temperature in the storage tank.

Where the storage tank is operated above atmospheric pressure, such a depressuring system can be readily tested, all the way from manual activation of the system to verification of flow through the discharge line. Testing of the automatic actuation mode requires more sophisticated checking procedures, such as the manual loading of a signal from each sensor and detection of the trip signal, with separate checks for proper functioning of each sensor in the system.

5.8.5. Emergency Relief Configuration

Emergency relief systems consist basically of one or more emergency relief devices, relieving either directly to the atmosphere "in a safe location" or by way of a relief header to a means of treating or handling the discharge, which can range from a stack to a combination of treatment devices in series.

When considering installation of an emergency relief system not required by codes, the need for the emergency relief system must be weighed against the consequences of an emergency relief. The risk involved in a relief device discharge of reactive material (either due to a genuine demand on the system or an inadvertent opening of the relief device) might actually be greater in some cases than the risk of containment rupture without the relief device, particularly if the tanks, vessels, piping, and connections are designed to fully contain most overpressure scenarios. However, overpressure containment is less practical for reactive chemical scenarios than, for example, for vapor-phase deflagration scenarios. The sizing basis for the emergency relief devices, along with general design considerations for emergency relief headers and treatment systems for reactive materials, are described in the following sections.

Rupture Disks, Relief Valves, and Fusible Plugs

As represented in **Summary 30** (page 317), different configurations of emergency relief devices are used in reactive materials storage and handling operations. Although relief valves are commonly used on hoppers, storage tanks, etc., relief valve/rupture disk combinations may be even more common, particularly for polymerizing materials. Rupture disks by themselves appear to be somewhat more commonly employed when handling pyrophoric, peroxide-forming, and water-reactive materials; they are also used alone where a relief valve does not have adequate discharge area or fast enough response time. Fusible plugs, designed to melt and provide relief in a fire situation, are used on some reactive material transport and storage cylinders. The advantages and disadvantages of rupture disks versus relief valves are described in detail in *Guidelines for Engineering Design for Process Safety* (CCPS/AIChE, 1993, 420–426).

Rupture Disk/Relief Valve Combination

The advantages of using a rupture disk and a relief valve together are twofold. First, plugging of the relief valve inlet line is less likely with a rupture disk underneath it, and the rupture disk can be installed nearly flush with the inside of the storage tank or with an air or nitrogen sweep to further minimize the plugging potential. Second, since many reactive materials are also highly corrosive, it may be necessary to use a relatively expensive relief valve fabricated from corrosion-resistant materials of construction. The use of a corrosion-resistant rupture disk under the relief valve may allow a much less expensive relief valve to be used.

If such a configuration is used, it is essential that the space in between the rupture disk and relief valve is monitored for leakage past the rupture disk. Since rupture disks work on the basis of differential pressure across the disk, the tank pressure would need to be up to twice as high as normal before bursting of the disk if the space downstream of the rupture disk is pressurized. Also, leakage of reactive material past the disk may result in physical blockage of the line and/or corrosion of the relief valve.

Several configurations are used to monitor the intermediate space, the most common of which is a local pressure gage that is checked as part of routine operator patrols. A small bleed line may also be added if the material is not overly toxic or corrosive. However, if the relief system is at an elevated or difficult-to-access location, the checking of the pressure gage might be neglected over time. Other configurations include a high pressure switch with control room alarm and the addition of a compatible pad gas in the intermediate space with both high and low pressure alarms.

5.8.6. Emergency Relief Sizing Basis

As can be seen in **Summary 31** (page 318), a uniform means does not exist for determining the emergency relief sizing basis in reactive chemical storage and handling facilities. However, some common threads emerge from the practices of different companies:

- The most prevalent scenarios used for sizing the emergency relief systems are external fire exposure and runaway reaction (polymerization, thermal decomposition, etc.). Consideration should also be given to incompatible mixing, such as the inadvertent addition of water to a water-reactive material or vice versa.
- Industry recommended practices are commonly used for sizing of relief devices where fire scenarios and single-phase flow are anticipated.
- The potential for two-phase flow through the relief system is commonly but not universally recognized. This potential is especially important to consider in reactive systems. The methodology developed by the Design Institute for Emergency Relief Systems (DIERS) is used by many companies where two-phase venting with or without reaction is possible (DIERS, 1992).
- Small-scale testing is often required to determine various parameters necessary for relief system sizing. Test apparatuses in use include the Vent Sizing Package (VSP), Reactive System Screening Tool (RSST), and Accelerating Rate Calorimeter (ARC).
- Some companies have internal corporate standards or guidelines for determining the emergency relief sizing basis and performing the sizing calculations.

Further discussion of the sizing of emergency relief systems is given in *Guidelines for Engineering Design for Process Safety* (CCPS/AIChE, 1993, 428–440).

5.8.7. Emergency Relief Headers

The purpose for having headers as part of an emergency relief system is to collect emergency relief discharges and direct them to a stack, flare, scrubber, or other treatment system. As represented in **Summary 32** (page 319), some companies use relief headers in their reactive material storage and handling facilities and some do not use them or even consciously avoid using them.

Relief headers designed to handle emergency discharges are more difficult to design for safe operation than headers capturing normal process vents such as for vapor emission control. Some basic considerations that must be included in the design of emergency relief headers include:

- Sizing of the relief header and downstream treatment system to handle multiple simultaneous ventings, such as in the case of a massive fire engulfing several tanks.
- Sizing of the headers such as to prevent backpressure in excess of 10% on any individual relief device (API RP 520, 1988, 1993). If balanced-bellows relief valves are used, the valve capacity is not significantly reduced until the backpressure exceeds 30%.
- Specification of the header system material of construction for adequate corrosion resistance.
- Effect on the material of construction of extremely cold temperatures generated by flashing liquid discharges involving low-boiling-point materials.
- Intercompatibility of all streams potentially venting into the header simultaneously.
- Consideration of drainage from each discharge line and header, cleanout and maintenance, potential for plugging or freezing, and avoidance of pockets and dead legs.
- Inclusion of flame arresters and/or header system purging to avoid flashback where flammable vapors may be present.
- Inclusion of one or more knockout drums to allow disengagement of the liquid and vapor phases where two-phase venting may occur.
- Supporting of the entire header system for both emergency relief reaction forces and the weight of liquid if the potential exists for liquid to discharge or overflow into the header system.

Design and installation of the header system may be covered by various regulations or recommended practices, such as ASME B31.3 and API RP 520, Part II. As can be seen from the above, design of an emergency discharge header system for safe operation is by no means trivial. The final design of such a system warrants careful review such as by a process hazard analysis team review.

5.8.8. Emergency Relief Treatment Systems

To avoid discharge of reactive materials directly to the atmosphere in the event of an emergency venting, several types of treatment systems are in use that remove, recover, separate, neutralize, dilute, react, or thermally destroy the reactive material in the emergency effluent stream. The various treatment systems in use and how they may be combined are described in *Guidelines for Vapor Release Mitigation* (CCPS/AIChE, 1988, 33–47), *Guidelines for Engineering Design for Process Safety* (CCPS/AIChE, 1992, 445–486), and *Guidelines for Handling of Emergency Relief Effluents* (CCPS/AIChE, in progress).

The information in **Summary 33** (page 319) indicates that most companies do not employ effluent treatment systems for reactive material storage/handling emergency relief discharges. The treatment systems that are used range from discharge of thermally decomposing material from a relief valve into a dedicated dumpster to allow decomposition to occur outside of the storage containment, to scrubbing or flaring of vented gases. In general, where emergency relief treatment systems are used, combustible and/or toxic effluents are sent to a knockout pot or blowdown tank if necessary for phase separation and depressurization, with vapors then being flared (if combustible and if a flare is available) or neutralized in a water or caustic scrubber. Relatively nonhazardous releases are more commonly vented to the atmosphere either directly or through a vent stack. As for header systems, significant precautions must be built into such systems to ensure safe operation, such as the need for inert gas purges where flammable vapor–air mixtures can be generated inside the equipment during emergency venting.

5.8.9. Explosion Suppression

As indicated in **Summary 34** (page 320), the only "explosion suppression" systems indicated as being used in reactive chemical storage and handling equipment are deflagration arresters used in a minority of applications.

5.9 Passive Mitigation

Mitigation refers to reducing the severity of consequences of a fire, explosion, toxic release, or other accidental event involving loss of containment of process material or energy. *Passive mitigation systems* are designed to mitigate accidental events without active operator intervention or the necessity for engineered devices such as interlocks or valves to work properly. Industry practice regarding the use of passive mitigation systems with reactive chemical storage and handling operations are indicated in **Summary 35** (pages 320–321) and discussed below.

5.9.1. Flow-Limiting Orifices

A flow restriction such as a spool piece, a control valve with a minimum C_V, or a restricting orifice can be intentionally included in the design of a reactive materials piping system or tank connection, in order to limit the release rate if an opening to the atmosphere or other unprotected location inadvertently occurs downstream of the restriction. Since the restriction requires no active components to function properly in order for the release rate to be reduced, it qualifies as a passive mitigation device. Such a restriction can be used in essentially any type of reactive material operation, although if a restricting orifice is used, it must be inspected or replaced regularly to ensure the orifice diameter is not enlarged over time by corrosion or erosion and thus defeat the benefit of the safeguard. An additional benefit of using flow-restricting devices is that a larger line size may be able to be used, for purposes of increased mechanical strength (and perhaps reduced overall friction losses as well), and still limit the maximum flow if an accidental release occurs. Such devices are not able to be used in all installations, however, due to plugging or backpressure considerations.

5.9.2. Fire-Resistant/Explosion-Resistant Construction

Fire/explosion-resistant construction is used occasionally with the storage and handling of all types of reactive materials. Fire-resistant construction commonly includes insulation of major structural steel to allow it to maintain its structural strength during a fire, and insulation of storage tanks to reduce the rate of heat input to the tank contents in the event of an external fire. Protection of instrument and electrical cable trays from fire exposure can also be an important passive mitigation safeguard. Cable trays can be protected by either having the cable raceway buried below grade in a trench, or using a 14- to 16-gauge metal sheet below the tray with either fireproofing on the metal sheet or a water spray directed on the top side of the tray (AIChE, 1994a). Explosion-resistant design includes the use of reinforced concrete or reinforced block wall construction for area buildings; this is also a fire-resistant feature.

5.9.3. Weak Seams and Explosion Venting

Storage tanks with weak-seam roofs are used primarily with polymerizing materials in steel tanks. Due to the rapidity of some runaway polymerization reactions, it is difficult to size such a storage tank with a sufficiently large rupture disk and relief system; hence, to avoid a totally uncontrolled rupture of the storage vessel in a runaway situation, either the vessel must be designed with a weak-seam roof or the reactive material must be stored in smaller tanks where stronger construction and emergency venting are feasible.

5.9.4. Bunkers, Blast Walls and Barricades

Structural design of storage areas for passive explosion mitigation is more commonly used in process areas than storage areas, except for the handling of shock-sensitive materials in relatively large quantities, where it is common to use such mitigation features for both processing and storage. Bunkers are commonly constructed at least partially below grade level and use mounding of dirt or other homogeneous material for attenuation of explosion blast effects. Blast walls and barricades, on the other hand, are generally constructed of reinforced concrete. A common configuration is to have the storage area enclosed on three sides, with a blow-out roof and one blow-out wall (such as of corrugated fiberglass panel construction), and with the access to the area outside of the blow-out wall strictly controlled. TM 5-1300 (1990) is a widely used reference in the design of blast-resistant structures.

5.9.5. Secondary Containment

The most widely used form of secondary containment, for all types of reactive chemicals, is diking around storage tanks. This provides a means of capturing liquid leaks to prevent spreading and seepage into the ground or nearby rain water runoff drainage, and is required by regulation or standard in most locations. Storage tanks handling unstable liquids that are also flammable must be placed in individual dikes separated by drainage channels or by curbing at least 18 in (0.46 m) high, instead of large common dikes, for segregation of materials per NFPA 30 (1993). The definition of an *unstable liquid* for this purpose is "a liquid that, in the pure state or as commercially produced or transported, will vigorously polymerize, decompose, undergo condensation reaction, or become self-reactive under conditions of shock, pressure, or temperature" (NFPA 30, 1993).

Sloping of the dike floor to a sump can minimize the surface area for evaporation. Sloping of the dike floor away from storage vessels provides a degree of protection when flammable liquids are involved, especially if a fire wall is placed in between the tank and the collection location (Englund, 1991). The preferred drainage configuration for flammable reactive liquids is remote impoundment per NFPA 30 (1993).

Double-walled piping and double-walled tanks do not appear to be significantly used for reactive chemical storage and handling facilities. Covered dikes and ducting around drainage piping are used in some facilities handling water-reactive materials.

5.9.6. Separation Distances

Distance can be the most effective passive mitigation strategy, since the severity of effect of most process accidents decreases rapidly with distance from the source. (The decrease is sharper with explosions and fires than with toxic vapor releases; however, distance also gives time for warning, sheltering, donning of escape masks or respirators, evasion, and/or evacuation in the event of toxic vapor releases.) In storage and handling areas, the benefit of distance is most commonly achieved by separation of reactive material storage tanks from each other, from storage of other materials, from nearby personnel, and/or from handling or transfer areas. Storage quantity and configuration are discussed in Section 5.4.

5.10. Detection, Warning and Isolation

In addition to passive mitigation measures, it is prudent for even the best-run facilities to have preestablished means of actively mitigating, or reducing the severity of consequences of, a process accident if one were to occur. This section addresses three parts of an effective mitigation strategy that are critical in the early stages of active mitigation; namely, loss-of-containment detection, warning systems, and source isolation.

5.10.1. Release Detection

Before active mitigation measures can be initiated, the loss of containment of reactive material or energy must first be detected. Accidental explosions are generally self-evident, but fires and especially reactive material releases may take some time to detect. Means of detection can range from operator patrols to highly sophisticated electronic sensors. **Summary 36** (page 322) shows some industry practices related to detecting reactive material releases from storage and handling facilities.

Routine Operator Patrols and Area Security

The most common means employed for detecting loss of containment are by routine operator patrols and/or area security monitoring. Operator patrols are typically made at least twice a shift. However, even if patrols are as frequent as once every 2 hours, releases from storage areas that are away from normal site traffic would take on the average 1 hour to detect (i.e., half of the time interval between patrols). This may be unacceptable for major releases of some reactive materials, particularly those that do not have good warning properties such as vapor cloud visibility or odor. For this reason, automatic sensing devices may need to be installed.

Fixed Concentration Monitors

A wide range of vapor detectors are available to the chemical industry. Some detectors are intended for use with flammable vapors and are calibrated to read out in percent of the lower flammable limit (LFL), whereas others are specific to a given chemical or class of chemicals. Typical placement of concentration monitors include likely leak locations (such as pumps and unloading stations), outlying facilities (such as remote storage tanks), and areas where released material might accumulate (such as inside a sump or diked area). In order to be effective, sensors must be placed at a height appropriate to the density of the vapors to be detected, close enough together that the vapors are not likely to pass between two adjacent sensors without being detected, and must be tested and calibrated frequently. CCPS/AIChE (1988, 77–81) discusses types, response times, and positioning of vapor detectors.

Vapor detectors are often integrated into alarm systems that typically warn when a concentration of 10% or 25% of the LFL is detected for flammable vapors, or at a comparable warning level for toxic or other reactive chemical vapors. Some detectors are also incorporated into systems that will shut down and/or automatically isolate a handling/storage arrangement at a higher vapor concentration or upon detection of vapors by two or more sensors. A water deluge or other active mitigation system might likewise be actuated by fixed concentration monitors.

Remote Sensors

Equipment for sensing a release from a remote location are not commonly used in reactive materials storage/handling installations. Some facilities use television cameras, with displays in the control room, to monitor remote storage facilities or likely release locations such as unloading areas. It is possible to install contrast alarms on television monitors that will warn of a sudden change in contrast such as the formation of a vapor cloud.

Other Detection Systems

In addition to the above-described approaches, it may also be possible to detect large releases by process instrumentation, such as a low level or low pressure measurement that is indicated in the control room. Fires and visible plumes are commonly detected and reported by personnel in adjacent operating units or even off-site populations. Heat sensors can also be used to detect an area fire where thermally sensitive materials are stored and handled; the sensors may be tied into mitigation systems such as water deluge sprays.

5.10.2. Release Warning

Also essential to mitigation of a storage and handling incident's effects is the warning of site personnel of the dangerous situation, so that proper pre-planned action can be taken. Summary 37 (page 323) indicates that both control room alarms and plant-wide warning sirens are used commonly in industry as warning systems for reactive chemicals storage and handling facilities.

Control Room Alarms

Control room alarms, particularly those that are hard wired, identified as safety critical, and not susceptible to multiple false trips, are an effective means of bringing a dangerous condition to the immediate attention to the control room operator or operators. Alarms may be triggered by one or more of three different types of input signals:

- From direct detection of a release, such as a concentration monitor detecting a vapor concentration above a predetermined action level or an infrared detector recognizing the location of a release or fire.
- From indirect detection of a release, such as a pressure transmitter detecting loss of pressure in a transfer line or a level measurement device detecting loss of level in a storage tank. Note that indirect detection often requires alarm bypassing at some stage in the operation of the facility, e.g., low pressure or low level may be a normal process state during part of a transfer or unloading operation. This makes indirect detection potentially less reliable, since administrative controls are necessary to ensure alarms are not allowed to remain in the disabled or bypassed state.
- From inferred detection of a release, such as by a calculated excess change in level or pressure. This avoids some of the difficulties of indirect detection, such as needing to bypass alarms on start-up or shutdown; however, functional testing of rate-of-change alarms may be more difficult.

Note that it is imperative for the control room to be continually manned and the operators to be trained in proper and rapid response to such alarms in order for mitigation to be effective.

Plant-Wide Warning Siren

A second level of warning is provided on most sites by some form of plant warning system. This can take the form of a warning siren and visible alarm that can be heard and seen at any location in the operating areas over ambient noise and light levels. For larger sites, a plant-wide warning siren with separate warning devices or signals for each operating area may be desirable. For relatively small fires or releases that are not expected to have an effect beyond the immediate operating area, the local warning signal can be sounded; whereas for major accidents potentially affecting multiple areas, the plant-wide warning signal would be used. The plant-wide signal can also be used to warn neighboring plants, residential areas, etc. of a major incident.

Warning sirens or other warning signals are most effective when combined with a means of public-address or telephone announcement of the location and nature of the problem and whether evacuation, sheltering-in-place, or other evasive action is necessary. Obviously, the signals from such warning systems must be understood and remembered (or the location of immediately available posters, wallet cards, or other signal explanations be known) by all persons within the potentially affected area. Emergency response drills are particularly important in increasing the effectiveness of warning systems, thereby reducing the severity of consequences of potential reactive chemical accidents.

At some facilities, detection devices, such as hydrocarbon detectors, may need to be directly tied into an area-wide warning system in order to meet regulatory requirements. However, the potential for false alarms and the necessity for knowing what set off the warning system in order to take the proper action must be considered. A system with a sufficient number and reliability of detectors may be designed such that the tripping of, e.g., two adjacent detectors would almost certainly indicate a major release that would warrant immediate site-wide or area-wide warning. In any case, an unambiguous control room indication should exist as to what tripped the warning system.

5.10.3. Release Isolation

Another important means of mitigating a reactive chemical loss-of-containment incident is to isolate the source of the release, in order to stop the release as soon as possible and to reduce the total quantity of material released. Testing and maintenance of all release isolation devices is necessary to ensure their continued reliability. Some industry practices regarding the various types of valves used for release isolation are represented in **Summary 38** (page 324) and discussed in this section.

Excess Flow Valves

An excess flow valve is designed to close automatically if the flow through the valve exceeds a predetermined rate, such as if a line ruptures downstream of the valve. Excess flow valves are used selectively in installations where a high vapor pressure or other pressure driving force exists. They have the advantage of immediate leak isolation without any operator intervention and can be installed inside a tank at the outlet nozzle so as not to be vulnerable to external forces. However, drawbacks include a tendency to inadvertently close when not desired, such as at the beginning of a transfer operation if a manual valve is opened too rapidly; and difficulties in checking and maintaining the excess flow valves, particularly if they are installed inside a storage tank or rail car. Excess flow valves may also be ineffective with reactive materials that tend to polymerize or otherwise cause plugging at line restrictions. Another cause of an excess flow valve to fail to function properly is to have the valve installed at the upstream end of a long pipeline. If the leak occurs at the downstream end of the line, the friction in the line may limit the flow to less than the valve setting, but not enough to mitigate the hazard.

Emergency Block Valves

Remotely actuated valves, referred to as emergency block valves or remote block valves, are the most widely used devices for isolating the source of reactive material releases from storage and handling facilities. Control valves can usually be closed from the control room to isolate releases; however, it is more desirable to have independent, remotely actuated block valves for this purpose in order to avoid common mode failures. For example, if the positioner on a pneumatically operated control valve fails in such a way as to cause an overpressure incident and initiate a release, then the same failure would also prevent the control valve from closing. The block valves should be located as close to the storage tank or supply of reactive material as possible, to be most effective at minimizing the release quantity.

Block valves do not always need to be actuated from the control room for them to work effectively. It is becoming more common at unloading facilities handling flammable or reactive liquids to have a means of automatically closing the valve if a release and/or fire occurs. Fire-rated or fire-tested valves with fusible links can be employed to automatically close upon being engulfed in an area fire. Some isolation arrangements may require the proper functioning of instrumentation, such as a heat-sensing trip switch. Other creative arrangements are also being used in industry.

Manual Isolation Only

Release isolation using manual valves is occasionally employed with all types of reactive materials except perhaps shock-sensitive materials. Manual isolation without any remote isolation capability should be used only in situations where a release can be effectively and rapidly stopped without endangering the personnel that must close the manual valve or valves. More time may be required for manual isolation, if it is necessary to don protective equipment before isolation can be performed. Situations where manual isolation might be practical include very small lines and easily accessible valves where the location of the valve is not likely to be affected by the incident.

5.11. Fire Prevention and Protection

Technically speaking, fire protection systems such as fire water monitors and sprinkler systems are mitigation systems in the sense of reducing the severity of a process accident once the release or fire has already started. It is, of course, more preferable to prevent a fire than to have to mitigate it once started. Summaries 39 and 40 (pages 325 and 326) give industry practices related to fire prevention and mitigation.

5.11.1. Ignition Source Control

One means of preventing a fire, even after loss of containment occurs, is to eliminate possible ignition sources from the area. This will not be effective for pyrophoric materials, which require no ignition source for combustion to begin, but can be very important for other reactive materials that are also flammable or combustible.

Nearly all sites handling significant quantities of flammable, shock-sensitive, or thermally sensitive materials have rigorously enforced restrictions on nonessential ignition sources on-site such as matches, pocket cigarette lighters, or even automobile cigarette lighters. Hot work is only done by permit, with appropriate precautions being taken. The design of fired heaters can also include provisions to minimize their likelihood of being ignition sources for accidental releases, such as by having elevated air intakes. The use of specially designed forklift trucks and the routing of traffic away from potential leak locations can further reduce ignition potential. These and other potential sources of ignition such as mechanical sources and spontaneous ignition are identified in *Guidelines for Engineering Design for Process Safety* (CCPS/AIChE, 1993, Chapter 11).

Once open flames and other obvious ignition sources are controlled, the most common remaining source of ignition energy is the electrical instrumentation, switchgear, and equipment in process areas and adjacent buildings. Electrical ignition sources can be controlled by using classified electrical equipment. The most effective means of control is to use intrinsically safe equipment, where the instrumentation or other electrical devices do not contain or use sufficiently high energy levels to ignite flammable vapors, combustible dusts, etc. even if a failure occurs in the electrical equipment. Explosion-proof equipment can be nearly as effective, since it does not require active purging or ventilation of equipment to be maintained in order for ignition to be prevented. **Summary 39** indicates widespread use of intrinsically safe/electrically classified and purged equipment in reactive chemicals storage and handling areas. NFPA 30 (1993) should be consulted for electrical and other ignition source control requirements for installations involving flammable and combustible liquids.

It should be noted that having areas designated by code requirements with electrically classified equipment and instrumentation is only protective against more common but relatively small releases, such as pump seal leaks. Major releases such as from line failures or vessel ruptures can result in flammable/combustible concentrations extending well beyond the electrically classified area.

5.11.2. Fireproofing and Insulation

Fireproofing and insulation can reduce the impact of a fire by prolonging the integrity of structural steel and reducing the rate of heat transfer to storage tank contents. Fire-resistant construction is discussed in Section 5.9.2 as a passive mitigation approach.

5.11.3. Extinguishing Systems

Several means used to attempt to extinguish a fire or limit its spread after ignition include water sprinklers, deluges, and sprays; carbon dioxide systems; and foam systems. Halogenated agents such as Halon 1301 have been used in some reactive chemical installations, but were not included in the industry survey as their use is being phased out. Compatibility of extinguishing agents with reactive materials should always be considered in storage and handling facilities; extinguishing agent compatibility is discussed in Section 5.2.8.

Water-Based Systems

Water sprinkler systems are widely used in all reactive materials storage and handling facilities except with water-reactive materials. Wet-pipe or dry-pipe building sprinkler systems are designed based on flow rate of water per square foot, as a function of storage density and the type of material being stored. Guidelines such as NFPA 13, *Standard for the Installation of Sprinkler Systems* (1994), give general design requirements for sprinkler installations. NFPA 43B (1993) gives sprinkler system requirements that are specific to organic peroxide storage facilities.

Water spray is also used to cool and/or extinguish fires in many other configurations. Most notably, water-based systems include deluge sprays for external tank cooling, fixed fire monitor nozzles (operated either manually or, in some installations, remotely), and standard fire hoses. The use of water curtains, firewater monitors, and fire hoses with fog nozzles for vapor release mitigation are discussed in Section 5.12.1.

For water-reactive materials, a decision should be made ahead of time whether or not to use water-based systems. When used, such systems generally employ deluge quantities of water. Diking and containment must be designed to handle the large quantities of water that may be used.

Carbon Dioxide Systems

Carbon dioxide is used in only a few reactive chemical installations as a fire extinguishing agent. It has the advantages of being applied as a gas instead of a liquid, leaving no residue, and being nonconductive to electricity. It should be considered as an alternative to water spray systems in installations handling water-reactive materials. Carbon dioxide is not effective on reactive materials such as cellulose nitrate, organic peroxides, and most shock-sensitive materials that have available oxygen within the reactive molecule or do not require atmospheric oxygen to self-react; it should also not be used on metal hydrides or metals such as sodium or magnesium that will react with carbon dioxide. Consensus guidelines such as NFPA 12, *Standard on Carbon Dioxide Extinguishing Systems* (1993), have been developed for carbon dioxide-based installations.

Foam Systems

The use of various foam-producing systems have proven to be very effective in controlling and extinguishing many types of fires, including those involving flammable or volatile materials. Placement of a foam layer on the top of a burning liquid surface has a multiple effect in subduing a fire: exclusion of oxygen from the burning surface, suppression of liquid vaporization, and the cooling effect of the water content of the foam as it is turned to steam. Foam can be applied in a variety of ways, such as by portable foam generators, dedicated foam trucks, foam nozzles on water hoses with eduction tubes into containers of foaming agent, subsurface foam injection, and sprinkler system/foam system combinations. The use of foam to extinguish fires that are contained within diked areas can be particularly effective.

Foam systems do not appear to be as widely used in reactive chemical installations as in petrochemical installations, although they may be used in one way or another in the majority of large-scale storage and handling of polymerizing materials (most of which have significant flammability properties). Foams must be selected to be compatible with the reactive material being handled; water-based foams must, of course, be avoided when handling water-reactive materials.

5.12. Postrelease Mitigation

Besides standard fire prevention and protection systems, other measures can be employed to reduce the impact of a reactive chemical incident once it is detected and a plan of action is decided upon. Most of these measures, especially those such as water curtains requiring hardware installations or escape respirators that require pre-purchased equipment, require foresight and preplanning.

5.12.1. Release Countermeasures

The best ways to mitigate an accidental release of reactive material is highly dependent on the nature and properties of the material. For example, responding successfully to a spill of a solid, granular material onto a nonreactive surface requires much less pre-planning and immediate action than responding successfully to a large release of a highly reactive, volatile liquid that may also be flammable and toxic. Industry usage of various postrelease countermeasures is represented in **Summary 41** (pages 327–328) and discussed in this section.

Water Curtains

A fixed-pipe water spray system, installed on one or more sides of a tank or other equipment to provide a "wall" of water spray from a row of nozzles, is called a *water curtain.* Such an installation can mitigate a vapor or gaseous release by one or more of several mechanisms: absorption of the vapors in the water spray and/or heating of the vapor cloud by heat of absorption if the vapors are hygroscopic; condensation of the vapors if the water temperature is below the boiling point of the material; dilution of the vapor with entrained air; reducing the likelihood of flammable vapor ignition; cooling of equipment if involved in a fire; or destruction of the vapors if water-reactive.

Water curtains have the advantage of being rapidly actuated by the manual or automatic opening of the water valve. Potential disadvantages of water curtains include knocking out only a fraction of the vapor passing through it (depending on the water flow density), only working if the wind is blowing toward the water curtain (unless totally surrounding the unit being protected), and needing to deal with the dilute material resulting from the actuation of the curtain and absorption or condensation of the reactive material. Closely associated with water curtains are water deluge systems that are designed to keep storage tanks cool in the event of an external fire. Further details and recent research on water curtains and related mitigation measures can be found in *Guidelines for Postrelease Mitigation* (CCPS/AIChE, 1995).

Firewater Monitors

Fixed-location firewater monitors are commonly used where the potential for large fires exist. The monitors allow the directing of substantial water flow to where it is needed the most, whether to knock down a vapor cloud, dilute a spill, extinguish a fire, or (most often) cool adjacent vessels and equipment close to a fire. Most firewater monitor installations are activated and controlled manually; however, some can be remotely activated and controlled, and one company indicates the use of water cannons activated automatically upon release of monomer at rail car unloading stations. When used on spills of slightly to moderately water-reactive materials, sufficient firewater must be applied to overcome the heat of reaction and knock down or dilute any reaction products generated.

Fire Hoses with Fog Nozzles

While firewater monitors might be used primarily to cool nearby tanks and equipment close to a fire, fire hoses with fog nozzles can more effectively knock down and/or disperse hazardous material vapor clouds due to the fine spray and thus greater surface area for vapor/water contact. The mitigation of vapor clouds with fog spray is far more effective with hygroscopic materials such as acid gases and ammonia vapors than with materials that do not have a strong affinity for water. Fire hoses without fog nozzles are of course used for other mitigation purposes, such as directly fighting a fire, keeping adjacent tanks and equipment cooled down, and flushing or diluting spilled material.

Spill Absorbents

The use of spill absorbents is especially important for dealing with small spills of highly reactive liquids. The spill absorbent will tend to dilute and desensitize spills of shock-sensitive or thermally sensitive materials, and will allow cleanup without water for pyrophoric or water-reactive materials. Spill absorbents in various forms can be used both to limit the spread of a liquid spill and absorb the liquid for subsequent disposal. The compatibility of the absorbent with the spilled material should be established before use, and spill absorbents should be included in a chemical interaction matrix (Section 4.1.3) for identification of incompatibility hazards. Some absorbents can represent a reactivity hazard such as catalyzing polymerization.

5.12.2. Reactive Chemicals Personal Protective Equipment

Personal protective equipment (PPE) is important in mitigating reactive chemical incidents by reducing personnel exposure to spilled or splashed material, hazardous vapors or combustion products, fire thermal radiation, and/or flying debris from an explosion, given that a reactive material incident does occur. Industry practice with respect to the use of personal protective equipment in reactive chemical storage and handling facilities is represented in **Summary 42** (page 329).

Personal protective equipment is likely to be different for normal operations, high-hazard operations, and emergency operations. Equipment to be used during normal operations, such as eye and face protection, gloves, and protective footwear, should be specified in the standard operating procedures for the facility, and rigorously enforced. The material safety data sheet (MSDS) for each reactive chemical handled in the process should indicate appropriate personal protective equipment to be used during normal handling. During high-hazard operations, such as unloading of a highly reactive material from a tank car or tank truck or initial line breaks for maintenance, additional protection may be warranted such as an air-line respirator and/or a protective suit.

Emergency operations are more likely to require full protection, such as a full encapsulation suit and self-contained breathing equipment. The type of protective equipment to be used in emergency operations should be specified both in the emergency procedures for the facility and in the site emergency response plan. Training for and use of the full-protection equipment are essential parts of emergency preparedness and realistic emergency response drills.

Respiratory Protection

The use of full respiratory protection such as a self-contained breathing apparatus (SCBA) is fairly common in reactive chemical installations. Some companies have a policy to use full respiratory protection whenever dealing with a chemical spill; others limit its use to situations involving materials that are both volatile and toxic enough to pose an inhalation toxicity hazard, either under normal or emergency conditions.

Escape respirators of various designs are in fairly common use in reactive chemical handling, particularly where a release may generate a toxic vapor cloud.

Caution should be exercised in specifying and using canister-type escape respirators, since the canisters are chemical-specific and will only protect up to a given maximum vapor concentration and duration. Escape respirators having over-the-head masks and a supply of compressed breathing air can be used to escape from higher-concentration situations and are not chemical-specific.

Protective Clothing

Besides normal protective clothing used in chemical facilities, several types of special protective suits are in use, having names such as "bunker gear" and "slicker suit." The type of suit to be used will depend on the nature of the hazard (fire thermal radiation versus chemical corrosivity, for example) and whether used during normal or emergency operations. Emergency response where a high hazard may exist will usually dictate the use of a full protective suit that is appropriate to the nature of the anticipated hazard.

5.12.3. Reactive Chemicals Emergency Response

In general, Summary 43 (page 330) indicates that emergency response planning for the storage and handling of reactive materials is not significantly different from planning for any other highly hazardous chemicals such as large quantities of flammable or toxic materials. Emergency response plans need to include the determination of possible reactive chemical accident scenarios, a means of evaluating the potential range of effects (either real-time or predetermined reasonable-worst-case consequence calculations), development of detailed action plans, assembling of all information and equipment needed to effectively respond to a reactive chemical emergency, notification of potentially affected populations, training of emergency responders, and the periodic conducting and evaluating of realistic drills. Certain chemicals such as water-reactive materials may require special training for emergency response personnel.

5.13. Hazard Reviews

Hazard reviews, or process hazard analyses, are now widely used to rigorously examine process operations to determine what can go wrong, how likely such an event is to occur, how severe the consequences could be, and whether adequate safeguards exist to protect against all identified potential accident scenarios. For new facilities, detailed hazard reviews are generally performed at the final design stage. However, significant benefit can be obtained by having hazard reviews conducted from the very earliest stages of the design, when inherent safety and facility siting concepts can most effectively be addressed.

5.13.1. Hazard Severity Categories

Summary 44 (page 331) gives industry responses to the question of what hazard severity categories are used in hazard reviews. The most common responses were (a) the use of three severity levels and (b) that severity levels were not used or not differentiated. The number of severity levels ranged from one to five, with an average of 2.5 and a median of 3. Qualitative category descriptions such as high, medium, and low were most frequently used to differentiate the hazard categories. Some of the categorizations are based on the nature and severity of the underlying hazard (e.g., high reactivity, medium reactivity, and only combustible), while others are based on the severity of potential consequences, either in qualitative terms (e.g., catastrophic, serious, major, moderate) or in more specific personnel-injury-potential terms (e.g., minor injury, serious injury, fatality). Events with the potential for off-site consequences and those having only the potential for on-site effects were also differentiated.

5.13.2. Reactive Chemicals Hazard Reviews

Summary 45 (pages 332–334) outlines the types of hazard reviews being conducted for the different hazard severity categories used by various companies. The most common review types appear to be design reviews, process hazard analyses, personnel safety reviews, and pre-startup safety reviews.

Design Reviews

Design reviews are most commonly conducted on all new process designs without regard to hazard severity. One company conducts no design reviews on low-hazard processes and conducts reviews only of initial designs and major modifications for medium- and moderate-hazard processes.

Instrument Redundancy

Many different approaches are taken for reviewing a process for adequate safety instrumentation redundancy. These range from following corporate guidelines on alarm/interlock design to inclusion as part of quantitative risk analyses. Instrument redundancy reviews are often reserved only for high-hazard and some medium-hazard processes. A methodology for addressing instrumentation redundancy is included in *Guidelines for Safe Automation of Chemical Processes* (CCPS/AIChE, 1994b).

Process Hazard Analyses

Process hazard analyses, or PHAs, are required to be conducted for all processes coming under the U.S. OSHA process safety management standard 29 CFR 1910.119 and under the U.S. EPA risk management program rule 40 CFR Part 68. However, these regulations do not specify what PHA methodology is to be used. PHAs appear to be most often conducted by hazard and operability

(HAZOP) study, particularly for high- and medium-hazard processes. What-If, Checklist, and combined What-If/Checklist reviews are also common, although they are often reserved for lower-hazard processes. Failure modes and effects analyses, quantitative risk analyses, consequence analyses, fault tree analyses, and event tree analyses appear to be used more selectively. One company handles all PHAs the same except to vary the frequency of reevaluations after PHAs are completed. Another company conducts HAZOP studies every 2, 3, or 5 years for processes with high, medium, and low hazard, respectively.

Other Reviews

Reviews for operability, human factors, and facility siting are often addressed as part of a process hazard analysis, often within the context of a HAZOP study; these reviews are commonly reserved for higher-hazard processes. Personnel safety reviews may be included as part of overall safety reviews or Job Safety Analyses. Pre-startup safety reviews are also required by the OSHA standard mentioned above, and are usually conducted on all processes without regard to hazard severity category. Some facilities may also conduct maintenance safety reviews; management of change reviews; safety, health and environmental reviews; reactive chemical reviews; and/or product integrity reviews.

5.14. Codes and Standards

Industry-consensus codes and standards embody good industry practices that have developed over time based primarily on historical experience. **Summary 46** (page 335) lists codes and standards used in the design of reactive chemical storage and handling facilities. This list is obviously dominated by American codes and standards since CCPS's sponsors are predominately U.S. companies.

The codes used most commonly in U.S. reactive chemical storage and handling installations are those put out by the National Fire Protection Association (primarily dealing with fire prevention and protection design), the American Society of Mechanical Engineers (primarily dealing with mechanical integrity of equipment), and the American Petroleum Institute (dealing with many aspects of petrochemical and hydrocarbon processing facilities). Suppliers' recommendations, state and local ordinances, and insurance industry guidelines are commonly applicable. The Uniform Building Code (ICBO, 1994), Southern Building Code (SBCCI), and Building Officials and Code Administrators (BOCA) codes are also widely used for structural enclosures. Several companies have internal corporate engineering guidelines that have aspects pertaining to reactive chemical storage and handling. Other countries' codes and standards may also apply for companies having international operations.

Bibliography

Center for Chemical Process Safety, *Guidelines for Engineering Design for Process Safety,* American Institute of Chemical Engineers, New York, 1993.

Center for Chemical Process Safety, *Guidelines for Storage and Handling of High Toxic Hazard Materials,* American Institute of Chemical Engineers, New York, 1988.

Center for Chemical Process Safety, *Guidelines for Vapor Release Mitigation,* American Institute of Chemical Engineers, New York, 1988.

Center for Chemical Process Safety, *Guidelines for Safe Automation of Chemical Processes,* American Institute of Chemical Engineers, New York, 1994.

References

AIChE (1994). *Dow's Fire & Explosion Index Hazard Classification Guide,* 7th Ed. American Institute of Chemical Engineers, New York.

API RP 520 (1988). *Sizing, Selection, and Installation of Pressure-Relieving Devices in Refineries,* Part II—Installation, Third Edition, American Petroleum Institute, Washington, D.C., November.

API RP 520 (1993). *Sizing, Selection, and Installation of Pressure-Relieving Devices in Refineries,* Part I—Sizing and Selection, Sixth Edition, American Petroleum Institute, Washington, D.C., March.

BOCA. Building Officials and Code Administrators International, Inc., 4051 W. Flossmoor Rd., Country Club Hills, Illinois 60478-5795.

Bodurtha, F. T. (1980). *Industrial Explosion Prevention and Protection,* McGraw-Hill, New York.

CCPS/AIChE (1987). *Guidelines for Use of Vapor Cloud Dispersion Models.* American Institute of Chemical Engineers, New York.

CCPS/AIChE (1988). *Guidelines for Vapor Release Mitigation.* American Institute of Chemical Engineers, New York.

CCPS/AIChE (1989). *Guidelines for Chemical Process Quantitative Risk Analysis.* American Institute of Chemical Engineers, New York.

CCPS/AIChE (1993). *Guidelines for Engineering Design for Process Safety.* American Institute of Chemical Engineers, New York.

CCPS/AIChE (1994a). *Guidelines for Evaluating the Characteristics of Vapor Cloud Explosions, Flash Fire, and BLEVEs.* American Institute of Chemical Engineers, New York.

CCPS/AIChE (1994b). *Guidelines for Safe Automation of Chemical Processes.* American Institute of Chemical Engineers, New York.

CCPS/AIChE (1995). *Guidelines for Postrelease Mitigation.* American Institute of Chemical Engineers, New York.

DIERS (1992). *Emergency Relief System Design Using DIERS Technology, DIERS Project Manual,* Design Institute for Emergency Relief Systems, American Institute of Chemical Engineers, New York.

Englund, S. M. (1991). "Design and Operate Plants for Inherent Safety—Part 1." *Chem. Eng. Prog.* 87(3, March), 85–91.

FM (1993). "Criteria for Maximum Foreseeable Loss Fire Walls and Space Separation." *Loss Prevention Data Sheet 1-22* (periodically updated). Factory Mutual Engineering and Research, Norwood, Massachusetts.

ICBO (1994). *Uniform Building Code.* International Conference of Building Officials, 5360 Workman Mill Rd., Whittier, California 90601-2298, (310) 699–0541.

IRI (1992). "Plant Layout and Spacing for Oil and Chemical Plants," IRInformation Manual 2.5.2, Industrial Risk Insurers, Hartford, Connecticut, revised March 2.

Johnson, R. W. (1990). "Building Safety Layers Around Reactive Chemical Processes." AIChE Spring National Meeting, Paper No. 58d. Orlando, Florida, 21 March. American Institute of Chemical Engineers, New York.

NFPA (1994). *Fire Protection Handbook,* 17th Ed., National Fire Protection Association, Quincy, Massachusetts.

NFPA 12 (1993). *Standard on Carbon Dioxide Extinguishing Systems,* National Fire Protection Association, Quincy, Massachusetts.

NFPA 13 (1994). *Standard for the Installation of Sprinkler Systems,* National Fire Protection Association, Quincy, Massachusetts.

NFPA 30 (1993). *Flammable and Combustible Liquids Code,* National Fire Protection Association, Quincy, Massachusetts.

NFPA 43B (1993). *Storage of Organic Peroxide Formulations,* National Fire Protection Association, Quincy, Massachusetts.

NFPA 49 (1994). *Hazardous Chemical Data,* National Fire Protection Association, Quincy, Massachusetts.

NFPA 69 (1992). *Standard on Explosion Prevention Systems,* National Fire Protection Association, Quincy, Massachusetts.

NFPA 325M (1994). *Fire Hazard Properties of Flammable Liquids, Gases, and Volatile Solids,* National Fire Protection Association, Quincy, Massachusetts.

SBCCI. Southern Building Code Congress International, Inc., 900 Montclair Rd., Birmingham, Alabama 35213-1201.

TM 5-1300 (1990). *Structures to Resist the Effects of Accidental Explosions,* Joint Departments of the Army, the Navy, and the Air Force, TM 5-1300/NAVFAC P-397/AFR 88-22, Washington, D.C., 19 November 1990.

6

Process Safety Management of Reactive Material Facilities

Having a good design for a facility that will store and handle reactive materials is not sufficient for safe operation of the facility. It is essential to have a comprehensive system of process safety management to store and handle hazardous materials without incident. This is particularly true for reactive materials facilities, where uncontrolled reactions must be avoided as well as unplanned releases. *Process safety management (PSM)* has developed within the process industries as the set of management elements necessary to safely operate a facility handling hazardous materials.

Process safety management is at the heart of every well-run and incident-free operating facility. PSM emphasizes the containment and control of process materials and energies throughout the life cycle of a process by managing the technology, facilities, and personnel associated with the process. A comprehensive PSM system addresses all twelve of the following key elements (CCPS/AIChE, 1989a), although a given organization's PSM program may divide up what is covered by these elements in different ways:

- Accountability: objectives and goals
- Process knowledge and documentation
- Capital project review and design procedures
- Process risk management
- Management of change
- Process and equipment integrity
- Personnel training and performance
- Human factors
- Incident investigation
- Standards, codes, and laws
- Audits and corrective actions
- Enhancement of process safety knowledge.

All of the above elements play an important role in the safe operation of facilities that store and handle reactive materials. This chapter will emphasize features of a PSM system that are specifically related to operations storing and handling reactive materials. A guide produced by an Institution of Chemical Engineers Working Party (Barton and Rogers, 1993) briefly discusses several of these elements from a British perspective as they relate to reactive chemical processes.

6.1 Accountability: Objectives and Goals

This PSM element overlaps with the Management Leadership in Process Safety element that is part of the Responsible Care® code of management practices adhered to by chemical companies in many countries (CMA, 1990). Senior management in any company where hazardous materials are handled must communicate clear, explicit safety goals and objectives and commit the resources necessary to achieve them. A corporate and line management organization must be defined to carry out the goals and objectives, with roles and responsibilities clearly defined, and with a mechanism for checking performance and ensuring accountability.

Many changes will inevitably occur over time in every company. A series of management system issues must be addressed to ensure that the organization will be always accountable for the company's expectations, objectives, and goals that are related to process safety management. These management system issues are:

- accountability for PSM objectives and goals through shift changes, production changes, process upsets, and unplanned shutdowns
- accountability for PSM objectives and goals while facing limited personnel and funding resources
- accountability for PSM objectives and goals through plant and corporate organizational changes
- accountability for PSM as part of the quality process
- accountability for control of exceptions to established practice
- use of alternative methods of assuring accountability
- accountability of management to be accessible for PSM decisions.

None of these issues are unique to the storage and handling of reactive materials, but all are vitally important to developing PSM systems and maintaining them over the facility lifetime.

6.2. Process Knowledge and Documentation

A management system needs to be in place to capture and document the body of knowledge associated with a process as it develops through design and operational stages. This body of knowledge includes design criteria, chemical safety data, equipment design information, operating procedures and operational experience, protective systems criteria, and process risk management decisions.

In order to contain and control reactive chemicals safely, their behavior under a range of conditions must be determined with reasonable confidence and documented as part of the process knowledge. While material safety data sheets (MSDSs) may provide basic reactivity information, materials assessment including lab testing may be required to determine incompatibilities, reaction kinetics, and other important properties under specific planned and/or possible process conditions. A strategy for assessing and documenting this safety information is presented in Chapter 3 of this book.

When dealing with reactive chemicals, the section of the MSDS for each chemical that would typically address reactivity data (CCPS/AIChE, 1989a, 34) obviously becomes an important source of information. Therefore, MSDSs for reactive chemicals should be carefully reviewed and supplemented with additional information if necessary, to ensure that operating personnel have adequate safety information on the hazards of these materials. In addition, the sections for "Spill or Leak Procedures," "Special Protection Information," and "Special Precautions" should be likewise reviewed to ensure that they contain the best information concerning the material.

6.3. Capital Project Review and Design Procedures

Capital projects for facilities handling reactive materials usually involve significant changes such as new equipment, state-of-the-art controls, new operating procedures, and inventory increases. Standard practice for capital projects is a process safety review that addresses items such as materials of construction, redundant controls, alarms and interlocks, emergency relief system design, mitigation systems, and operator interfaces.

Operations in which reactive materials will be handled and stored need to go through detailed, systematic process safety reviews. In addition to checking to ensure the knowledge base of reactivity data and potential incompatibilities is complete, the process safety review should also, at the earliest opportunity, address issues of **inherent safety**. This includes a deliberate consideration of the following types of questions for each reactive material or class of reactive materials involved with the project:

- Is it essential to the project that this reactive material be stored and handled?
- Can an alternative material be substituted that would meet the project objectives but pose a lower overall risk of injury and loss in case of an operational accident?

- Can the reactive material be handled in a less hazardous form, such as at a lower concentration, pressure or temperature?
- Can the inventory and/or throughput of the reactive material be significantly reduced without a major increase in the complexity of the facility?
- Can a sufficiently large buffer zone be maintained such that no vulnerable populations will be affected by foreseeable accidental events involving the reactive material?

During the process safety reviews, special attention should also be paid to how the reactive material can affect materials of construction, and the adequacy of emergency relief systems.

Design considerations for facilities handling reactive materials will be dependent on the reactivity characteristics of the materials. Both the general and the specific design considerations given in Chapters 5 and 7 of this book, respectively, can be consulted to gain an appreciation for the range of factors that must be considered when designing such facilities.

6.4. Process Risk Management

Process risk management involves the systematic identification, evaluation, and control of potential losses that may arise in existing operating facilities from unplanned events such as fires, explosions, toxic releases, runaway reactions, or natural disasters. The practice of process risk management anticipates the possibility of such loss events and evaluates their potential impacts so they can be managed effectively (CCPS/AIChE, 1989a, 59), including the preparation of emergency response plans.

Because the potential for incompatibility reactions is important when handling most reactive materials, the first step in process risk management, which is the identification of hazards, should include the development of a chemical interaction matrix such as is described in Section 4.1.3 of this book. The identification of hazards for reactive materials obviously depends also on having a complete characterization of the reactivity of the materials.

When conducting a risk analysis of reactive material storage and handling operations, consideration should also be given to the risk associated with the on-site and off-site transport of the materials. The transportation risk, particularly with respect to off-site populations, is quite often greater than the on-site storage and handling risk. Guidelines are available for assessing both fixed-facility and transportation risks (CCPS/AIChE, 1989b; CCPS/AIChE, 1995). Part of a company's risk management strategy may also be to encourage customer and supplier companies to adopt similar risk management practices.

Emergency response plans should be carefully prepared when dealing with reactive materials. Special protective equipment may need to be used when handling a reactive chemical emergency (see for example NFPA 1991 and NFPA

1992), and personnel must be trained in its proper use. Some normal emergency response approaches may not be able to be used with certain reactive materials, such as the use of water to fight a fire involving water-reactive materials. Less-obvious situations should be identified as part of process hazard analyses, and should be incorporated into both the written plans and the actual drills conducted in preparation for dealing with a reactive chemical emergency.

6.5. Management of Change

A management system must be in place to control both temporary and permanent changes to the technology, facilities, and personnel associated with an operation involving hazardous materials (CCPS/AIChE, 1989a, 73). This management system will be similar regardless of the nature of the hazardous material. However, it should be ensured that the system will require a detailed safety review whenever the facility will go beyond its predetermined safe operating limits. For example, additional reactive materials testing may be required if the change will take the operation outside the composition range encompassed by previous testing, and incompatibility data may be needed if introducing a new material of construction or chemical into the operation. What may seem to be an unimportant change in composition, storage quantity, container size, temperature, etc. can cause a significant impact on the safety of a reactive chemical operation.

6.6. Process and Equipment Integrity

A management system is needed to ensure that all parts of a reactive materials containment system is constructed and maintained to a high degree of mechanical integrity, for obvious reasons. The components of such a management system include reliability engineering, materials of construction selection, fabrication and inspection procedures, installation procedures, preventive maintenance, inspection and testing, maintenance procedures, alarm and instrument management, and demolition procedures. These components are described by CCPS/AIChE (1989a, 85).

Because reactive chemicals tend to be more corrosive to materials of construction, industry-standard preventive maintenance programs may need to be adjusted to achieve high equipment reliability. Likewise, more rigorous administrative procedures may be needed for ensuring that the correct materials of construction are used for both initial fabrication and subsequent modifications.

In addition to more attention being given to maintaining process equipment, the handling and storage of reactive chemicals will almost certainly call for greater use and control of written procedures for performing specific inspection and maintenance activities. Process upsets, releases, or personnel exposures that are caused by maintenance errors or violation of safe work practices will generally have more severe consequences. Such written procedures may call for special personal protective equipment, more thorough purging of process equipment before being opened for maintenance, flushing of equipment with a material other than water for water-reactive materials, etc. Confined space entry procedures and demolition procedures will likewise need to incorporate special precautions depending on the nature of the reactive materials handled in the equipment being dismantled.

6.7. Human Factors

The design and operation of a facility should both anticipate and control the adverse impacts that could result from human actions and human/equipment interactions. Attention to human factors (ergonomics) and the assessment of human error potentials are necessary to achieve this (CCPS/AIChE, 1989a, 99). The necessity of wearing full-protection suits during some operations involving reactive materials carries with it the need to address the human factor issues associated with working with full-protection suits on (heat stress; restricted movement, hearing, and vision). Consideration should also be given as to the most effective means of providing cautions in procedures for safe handling of reactive materials. Training programs need to consider how to give operating personnel a proper respect for reactivity hazards without causing undue alarm or fear. Both designers and supervisors need to be particularly sensitive to human factor considerations and solicit feedback from operating personnel concerning such details as overly complex tasks, error-likely situations, infeasible time constraints, and strenuous work environments.

6.8. Personnel Training and Performance

The management system required to have an effective training program must include defining training requirements and job descriptions; determining minimum qualifications for new employees; selecting, developing, and revising training programs, including both initial and refresher training; interfacing with operating and maintenance procedures; qualifying and training the trainers, measuring the performance and effectiveness of training, and managing training records.

When the behavior of a reactive material is unique or difficult to predict, then a higher level of skill and additional training may be required in order to minimize the risk of handling the reactive chemical. It may be considered necessary to require a minimum level of understanding of chemistry for personnel to operate some facilities safely.

Extra training may also be required to ensure operating personnel are equipped to execute safely both normal and emergency procedures involving reactive materials. Training must include the proper and consistent use of all personal protective equipment required for safe handling of reactive materials.

6.9. Incident Investigation

The proper investigation of unplanned events with actual or potential injury/loss consequences presumes the accident or near miss had one or more management system failures as underlying causes (CCPS/AIChE, 1989a, 113f). Prompt and careful investigation of such incidents should identify the underlying management system failures, develop necessary corrective actions, and effectively communicate the results of the investigation to all affected parties and regulatory authorities as required by law. The CCPS/AIChE book *Guidelines for Investigating Chemical Process Incidents* (1992) can be consulted for details concerning incident investigation as an element of a process safety management system.

Some incidents involving reactive materials may require extensive analysis in order to determine the root cause of the incident. For instance, lab work may have to be performed to determine the behavior of the process and its contents under certain conditions that may have existed at a point in time during the accident sequence. Only after the mechanism of the incident is fully understood can effective actions be taken to prevent recurrence.

Management systems within an organization should also ensure that lessons learned from incidents are communicated to all affected facilities in the organization. The lessons learned should be incorporated into the process technology, so that they are not forgotten as soon as the affected persons move to new positions. Most companies also recognize the value of passing on safety-related information learned from incidents to the broader technical community via industry associations, conference presentations, technical articles, formal communications with suppliers and customers, and various other means of communication between organizations.

6.10. Standards, Codes, and Regulations

When designing facilities for storing and handling reactive materials, advantage can be taken of lessons learned from past experience and industry consensus practices by using standards and codes. Some will be mandatory for a given facility and others may be optional. Codes and standards issued by several organizations address the storage and handling of specific reactive materials or types of materials. The following list is an example of standards by one organization, the National Fire Protection Association, that relate to reactive materials:

NFPA 40	Storage and Handling of Cellulose Nitrate Motion Picture Film
NFPA 43B	Storage of Organic Peroxide Formulations
NFPA 49	Hazardous Chemicals Data
NFPA 51A	Acetylene Cylinder Charging Plants
NFPA 69	Explosion Prevention Systems
NFPA 325M	Fire Hazard Properties of Flammable Liquids, Gases, and Volatile Solids
NFPA 490	Storage of Ammonium Nitrate
NFPA 491M	Hazardous Chemical Reactions
NFPA 495	Explosive Materials Code
NFPA 651	Manufacture of Aluminum Powder
NFPA 1992	Liquid Splash-Protective Suits for Hazardous Chemical Emergencies
NFPA 1993	Support Function Protective Garments for Hazardous Chemical Operations

Other industry-consensus practices that may relate to reactive materials include procedures from the AIChE Design Institute for Emergency Relief Systems (DIERS) for emergency vent sizing; Health and Safety Executive guidelines and British Standards (United Kingdom); Loss Prevention Data Sheets issued by Factory Mutual Engineering & Research (Norwood, Massachusetts); publications related to acetylene cylinder and piping systems by the Compressed Gas Association (Arlington, Virginia); and state and local building codes. Some industries have industry-specific codes and standards. For example, pyrophoric gases such as silane are handled by the semiconductor industry, which has standards published by SEMI (1995).

Government regulations may be either prescriptive or performance-oriented. The following list of titles was extracted from one set of regulations issued by the U.S. Occupational Safety and Health Administration (OSHA). The list shows that some regulations apply to specific reactive materials or classes of materials, some apply to management systems, and others apply to specific facility operations.

29 CFR 1910.102	Acetylene
29 CFR 1910.109	Explosives and blasting agents
29 CFR 1910.119	Process safety management of highly hazardous chemicals
29 CFR 1910.120	Hazardous waste operations and emergency response
29 CFR 1910.1012	Ethyleneimine
29 CFR 1910.1017	Vinyl chloride
29 CFR 1910.1045	Acrylonitrile
29 CFR 1910.1047	Ethylene oxide

Regulations will vary depending on the locality and country in which the reactive material facilities are located. The above lists are intended only as examples of standards, codes, and regulations that may apply to a given situation.

The U.S. OSHA process safety management standard 29 CFR 1910.119 listed above is especially pertinent to the concepts of this chapter. Although this regulation only pertains explicitly to facilities handling greater than specified threshold quantities of highly hazardous chemicals, the performance requirements of this standard have had a significant effect on many companies worldwide. In comparison to the elements in the CCPS process safety management model, more specific emphasis has been placed in the OSHA standard on employee participation, the addressing of facility siting and human factors issues as part of process hazard analyses, pre-startup safety reviews, contractor safety, operating and maintenance procedures, safe work practices, and emergency planning and response. A similar proposed regulation by the U.S. Environmental Protection Agency (EPA), 40 CFR Part 68, will also require assessment of the potential for off-site and environmental consequences of worst-case accidental events, and the preparation of risk management plans that will be available to the public.

6.11. Audits and Corrective Actions

Auditing is a key management system by which periodic checks are made at various levels of depth to ensure ongoing conformity to company policies and procedures, and to develop corrective actions where deficiencies are found. Process safety management audits may need to check compliance with both internal company standards and external regulations related to management of hazardous materials operations. A process safety audit must look at all of the management systems described in this chapter (Accountability, Process Knowledge and Documentation, Capital Project Reviews, etc.), as well as how effectively they are being implemented on an ongoing basis. The CCPS/AIChE book *Guidelines for Auditing Process Safety Management Systems* (1993) contains details on the scope, frequency, protocol, staffing, techniques, reporting, and follow-up of PSM audits, as well as information on how to audit individual PSM program elements.

Audits of facilities that store and handle reactive materials will be essentially the same as process safety audits for other facilities handling hazardous materials. The scope of an audit for a reactive materials facility should include all storage and handling of the reactive materials. Less-obvious areas such as drum storage facilities are often overlooked during audits.

6.12. Enhancement of Process Safety Knowledge

Companies with outstanding process safety management programs contribute to advancing the state of the art of process safety by sharing nonproprietary results of internal process safety research, and supporting the process safety-oriented research and development programs of professional and trade associations or colleges (CCPS/AIChE, 1989a, 137). This is especially true in the reactive chemicals community, due to the more technical nature of the handling of reactive materials. Periodicals such as *Chemical and Engineering News* often have letters to the editor sharing incidents or near misses involving reactive materials, and technical sessions such as the triennial international symposium on Loss Prevention and Safety Promotion in the Process Industries, the annual AIChE Loss Prevention Symposium, and the Institution of Chemical Engineers' section meetings have historically had many presentations related to reactive materials. The proceedings of the CCPS International Symposium on Runaway Reactions (CCPS, 1989c) has a wealth of information on hazard assessment and emergency relief design.

6.13. Other Elements Required by Regulatory Authorities

The preceding sections are not exhaustive with respect to what may be required in a given process safety management system. Facilities handling greater than certain threshold quantities of hazardous materials in the U.S., for example, must conform to the performance requirements of 29 CFR 1910.119, "Process Safety Management of Highly Hazardous Chemicals." This regulation includes additional elements related to employee participation, pre-startup safety reviews, contractors, safe work practices, and trade secrets. Several of the individual program elements have additional requirements, also. For example, process hazard analyses must explicitly address facility siting and human factors issues, and the mechanical integrity program must include written maintenance procedures.

40 CFR Part 68, "Risk Management Programs for Chemical Accidental Release Prevention," promulgated by the U.S. Environmental Protection Agency (EPA), requires a hazard assessment and the preparation of risk management plans in addition to process safety management. Other requirements such as the Seveso Directive in Europe have less emphasis on management systems.

Bibliography

Center for Chemical Process Safety, *Guidelines for Technical Management of Chemical Process Safety,* American Institute of Chemical Engineers, New York, 1989.

Center for Chemical Process Safety, *Plant Guidelines for Technical Management of Chemical Process Safety,* American Institute of Chemical Engineers, New York, 1992.

Center for Chemical Process Safety, *Guidelines for Implementing Process Safety Management Systems,* American Institute of Chemical Engineers, New York, 1994.

References

Barton, J. A., and Rogers, R. L., eds. (1993). *Chemical Reaction Hazards—A Guide.* Institution of Chemical Engineers, Rugby, Warwickshire, U.K.

CCPS/AIChE (1989a). *Guidelines for Technical Management of Chemical Process Safety.* American Institute of Chemical Engineers, New York.

CCPS/AIChE (1989b). *Guidelines for Chemical Process Quantitative Risk Analysis.* American Institute of Chemical Engineers, New York.

CCPS/AIChE (1989c). *International Symposium on Runaway Reactions.* Proceedings of Center for Chemical Process Safety conference, Cambridge, Massachusetts. American Institute of Chemical Engineers, New York, March 7-9.

CCPS/AIChE (1992). *Guidelines for Investigating Chemical Process Incidents.* American Institute of Chemical Engineers, New York.

CCPS/AIChE (1995). *Guidelines for Chemical Transport Risk Analysis.* American Institute of Chemical Engineers, New York.

CMA (1990). "A Resource Guide for Implementing the Process Safety Code of Management Practices." Chemical Manufacturers Association, Washington, D.C., October.

NFPA 1991 (1990). *Vapor-Protective Suits for Hazardous Chemical Emergencies.* National Fire Protection Association, Quincy, Massachusetts.

NFPA 1992 (1990). *Liquid Splash-Protective Suits for Hazardous Chemical Emergencies.* National Fire Protection Association, Quincy, Massachusetts.

SEMI (1995). *SEMI International Standards.* Semiconductor Equipment and Materials International, 805 East Middlefield Road, Mountainview, California 94043, (415) 964-5111.

7

Specific Design Considerations

This chapter gives design considerations for a few specific reactive materials, to illustrate the application of safe storage and handling practices to commercially available materials in each reactivity category. The design considerations are presented in the same order for each chemical as the topics in Chapter 5 for general design considerations, so the user can cross-reference both Chapter 5 and the tables in Appendix B when using the information in this chapter.

Specific design considerations for selected reactive materials in each category apply to that reactive material only. These considerations do not necessarily apply to other reactive materials in the same category.

Sources of information are also referenced for each material throughout the chapter, with references listed at the end of the chapter. ***It should be emphasized that the design features listed in this chapter are intended to be considerations only, and not design standards. Because these considerations were taken from many sources and company practices, it is possible that some considerations may be contradictory or may not be consistent with current regulatory requirements for a given location.*** It should also be noted that some of the information in this chapter was extracted from publications now out of print and from unpublished sources. In all cases, users should contact suppliers and manufacturers of each chemical of interest for current material safety data sheets, data verification, and any other more detailed and up-to-date information pertaining to the safe storage and handling of the specific chemical of interest.

Monomers in total are the most prevalent reactive materials in terms of volume production, based on 1993 U.S. data (C&EN, 1994). Specific design considerations for two monomers, acrylic acid and styrene, are given in Sections 7.1 and 7.2, respectively.

Organic peroxides are a special class of reactive materials that are thermally sensitive, with some formulations being shock-sensitive also. Some specific design considerations that are commonly applicable to most organic peroxides are given in Section 7.3. Considerations pertaining to two specific organic peroxide formulations, benzoyl peroxide and methyl ethyl ketone peroxide, are tabulated in Sections 7.4 and 7.5.

Ethylene oxide can decompose energetically under certain conditions, in addition to its ability to polymerize. Specific design considerations for safe storage and handling of ethylene oxide are given in Section 7.6.

Design considerations for pyrophoric aluminum alkyls and the peroxide-forming 1,3-butadiene are tabulated in Sections 7.7 and 7.8, respectively. Considerations for sodium metal and chlorosulfonic acid as examples of water-reactive materials are given in Sections 7.9 and 7.10.

7.1. Polymerizable Materials: Acrylic Acid

$$CH_2{=}CH\overset{\overset{O}{\|}}{C}OH$$

Acrylic acid (propenoic acid) is produced by air oxidation of propylene. It is used in the manufacture of acrylic plastic products, leather treatments, and paper coating. The purity of glacial acrylic acid, which is the grade available in bulk for commercial applications, is 98.0-99.5%. Technical-grade acrylic acid is 96.0% with the remainder being dimer and higher "mers."

Glacial acrylic acid is a colorless, combustible liquid with a freezing point of 55°F (13°C) and a flash point of 130°F (54°C). It has a sharp, acrid odor; the odor threshold of 1 ppm provides good warning properties for avoidance of exposure to dangerous vapor concentrations. Acrylic acid is strongly corrosive to skin and eyes. Vapor inhalation should be avoided by local exhaust ventilation and/or respiratory protection.

The most important aspect of acrylic acid storage is the use of a polymerization inhibitor. An investigation of an uncontrolled polymerization (Levy and Penrod, 1989) has shown that even low concentrations of a potent inhibitor such as 10 ppm of phenothiazine can delay uncontrolled polymerization long enough at elevated temperatures so that a significant portion of monomer will have undergone a secondary, less exothermic dimerization reaction by the end of the induction period, thereby reducing and slowing down the energy release from free-radical polymerization. The most widely used inhibitor for acrylic acid is monomethyl ether of hydroquinone (MEHQ) at a level of about 200 ppm.

The uncontrolled polymerization of acrylic acid is more rapid and violent than that of most commercially available vinyl-type monomers and evolves considerable heat and pressure-generating gases. Acrylic acid users have collectively experienced about one serious incident per year.

Overheating is by far the most common cause of uncontrolled polymerization. In one incident, a 17-m^3 tank trailer of acrylic acid was being warmed by internal heating coils to prevent the acrylic acid from freezing in subzero temperatures. The heat supplied to the coils from an unmonitored steam and water mixer was sufficient to initiate a runaway polymerization that led to an extremely violent explosion (Bretherick, 1990, 360). Another recent storage tank explosion resulted in one fatality and 13 injuries and destroyed the storage unit, a nearby manufacturing unit, and a warehouse (TCE, 1994). The freezing of acrylic acid due to a power failure led to blockage of a recirculation pump discharge line leading to a heat exchanger. This caused the pump to continuously recirculate acrylic acid directly back to the tank through a bypass line (also partially blocked with frozen acrylic acid). Due to the lack of heat exchange and lack of tank temperature monitoring, the heat input from the pump raised the temperature of the tank contents to the point where a runaway polymerization was initiated.

One important overheating mechanism involved in many acrylic acid incidents is the improper heating of a tank shell or other container to thaw acrylic acid that has become frozen. The uninhibited solid phase is susceptible to uncontrolled polymerization if improperly thawed (BASF et al., 1993), since the inhibitor is no longer properly distributed throughout the mass of material. Acrylic acid incidents have also been caused by contamination with other chemicals and by depletion of the oxygen necessary to activate the polymerization inhibitor (BASF et al., 1993).

Specific design considerations for the safe storage and handling of acrylic acid, *in addition to general precautions to be taken with combustible liquids and highly corrosive materials,* are listed in the "Feature" column of Table 7.1. The reason or explanation for each feature is given in the "Purpose" column of the table, with references given in the last column if further information is needed. Note that the sections of Table 7.1 (and the remaining tables in this chapter) correspond to the sections in Chapter 5, General Design Considerations, and to the topics covered by the industry survey documented in Appendix B.

The specific design considerations in Table 7.1 for acrylic acid do not necessarily apply to other polymerizable chemicals. In all cases, whether designing or evaluating a storage and handling arrangement for acrylic acid or for any other polymerizable chemical, suppliers and manufacturers should be contacted for current MSDSs and design considerations.

In brief, the most important requisites for safely storing and handling acrylic acid are

(a) preventing polymerization by maintaining an active inhibitor and proper dissolved O_2 levels,
(b) avoiding contamination under normal storage and handling conditions,
(c) avoiding unintentional freezing of the acid and using extreme care in thawing frozen acrylic acid,
(d) protecting against overheating or fire, and
(e) consider mitigating the consequences of a runaway reaction or accidental release by providing last-resort safety systems and personnel protection.

The combination of acrylic acid's corrosivity, relatively high freezing point, combustibility, and the possibility of uncontrolled, violent polymerization require very careful attention to both the design of acrylic acid piping and equipment and the procedures used for safely storing and handling acrylic acid. Safety and handling information is available from producers of acrylic acid (BASF, Hoechst Celanese, Rohm and Haas, and Union Carbide, 1991, 1993); the producers should be consulted for special design features.

TABLE 7.1
Specific Design Considerations for Acrylic Acid

FEATURE	PURPOSE	REFERENCE
1. Compatibility		
1.1 Separate acrylic acid storage and handling areas from, and take particular precautions to avoid contact with compounds such as: acids, oxidizing materials, 2-aminoethanol, ammonium hydroxide, chlorosulfonic acid, ethylenediamine, ethyleneimine, ethylidenenorbornene and other oxygen scavengers, iron salts, oleum, peroxides, peroxide forming and free radical forming compounds (aldehydes, ethers, azides, etc.)	The rate of acrylic acid polymerization violently increases with many chemicals; this hazard is reduced if the acrylic acid is properly inhibited (see 2.1 below), but oxygen scavengers can deplete the available oxygen that is necessary to activate the inhibitor	BASF et al., 1991, 1993; Bretherick, 1990, 360; DOT, 1984; NFPA 491-M
1.2 For tank storage, use dedicated loading and unloading lines with no cross-connections to other tanks or materials	Dedicated equipment reduces the likelihood of dangerous contamination, including from cross-connections via vapor return, vent, and transfer lines	BASF et al., 1993
1.3 Avoid contact of acrylic acid with carbon steel, and do not use carbon steel as a material of construction (see also 3.3 and 6.1 below)	Acrylic acid is corrosive to carbon steel	BASF et al., 1991; Nemec and Bauer, 1993; Niehaus, 1971
1.4 Do not use copper or copper-containing alloys in any equipment or part that may come into contact with acrylic acid	Copper causes discoloration of acrylic acid and may introduce copper ions and salts that could adversely affect its end use	Niehaus, 1971
2. Storage Time/Shelf Life		
2.1 During storage and handling of acrylic acid, maintain the concentration of an inhibitor as specified on the manufacturer's material safety data sheet	Both MEHQ and phenothiazine inhibit the free-radical polymerization of acrylic acid; 200 ppm MEHQ is most commonly used for bulk storage	Bretherick, 1990, 360; Kirch et al., 1988; Nemec and Bauer, 1983; Niehaus, 1971; Sabaa, 1990

TABLE 7.1
Specific Design Considerations for Acrylic Acid

FEATURE	PURPOSE	REFERENCE
2.2 Analysis of acrylic acid for the inhibitor MEHQ should be carried out chromatographically rather than by nitrite-colorimetry	The nitrite-colorimetric method erroneously identifies MEHQ degradation products (not active inhibitors) as MEHQ	BASF et al., 1991
2.3 If stored in 55-gal drums as received from the manufacturer, consult the time-temperature recommendations listed on the material safety data sheet	The polymerization inhibitor may be gradually depleted, and dimer formation reduces quality over time (e.g., 1 to 1.5% dimer is formed after one month at 85-90 °F (29-32 °C)	BASF et al., 1991; Wampler, 1988, 189
3. Storage Quantity and Configuration		
3.1 Minimize stored inventories of acrylic acid, especially in hot weather	Reducing the quantity of acrylic acid stored on-site makes the facility inherently safer in the event of loss of containment/control or an area fire	
3.2 Tank cars and tank trucks are not recommended alternatives to fixed tank storage	Temperature and inhibitor level control, mixing, last-resort safety features, mitigation systems, etc. are not as effective as with fixed tank storage with dedicated safeguards	BASF et al., 1991
3.3 Construct bulk storage tanks of stainless steel type 304 or 316; aluminum 3003 and series 5000 may also be able to be used with glacial acrylic acid (see also 1.3 above)	Acrylic acid is corrosive to carbon steel	Niehaus, 1971; BASF et al., 1991
3.4 Provide for mixing of storage tank contents; for example, jets or eductors on the tank inlet line have been found to provide effective mixing	Mixing is necessary, at least on a periodic basis, to allow exposure to atmospheric oxygen and to prevent stratification of the tank contents	BASF et al., 1991, 1993
4. Air and Moisture Exclusion		
4.1 Do not exclude air from the storage containment and do not inject nitrogen or other inert gas into the containment such as by a bubble pipe level indication system	MEHQ and other phenolic inhibitors require some oxygen for activation; acrylic acid should always be stored under an atmosphere containing 5-21 vol% oxygen; inert blanketing is not necessary to keep acrylic acid vapors out of the flammable range at ambient temperatures	BASF et al., 1991, 1993; Nemec and Bauer, 1983; Nicolson, 1991, 182
5. Monitoring and Control		
5.1 Store glacial acrylic acid above 59 °F (15 °C) to prevent freezing, whether in tank storage or inside temperature-controlled rooms or buildings	Freezing of acrylic acid in bulk storage should be avoided, since thawing presents a risk of overheating and thereby initiating uncontrolled polymerization	BASF et al., 1991

TABLE 7.1
Specific Design Considerations for Acrylic Acid

FEATURE	PURPOSE	REFERENCE
5.2 Store glacial acrylic acid below 77 °F (25 °C) to minimize dimerization, whether in tank storage or inside temperature-controlled rooms or buildings; do not exceed maximum storage temperature recommended on the material safety data sheet	Dimer formation increases, and time to depletion of inhibitor and onset of polymerization decreases, with increasing temperature	BASF et al., 1991; Nemec and Bauer, 1983
5.3 Insulate all outdoor tanks containing acrylic acid	Proper tank insulation can delay freezing of acrylic acid when the ambient temperature falls below its 55 °F (13 °C) freezing point	BASF et al., 1991; Niehaus, 1971
5.4 Provisions should be made to automatically control the acrylic acid temperature within the desired range and to automatically alarm and shut down the cooling and/or heating devices at preset over- and under-temperature points. Tempered water is recommended for controlling the storage temperature; pumps, pipes, and valves, etc. should also be traced with tempered water. The temperature of this water should not exceed 113 °F (45 °C); do not use steam under any circumstances	Temperature is the key process parameter for safe storage of acrylic acid. The tempered water system is primarily for the prevention of freezing and for thawing frozen acid. It may be used for cooling in extremely hot weather to control dimer formation. The cooling capabilities are generally limited, however, by the temperature of the available water	BASF et al., 1991, 1993; Niehaus, 1971; Current industry practice
5.5 Provide storage tank/tempered water systems with redundant temperature sensors; install high and low temperature alarms and cooling/heating system shutdowns on both control and redundant temperature sensors with setpoints as recommended on the material safety data sheet and with the ability to detect an abnormal temperature	Redundant sensors are installed in case of failure of the primary sensors used for temperature control; an unexpected temperature rise can signal the onset of an uncontrolled reaction initiated by heat, inhibitor depletion, lack of oxygen to activate the inhibitor, or acrylic acid contamination	BASF et al., 1991, 1993
5.6 Drums containing frozen acrylic acid must not be thawed with direct steam impingement or electrical resistance heating elements	Excess heat can lead to uncontrolled polymerization	BASF et al., 1991
5.7 When melting frozen acrylic acid, heat gradually and thaw slowly. Do not use steam. Ensure that the acrylic acid is completely thawed and mixed before withdrawing any liquid	When freezing occurs, the first crystals (low in inhibitor) will form along the outer wall of the container, and polymerization along the walls can be initiated at steam temperatures	BASF et al., 1991, 1993; Bretherick, 1990, 360; Niehaus, 1971; Nemec and Bauer, 1983; NFPA 49, 35; Sabaa, 1990; Johnson et al., 1961

TABLE 7.1
Specific Design Considerations for Acrylic Acid

FEATURE	PURPOSE	REFERENCE
6. Handling and Transfer		
6.1 Construct all pumps, pipes, and valves of stainless steel; some accessory equipment may be constructed of polyethylene, polypropylene, or fluoropolymers	Acrylic acid is corrosive to carbon steel	BASF et al., 1991
6.2 Maintain a closed system when transferring acrylic acid	Vapors displaced from the storage tank by incoming liquid should be vented into the tank being emptied rather than into the atmosphere	Current industry practice
6.3 Design and locate all piping, valves, gages, vents, flame arresters, tank openings, emergency relief systems, and engineering controls such that they may be readily and periodically inspected and cleaned	Acrylic acid tends to form polymers that can plug equipment, particularly where vapors can get to, since acrylic acid vapors are uninhibited when, e.g., MEHQ is used as an inhibitor	Bretherick, 1990, 360; NFPA 49, 35
6.4 Trace pumps, pipes, valves, etc. with tempered water with a redundantly controlled maximum water temperature of 113 °F (45 °C) or heating devices with high-temperature shutdown systems; do not use electric resistance heaters	Tracing is necessary to prevent freezing of acrylic acid inside equipment; tempered water has the advantage that it can be used to cool the acrylic acid in very hot weather to control dimer formation	BASF et al., 1991, 1993; Current industry practice
6.5 Use centrifugal pumps for the recirculation and transfer of acrylic acid. Sealless pumps can be used provided care is taken to not overheat the acid in internal passageways	Residual heat may be present in canned pumps after use, causing polymer formation in the pump	BASF et al., 1991, 1993
6.6 Reliably protect against dead-head pumping of acrylic acid by multiple levels of design/operating safeguards such as redundant high temperature sensors and switches on the pump discharge or recirculation lines	Dead-head pumping, such as against a closed valve, quickly generates heat and may lead to polymerization in the pump	BASF et al., 1993
7. Last-Resort Safety Features		
7.1 For tank storage, immediately inject phenothiazine (PTZ) in solution via a restabilization (shortstop) system at the suction side of the recirculation pump if a runaway polymerization onset is detected by • a liquid temperature measurement of 122 °F (50 °C) and rising in a system without temperature control, • two temperatures of 113 °F (45 °C) and rising in a system with temperature control, or • a temperature rise greater than 18 °F (10 °C) in one hour or less. Consider injection of PTZ by gravity, such as from a small head tank on top of the storage tank	Addition of PTZ even after polymerization has begun, to a final mixed PTZ concentration between 200 and 1000 ppm in each affected tank, may stop or at least slow down a thermally initiated polymerization reaction	BASF et al., 1993; Levy and Penrod, 1989; Current industry practice

TABLE 7.1
Specific Design Considerations for Acrylic Acid

FEATURE	PURPOSE	REFERENCE
7.2 Pressure-relief manholes are recommended on outdoor tanks for emergency venting; indoor tanks must have rupture disk nozzles fitted with vent lines discharging outdoors or to a containment system if needed	Emergency venting capabilities reduce the risk of vessel failure in the event of an uncontrolled polymerization; two-phase venting will require a relatively large vent area	BASF et al., 1991; DIERS, 1992
7.3 Consider the use of water dilution into the tank as a last-resort safety feature in acrylic acid storage vessels	The vaporization of water can provide an efficient means of removing heat from acrylic acid that is above 212 °F (100 °C); however, the release of large volumes of steam/acid vapor and the possibility of tank overflow must be considered	BASF et al., 1993; Levy and Penrod, 1989
8. Passive Mitigation		
8.1 For indoor storage of drums, do not store more than two high and two wide in groups of 48; have aisles of 4 ft (1.2 m) minimum on all sides	Proper drum spacing reduces the likelihood of an incident with one group of drums initiating a follow-on effect with adjacent drums	BASF et al., 1991
8.2 For larger-volume storage of acrylic acid, consider the following: • using a low-design-pressure storage tank (with appropriate secondary containment) •using a weak-seam roof design	There is currently no known reliable method of relieving pressure from a "worst-case" runaway polymerization, and opening at a lower pressure would result in less energy release	BASF et al., 1991; Current industry practice
9. Detection, Warning, and Isolation		
No post-release detection, warning, or isolation systems unique to acrylic acid service		
10. Fire Prevention/Protection/Mitigation		
10.1 Cool all equipment handling acrylic acid (particularly storage tanks, drums, and containers) by water stream and/or fixed deluge protection if exposed to fire	Uncontrolled polymerization is initiated by elevated temperatures	BASF et al., 1991; NFPA 49, 35
10.2 Maintain maximum possible distance when fighting fires involving acrylic acid liquid, or fight fire from protected location	Initiation of acrylic acid polymerization by heat (particularly in closed drums, inadequately relieved storage tanks, and blocked-in liquid lines) can result in rupture of the containment with ensuing explosion blast, fireball, and missile effects	NFPA 49, 35
11. Post-Release Mitigation		
11.1 Avoid routing acrylic acid spills to an enclosed sewer system	The acrylic acid may polymerize exothermically in the sewer system and possibly cause ignition of vapors and/or containment rupture by pressure generation	Bretherick, 1990, 360; NJDOH, 1994
12. Codes and Standards		
ORGANIZATION	TITLE	REFERENCE
U.S. Occupational Safety and Health Administration	Flammable and Combustible Liquids	29 CFR 1910.106

7.2. Polymerizable Materials: Styrene

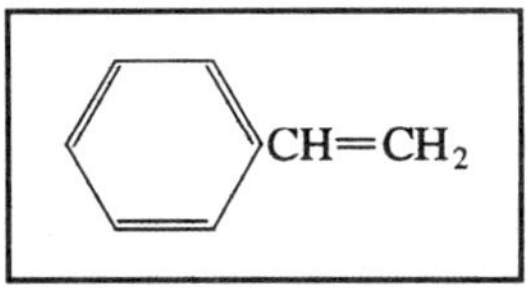

Styrene (styrene monomer or vinylbenzene), produced primarily by the dehydrogenation of ethylbenzene, is a commodity chemical used in various polymerization processes to make polystyrene and various copolymers. With an annual production of just over 10 billion pounds (4.6 million metric tons) in 1993, it ranked as the twenty-second highest-volume chemical in the U.S. (C&EN, 1994).

Styrene is a colorless, oily, flammable liquid that has an atmospheric boiling point of 293°F (145°C) and a flash point of 88°F (31°C). Its lower flammable limit is variously reported as 0.9 to 1.1% by volume, and its reported upper flammability limit likewise ranges from 6.1% to 7.0% (Dow, 1967, 3; MCA SD-37, 1971; NFPA 49, 1994; NFPA 325M, 1994). Styrene has a sweet, aromatic odor at low concentrations (odor threshold 0.3 ppm) and a sharp, penetrating, disagreeable odor at higher concentrations. Hence, it has good warning properties for avoidance of exposure to dangerous vapor concentrations. Although only mildly toxic to humans by inhalation, producing narcotic or anesthetic effects upon exposure to high concentrations, styrene can cause severe but reversible eye injuries. Since it is positive in mutagenicity tests (NJDOH, 1994) and a suspected human carcinogen, vapor inhalation should be avoided by local exhaust ventilation and/or respiratory protection.

Styrene is considered relatively safe to handle when properly inhibited, compared to many other flammable and reactive chemicals such as ethylene and vinyl chloride monomer. However, it has been involved in several plant-scale explosions. In 1966, styrene vapors ignited following a sight glass failure, with the ensuing explosion resulting in 11 fatalities, 10 injuries, and over $2 million in losses (Slater, 1978; Lees, 1980, 890). In another incident, accidental heating of styrene in a storage tank initiated a runaway polymerization reaction that resulted in the violent ejection of liquid and vapor from the tank (Bond, 1985, 25). The heat of polymerization of styrene is sufficient to raise the temperature of the polymer over 570°F (300°C) under adiabatic conditions (Dow, 1967, 10).

Specific design considerations for the safe storage and handling of styrene monomer that are *in addition to the usual precautions to be taken with flammable liquids* are listed in the "Feature" column of Table 7.2. The reason or explanation for each feature is given in the "Purpose" column of the table, with references given in the last column if further information is needed. Note that the specific design considerations in Table 7.2 for styrene do not necessarily apply to other polymerizable chemicals. In all cases, whether designing or evaluating a storage and handling arrangement for styrene or for any other polymerizable chemical, suppliers and manufacturers should be contacted for current MSDSs and design considerations.

In brief, the three most important objectives for the safe storage and handling of styrene (Shelly and Sills, 1969) are

(a) personnel protection,
(b) protection of the quality and monomeric status of the material under normal circumstances, and
(c) protection against fire.

Maintaining the integrity of the styrene containment can be considered an integral part of both (a)and (c).

TABLE 7.2
Specific Design Considerations for Styrene

FEATURE	PURPOSE	REFERENCE
1. Compatibility		
1.1 Separate styrene handling from, and take particular precautions to avoid contact of the styrene with: strong acids (sulfuric acid and oleum, phosphoric acid, hydrochloric acid, chlorosulfonic acid), oxidizing materials, peroxides and other polymerization initiators, and metal salts such as iron chlorides, aluminum chloride, and other metal halides	The rate of styrene polymerization increases significantly with many common reagents; this hazard is reduced if the styrene is properly inhibited (see 2.1 below)	DOT, 1984; Dow, 1967, 10; NFPA 49, 163; Sax and Lewis, 1989, 3128
1.2 Design and operate to avoid contact with pure oxygen	Styrene reacts with oxygen above 104 °F (40 °C) to form a thermally sensitive organic peroxide	Sax and Lewis, 1989, 3128
1.3 Do not use copper or copper-containing alloys in any equipment or part that may come into contact with styrene	Styrene is corrosive to copper and copper-containing alloys, causing greenish-blue coloration of the styrene and inhibiting polymerization	Dow, 1967, 8; Hagopian et al., 1983, 791; MCA SD-37, 5
1.4 Do not use rubber in any equipment or part that may come into contact with styrene	Rubber dissolves in styrene, causing discoloration of the styrene	Hagopian et al., 1983, 791
2. Storage Time/Shelf Life		
2.1 Maintain a concentration of 4-tert-butylcatechol (TBC) inhibitor of 10-50 ppm by weight during storage and handling of styrene monomer (Note: nitrogen-substituted aromatics and sulfur-based compounds have also been used as inhibitors; however, TBC is almost universally used for styrene in storage)	TBC is an antioxidant that inhibits the polymerization of styrene monomer, which may be initiated by elevated temperature, oxidizers, peroxides, metal salts, or sunlight; however, TBC concentrations greater than 40-50 ppm may increase color control problems while adding little to the storage stability	Hagopian et al., 1983, 785; Dow, 1966, 5-7; Shelly and Sills, 1969, 30

TABLE 7.2
Specific Design Considerations for Styrene

FEATURE	PURPOSE	REFERENCE
2.2 Check TBC concentration of styrene in storage tanks at least weekly if storage temperature is 70 °F (21 °C) or less and daily or as experience indicates if storage temperature is above 70 °F; do not allow the TBC concentration to fall below 10 ppm (see 2.5 below for drum storage)	TBC concentration can decrease over time as it performs its inhibiting function; TBC concentration can be measured spectrophotometrically by the formation of a red quinone when styrene is extracted with aqueous NaOH; polymer formation in the monomer can also be determined using ASTM D-2121	Dow, 1989, 7: Hagopian et al., 1983, 790; MCA SD-37, 6; Shelly and Sills, 1969, 30
2.3 Provide a means of pumping inhibitor solution to the storage tank and mixing it with the tank contents, such as injecting it into a pumped recirculation stream	Complete mixing of the inhibitor with the tank contents is essential for proper effecting of its inhibiting action; proper mixing can also ensure adequate dissolved oxygen; multiple injection points may be necessary for very large tanks	Current industry practice; Dow, 1967, 8
2.4 The relatively small quantities of TBC required to raise the inhibitor level in stored monomer can be easily added using a concentrated stock solution.	A TBC concentrate can be prepared by dissolving 704 grams of pure TBC in 1 gallon of styrene monomer; 1 cc of the concentrate will raise the inhibitor level 1 ppm in a 410 lb drum of monomer	Dow, 1989, 7
2.5 Expect the shelf life of styrene to vary from 1 month to over 1 year, depending on the storage temperature, TBC concentration, and oxygen concentration	With 12 ppm TBC, the expected shelf life is 3 mo at 85 °F (29 °C) and 6 mo at 60 °F (16 °C) if saturated with oxygen (only 4-5 days and 10-15 days, respectively, with less than 3 ppm oxygen); with 50 ppm TBC, the shelf life is 6 mo at 85 °F and over 1 yr at 60 °F if saturated with oxygen	Dow, 1989, 4; Dow, 1967, 8; Shelly and Sills, 1969, 30
3. Storage Quantity and Configuration		
3.1 Minimize stored inventories of styrene, especially in hot weather	Reducing the quantity of styrene stored on-site makes the facility inherently safer in case of loss of containment or control or an area fire	AIChE-CCPS, 1988, 15; Dow, 1967, 6
3.2 Design vertical storage tanks without internal bracing, such as an API 650 tank with a self-supporting dome roof. Inspect the roof, sides, and floor of bulk storage tanks at least annually for polymer buildup. One company reports painting the tops of tankage black to ensure high surface temperature to reduce condensation and polymerization sites (but see also 5.4 below)	Styrene vapor is uninhibited, and will condense and readily polymerize on all vapor space surfaces; the polymer "icicles" can cause off-specification product or even damage the roof or roof supporting structures by their weight	Current industry practice; Dow, 1967, 23; Shelly and Sills, 1969, 31
3.3 Smooth and grind all tank and nozzle welds; allow no rough surfaces; have a smooth coating such as a baked phenolic, modified epoxy, or catalyzed epoxy lining on the inside of the tank (except as mentioned in 3.5 below)	Rough surfaces will promote polymerization of styrene, particularly in the vapor phase	Current industry practice; Dow, 1967, 7; Shelly and Sills, 1969, 31

TABLE 7.2
Specific Design Considerations for Styrene

FEATURE	PURPOSE	REFERENCE
3.4 Use a rust-resisting inorganic zinc silicate material to coat tank and nozzle internals, particularly the inside bottom of the tank and the lower 2 ft of wall	The epoxy or phenolic internal tank lining is nonconductive; the coating allows static charges in the liquid to drain off through the tank ground	Current industry practice; Shelly and Sills, 1969, 31
3.5 Apply an epoxy-novolac coating with TBC inhibitor dissolved in methanol on the internal necks of the pressure-vacuum relief device, manhole, and rupture disks	The inhibitor formulated paint reduces the formation of polystyrene, which may block safety equipment	Current industry practice
4. Air and Moisture Exclusion		
4.1 Air cannot be totally excluded from the storage containment; styrene must be either stored with exposure to the atmosphere or air must be added periodically if nitrogen blanketing is used	The TBC inhibitor needs a minimum of 15 ppm O_2 to activate it (the equilibrium concentration of O_2 in liquid styrene is ~50 ppm at room temperature); however, styrene's flammable limits should be avoided	Dow, 1967, 4; Shelly and Sills, 1969, 30; Current industry practice
4.2 Rust and moisture should be excluded from styrene storage and handling facilities	Metallic salts and hydrates of TBC form highly colored compounds and can change the color of the monomer markedly	Dow, 1966, 7
5. Monitoring and Control		
5.1 Store at or below room temperature if possible; circulate through an external refrigeration unit to keep around 70-75 °F (21-24 °C) in areas where average daily temperatures above 80 °F (27 °C) are common	Maintaining the temperature of the styrene below its 88 °F (31 °C) flash point will prevent the vapors from being in the flammable range, and lower temperatures extend the shelf life of the styrene monomer	Shelly and Sills, 1969, 30
5.2 For storage tanks, indicate the styrene temperature at three to five points along the height of the liquid (see Figures 7.2 and 7.3), check the liquid and vapor temperatures at least daily, and alarm on high temperature	Generally a temperature above 150 °F (66 °C) is required to initiate a rapid polymerization reaction; however, this onset temperature depends on the concentration of inhibitor, degree of mixing, and heat transfer to the environment determined by such factors as the storage quantity and configuration and ambient temperature	MCA SD-37, 6; Dow, 1967, 18-19, 23
5.3 Do not permit drums of monomer to stand in the sun for more than a short period of time; drums can be cooled by water spray in hot weather or kept in a properly designed air-conditioned building	Drum contents can be heated by solar heating more rapidly than the contents of large tanks	Dow, 1967, 6, 20
5.4 Paint the outside of storage tanks white or aluminum; consider insulating storage tank; burial or shading of tanks may be practical for smaller tanks, although underground storage is strongly discouraged	Minimizing heat input to the tank will provide an additional safety factor against an uncontrolled temperature rise (but see also 3.2 above)	Current industry practice; Dow, 1967, 20
5.5 Do not use heater coils in styrene service under any conditions	Inadvertent heating of styrene monomer can rapidly lead to uncontrolled polymerization	MCA SD-37, 11

TABLE 7.2
Specific Design Considerations for Styrene

FEATURE	PURPOSE	REFERENCE
6. Handling and Transfer		
6.1 Wear protective equipment, stand to one side, and carefully vent any internal pressure when removing the bung from a drum of monomer	Pressure can be generated due to heating of the drum contents, if polymerization of the contents has begun to occur	MCA SD-37, 10
6.2 Do not use a rubber-type hose when a flexible connection is needed for unloading styrene	Rubber is dissolved by styrene; EPDM and fluoroelastomers have been found to be best overall with styrene monomer for flexible hoses; an aluminum pipe with a swing joint is lightweight and easily cleaned	Dow, 1967, 22; MCA SD-37, 10
6.3 Design and locate all piping, valves, gages, flame arresters, tank openings, emergency relief systems, vents, and engineering controls such that they may be readily and periodically inspected and cleaned	Styrene tends to form polymers that can plug equipment	MCA SD-37, 6
6.4 Drain transfer lines and blow clear when not to be transferring liquid for more than 2 weeks, or shorter in summertime unless lines are cold water traced, buried, or otherwise temperature controlled	Even in uninsulated, cold-water-traced lines, no oxygen is available except existing dissolved oxygen to activate the TBC inhibitor	Current industry practice; Dow, 1967, 21
6.5 Avoid venting of styrene tanks directly to the atmosphere with probable halogen vapor sources nearby	Styrene concentrations in air as low as a few ppb will form an extremely potent lachrymator (tear gas) when contacted by bromine in sunlight; chlorine contact requires only slightly higher concentrations	Shelly and Sills, 1969, 31
6.6 Piping systems containing styrene monomer should use plug valves rather than ball valves for all manually and intermittently operated valves	Ball valves can trap pockets of monomer between the ball and valve body when opened or closed; the monomer can polymerize and seize the ball, rendering the valve inoperable	Current industry practice
6.7 Reliably protect against dead-head pumping of styrene by multiple levels of design/operating safeguards	Dead-head pumping, such as against a closed valve, quickly generates heat and may lead to runaway polymerization	Dow, 1967, 20; MCA SD-37, 11
7. Last-Resort Safety Features		
7.1 Add TBC and aerate the monomer if TBC has been depleted and polymerization has begun; do not add solid TBC to polymerizing monomer	Addition of TBC even after polymerization has begun may stop or at least slow down the polymerization reaction	Dow, 1967, 6; Dow, 1989, 11
7.2 Dilute with ethylbenzene or toluene if runaway polymerization has begun	Addition of ethylbenzene or toluene reduces viscosity of the polymerizing liquid and slows polymerization by dilution	Dow, 1989, 11

TABLE 7.2
Specific Design Considerations for Styrene

FEATURE	PURPOSE	REFERENCE
7.3 Dump systems, if employed, should be operated under an inert gas atmosphere at all times	Introduction of polymerizing liquid styrene into a heated styrene vapor-air mixture can, despite the temporary cooling effect, cause ignition to occur under conditions where it would not normally do so; this effect has been noted in previous plant incidents	Bretherick, 1990, 743
7.4 Size the storage tank emergency relief system for runaway reaction with two-phase venting (Note: the manhole is often used for emergency relief venting; however, this may be vastly undersized for a runaway reaction in a large field tank)	Sizing the emergency relief for only fire exposure would be inadequate due to the heat of polymerization of styrene and the increased relief area required due to the potential for two-phase flow; in addition, styrene, even when inhibited, will always start to polymerize well below its atmospheric boiling point	Dow, 1967, 21; MCA SD-37, 6
8. Passive Mitigation		
No passive mitigation systems unique to styrene service		
9. Detection, Warning, and Isolation		
No detection, warning, and isolation systems unique to styrene service		
10. Fire Prevention/Protection/Mitigation		
10.1 Cool all equipment handling styrene monomer (especially storage tanks) by water spray if exposed to fire; full coverage of all exposed areas at a water rate of 0.25 gpm per ft^2 of equipment (10.2 L/min per m^2 of equipment) is recommended	Uncontrolled polymerization may occur under fire conditions	MCA SD-37, 9; NFPA 49, 162; Current industry practice
10.2 Consider foam injection in the design of styrene storage tanks; a suggested design for supplying liquid foam solution is at a rate of 0.16 gpm per ft^2 of liquid surface (6.5 L/min per m^2 of liquid surface)	Foam injection can rapidly extinguish a fire by excluding oxygen from the surface of the liquid; subsurface foam injection has maintenance, cost, and reliability advantages over foam chamber systems	Appendix B survey results; Shelly and Sills, 1969; Current industry practice
10.3 Maintain maximum possible distance when fighting fires involving styrene liquid, or fight fire from protected location	Initiation of styrene polymerization by heat, particularly in closed drums, inadequately relieved storage tanks, or blocked-in liquid lines, can result in containment rupture, with ensuing explosion blast and fireball effects	DOT, 1993; NFPA 49, 162
11. Post-Release Mitigation		
11.1 Absorb spills in pretested noncombustible material for proper disposal	Some absorbing agents such as untreated clays and micas will cause an exothermic reaction that might ignite the monomer, and use of combustible material for spill cleanup amplifies the fire hazard in case an ignition source is present	Dow, 1966, 19; NFPA 49, 162

TABLE 7.2
Specific Design Considerations for Styrene

12. Codes and Standards

ORGANIZATION	TITLE	REFERENCE
American Society of Testing and Materials	"Tentative Method of Test for Inhibitor, 4-Tertiary-Butylcatechol, in Styrene Monomer"	ASTM Method D 2120-62T

7.3. Organic Peroxides

The organic peroxides are a class of chemicals that have a peroxide (O–O) bond with organic substituents. Organic peroxides are commercially important because they are relatively unstable. They are readily decomposed by heat or chemical agents, yielding free radicals that function as polymerization initiators and curing agents. The same instability that makes organic peroxides useful requires that special precautions be taken in their storage and handling. If not handled properly, organic peroxides can be extremely hazardous. However, the degree of hazard differs from product to product. Some peroxides, such as methyl ethyl ketone (MEK) peroxide, are so reactive they are not commercially available in concentrations over 60% (Noury, 1985). Explosive decomposition of dibenzoyl peroxide can be initiated by friction if allowed to dry and be carelessly handled. However, many other organic peroxide formulations are no more hazardous than ordinary combustibles.

The primary hazards involved in the safe storage and handling of organic peroxides are thermal sensitivity and, in many cases, flammability or combustibility. In addition, the stability of organic peroxides can be significantly affected by various contaminants.

All organic peroxides should be considered flammable or combustible unless and until the specific flammability characteristics are known. Most will burn vigorously once ignited, and some are susceptible to detonation if not diluted (such as dry benzoyl peroxide) or if confined when decomposition occurs.

Organic peroxides also all exhibit sensitivity to heat, which increases with the reactivity of the peroxide. Organic peroxides are constantly undergoing decomposition, at a rate that is dependent on the temperature of the material. When thermal energy available from the heat of decomposition exceeds the rate of heat dissipation to the surroundings, the increasing temperature accelerates the decomposition of the remaining peroxide. This may proceed out of control in a characteristic "runaway reaction" decomposition. The danger of runaways is increased for some peroxides, such as cumene hydroperoxide, because the decomposition is autocatalytic; i.e., the decomposition products catalyze further decomposition and cause the reaction to self-accelerate.

Accelerated decomposition of organic peroxides can also be caused by chemical contamination. Certain chemicals initiate rapid decomposition at ambient temperatures. These hazards demonstrate the importance of safety in the storage and handling of organic peroxides (Noury, 1985).

NFPA 43B (1993) can be consulted for specific details of organic peroxide storage arrangements. This standard also gives a classification of organic peroxide formulations relative to their decomposition and flammability hazards (see Section 2.3.2).

Some design considerations for the safe storage and handling of organic peroxides that are *in addition to the usual precautions to be taken with flammable/combustible solids and liquids* are listed in the "Feature" column of Table 7.3. (Some of these considerations would not apply to NFPA Class V organic peroxides.) The reason or explanation for each feature is given in the "Purpose" column of the table, with references given in the last column if further information is needed. Additional design considerations for two specific organic peroxides, benzoyl peroxide and methyl ethyl ketone (MEK) peroxide, follow in Tables 7.4 and 7.5.

TABLE 7.3
Specific Design Considerations for Organic Peroxides

FEATURE	PURPOSE	REFERENCE
1. Compatibility		
1.1 Separate organic peroxide handling from, and take particular precautions to avoid contact of peroxides with incompatible materials. Strong acids, catalytic accelerators, and many organic and inorganic compounds are incompatible with organic peroxides; consult material safety data sheets and the organic peroxide supplier for compatibility with other compounds	Direct contact with incompatible materials or accelerators could result in a violent decomposition or fire	Hatayama et al., 1980; Luperox, 4; Noury, 1985
1.2 Do not use mild steel, copper, or rubber in any equipment or part that may come into contact with organic peroxides (note that there may be some exceptions such as cumene hydroperoxide regarding the use of mild steel equipment)	Contact with these materials can lead to decomposition of the peroxide	Luperox, 5
2. Storage Time/Shelf Life		
2.1 Rotate organic peroxide inventories to avoid shelf-life problems	While some refrigerated organic peroxides have a shelf life of only 1 month, most organic peroxide will last 6 to 12 months	Walrod, 2
3. Storage Quantity and Configuration		
3.1 When possible, store organic peroxides in their original containers; if repackaging is necessary, use only clean, new, compatible packaging and do not repackage in storage area	Storage in the original container prevents contamination of the peroxide	Noury, 1985

TABLE 7.3
Specific Design Considerations for Organic Peroxides

FEATURE	PURPOSE	REFERENCE
3.2 If windows are present in the storage area, they should be situated on the shaded side and be made of fire retardant plastic or reinforced glass; painting and/or shading of windows is also recommended	Containers of organic peroxides should not be directly exposed to sunlight; reinforced glass will reduce the risk of missile hazards	Noury, 1985
3.3 Locate storage areas away from manufacturing areas, in a secured and monitored remote location if possible	Remote storage minimizes the danger of fire, explosion, and health problems	Luperox, 3; Varjavandi and Mageli, 1971
3.4 Do not use organic peroxides in pressurized vessels or other confinement unless adequate provision is made to safely relieve any sudden internal pressure buildup	A pressure surge may develop if the peroxide is decomposed by heat or contamination	SPI, 1978
3.5 Arrange storage to facilitate manual access and handling, to maintain pile stability, to minimize breakage and spillage, and to promote good housekeeping; maintain minimum spacing requirements as specified in NFPA 43B	Proper storage arrangement reduces the risk of damage to the organic peroxide containers	NFPA 43B, 7
4. Air and Moisture Exclusion		
No air and moisture exclusion considerations unique to organic peroxide service		
5. Monitoring and Control		
5.1 Know the maximum storage temperature and quantity for each organic peroxide formulation to be stored or handled, and keep material in storage below the maximum storage temperature and quantity at all times	Maintaining the organic peroxides at lower temperatures extends the shelf life of the material; most organic peroxides can be stored at room temperature, 90 °F (30 °C); for long-term storage or in areas where excessive temperatures may occur, a refrigeration unit is recommended to maintain the temperature below 77 °F (25 °C); some formulations require cold storage in a freezer compartment	NFPA 43B, 5; Noury, 1985
5.2 Paint the outside of storage units with a reflective paint, usually white; in warmer climates, the roof can be sprayed with water or a sunshield may be installed	Minimizing heat input to the storage facility will provide an additional safety factor against an uncontrolled temperature rise	Noury, 1985
5.3 Heater coils should generally not be used in organic peroxide service; some exceptions exist, such as the heating of dicumyl peroxide, which has a melting point of 104 °F (40 °C), for transfer purposes	Inadvertent heating of organic peroxides can lead to explosive decomposition	SPI, 1978
6. Handling and Transfer		
6.1 Unused organic peroxides should not be returned to their original containers	This protects against contamination of the remaining material in the container	Luperox, 5; Noury, 1985

TABLE 7.3
Specific Design Considerations for Organic Peroxides

FEATURE	PURPOSE	REFERENCE
6.2 Material and equipment used in the handling and transfer of organic peroxides should be dedicated for their use only	The use of dedicated equipment reduces the likelihood of contamination by an incompatible material through an operational or mechanical failure	Varjavandi and Mageli, 1971; Noury, 1985
6.3 Transfer liquid organic peroxides by gravity flow when feasible; if gravity flow is not an option, centrifugal pumps, diaphragm pumps, or inert gas pressurizing methods can be employed. Pumps, when used, should be instrumented with cutoffs for low flow or high temperature	Transfer of organic peroxides is best accomplished by gravity flow due to the avoidance of potentially dead-heading a transfer pump and initiating a decomposition reaction	Luperox, 5; Noury, 1985; Current industry practice
7. Last-Resort Safety Features		
7.1 Use dry ice to cool small containers of organic peroxide that are in danger of overheating	The use of dry ice to externally cool small containers may prevent or at least reduce the possibility of decomposition by overheating, such as by temporary loss of refrigeration for formulations needing to be refrigerated	ATO, 1993
8. Passive Mitigation		
8.1 For large-quantity storage buildings, roof construction should be light, allowing for the release of excess pressure that might develop in an emergency	It is preferable to have a known failure point in case of a peroxide explosion than to have building structural failure	Noury, 1985
8.2 Buildings containing organic peroxides should be of noncombustible construction unless sprinkler protection is provided	Fireproof construction minimizes the danger of an organic peroxide fire spreading to other buildings	Luperox, 3; Varjavandi and Mageli, 1971; Noury, 1985
9. Detection, Warning, and Isolation		
9.1 Have operations personnel routinely patrol organic peroxide storage areas	Routine patrols can give an early detection of loss of containment	Current industry practice
10. Fire Prevention/Protection/Mitigation		
10.1 Keep organic peroxides away from all sources of heat such as steam pipes, space heaters, hot water pipes, radiators, and hot surfaces; protect against direct sunlight	Organic peroxides are heat-sensitive, and an excess of heat could lead to thermal decomposition	Luperox, 4; Noury, 1985
10.2 Maintain maximum possible distance when fighting organic peroxide fires, or fight fire from protected location; evacuate all personnel to safe area	Heat from the fire could cause decomposition of the peroxide and result in rupture of the containment with ensuing explosion blast, missiles, and fireball effects	Noury, 1985
10.3 Cool all equipment handling organic peroxides (especially storage containers) by water stream or fog if exposed to fire	Decomposition and explosion may occur under fire conditions	Luperox, 12; Noury, 1985; SPI, 1978

TABLE 7.3
Specific Design Considerations for Organic Peroxides

FEATURE	PURPOSE	REFERENCE
11. Post-Release Mitigation		
No post-release mitigation design considerations unique to organic peroxides		
12. Codes and Standards		
ORGANIZATION	TITLE	REFERENCE
National Fire Protection Association	Storage of Organic Peroxide Formulations	NFPA 43B, 1993
United Nations	United Nations Recommendation on Transport of Dangerous Goods, Tests and Criteria	UN, 1990

7.4. Organic Peroxides: Dibenzoyl Peroxide

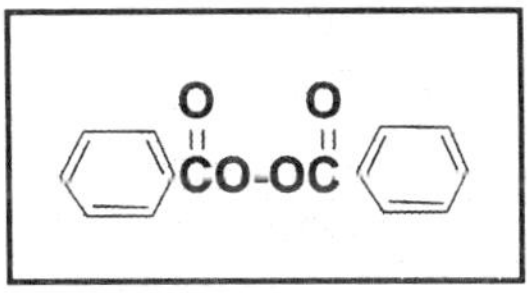

Dibenzoyl peroxide (benzoyl peroxide or BPO), produced by the reaction of benzoyl chloride with hydrogen peroxide or sodium peroxide, is used mainly as a polymerization initiator. It is also used in the vulcanization of rubber, as a drying agent for ink, as a bleaching agent, and in pharmaceutical applications.

Pure dibenzoyl peroxide is a white, granular, tasteless, odorless powder at room temperature. It has low toxicity, but is a suspected human carcinogen (NJDOH, 1994; Sax and Lewis, 1989) so dust inhalation should be avoided by local exhaust ventilation and/or respiratory protection. Dibenzoyl peroxide may decompose explosively above its melting point of 217-223°F (103-106°C), and has an self-accelerating decomposition temperature (SADT) of 155°F (68°C) for a standard packaging arrangement (Sax and Lewis, 1989).

In the pure, dry state, dibenzoyl peroxide is sensitive to shock and friction. It has been involved in several industrial fires and explosions. An explosion took place in a polyvinyl acetate plant, killing an operator, when dibenzoyl peroxide was contaminated with residual vinyl acetate and decomposed on a heated funnel. In another incident, a serious fire started when dibenzoyl peroxide was ignited by friction generated by sweeping the floor with a broom. Several other incidents involving dibenzoyl peroxide in fires and/or explosions have been recorded.

Because of its sensitivity to shock and friction, dibenzoyl peroxide is generally combined with a diluent in commercial formulations. Common forms include granular mixtures with up to 30 wt% water, pastes or slurries made with various plasticizer, water, and/or silicone oil combinations, and powders with 63 to 67 wt% wheat starch or dicalcium phosphate.

In summary, the four most important objectives for the safe storage and handling of dibenzoyl peroxide are

(a) protection against drying out or otherwise separating the pure dibenzoyl peroxide,
(b) protection against shock and friction in case the dibenzoyl peroxide does dry out or separate from its diluent,
(c) protection against contamination, and
(d) protection against fire.

Specific design considerations for the safe storage and handling of dibenzoyl peroxide that are *in addition to the precautions to be taken with organic peroxides that are listed in Table 7.3* are listed in the "Feature" column of Table 7.4. The reason or explanation for each feature is given in the "Purpose" column of the table, with references given in the last column if further information is needed. Specific considerations may vary depending on the concentration of dibenzoyl peroxide being stored and handled. Note that the design considerations in Table 7.4 for dibenzoyl peroxide do not necessarily apply to other organic peroxides. In all cases, whether designing or evaluating a storage and handling arrangement for dibenzoyl peroxide or for any other organic peroxide, suppliers and manufacturers should be contacted for current MSDSs and design considerations.

TABLE 7.4
Specific Design Considerations for Dibenzoyl Peroxide

FEATURE	PURPOSE	REFERENCE
1. Compatibility		
1.1 Separate dibenzoyl peroxide handling from, and take particular precautions to avoid contact of dibenzoyl peroxide with: acids, alcohols, amines, ethers, metallic naphthenates, polymerization catalysts, reducing agents, aniline, N-bromosuccinimide + 4-toluic acid, carbon tetrachloride + ethylene, charcoal, N,N-dimethylaniline, dimethyl sulfide, lithium tetrhydroaluminate, methyl methacrylate, and vinyl acetate	Dibenzoyl peroxide reacts violently and/or forms explosive mixtures with many common chemicals	Bretherick, 1990, 868; NFPA 491M, 31; Sax and Lewis, 1989, 393
1.2 Do not allow dry dibenzoyl peroxide to come into contact with combustible materials such as wood or paper	Dibenzoyl peroxide is a strong oxidizer (however, note that paper is sometimes used as a packaging material for dibenzoyl peroxide formulations)	NJDOH, 1994

TABLE 7.4
Specific Design Considerations for Dibenzoyl Peroxide

FEATURE	PURPOSE	REFERENCE
2. Storage Time/Shelf Life		
2.1 Store dibenzoyl peroxide at less than 86 °F (30 °C) in order to obtain optimum shelf life; dry dibenzoyl peroxide lasts approximately 12 months, depending on the storage temperature, but dibenzoyl peroxide pastes have a shelf life of 3 to 6 months	Reduced shelf-life of pastes is due to slight solubility of dibenzoyl peroxides in the carrier resulting in lower thermal stability	ATO, 1992; Luperox, n.d., 9
2.2 Dispose of too-old dibenzoyl peroxide by hydrolysis, which can be accomplished by treatment with ten times its weight of 10% sodium hydroxide solution	Burning is not recommended, as dibenzoyl peroxide may burn or decompose in an uncontrolled manner	Noury, 1985
3. Storage Quantity and Configuration		
3.1 Protect storage containers of dibenzoyl peroxide from shock and/or physical damage	Dry dibenzoyl peroxide is sensitive to shock and friction	NFPA 49, 1991, 114
3.2 Do not store bags, drums, and other packages of dibenzoyl peroxide more than 6 ft (1.8 m) high and do not exceed a pile width of 4 ft (1.2 m); there should be a main aisle in the storage facility of at least 8 ft (2.4 m), and all other aisles should be at least 4 ft (1.2 m) wide	Proper storage spacing allows good sprinkler coverage	NFPA 43B, 1993, 7
3.3 Dibenzoyl peroxide that is diluted with water should remain sealed in the original container until used	Opening the container to air leads to evaporation of the water to expose the more hazardous dry peroxide	
4. Air and Moisture Exclusion		
4.1 When mixed with a large surplus of water (~30%), the hazards of dibenzoyl peroxide are significantly reduced	Dibenzoyl peroxide is often sold commercially in diluted form for use in water-tolerant applications	Bretherick, 1990, 869
5. Monitoring and Control		
5.1 Monitor temperatures inside storage building; alarm on high temperature	Dibenzoyl peroxide may decompose explosively when overheated under confinement	Varjavandi and Mageli, 1971; Sax and Lewis, 1989, 392
6. Handling and Transfer		
6.1 Do not transfer into rigid containers with tight or screw-on closures	Frictional heat developed in the process of closing the containers may cause rapid decomposition of material trapped under the closure	Catalyst Resources, 1983
7. Last-Resort Safety Features		
No last-resort safety features unique to dibenzoyl peroxide		

TABLE 7.4
Specific Design Considerations for Dibenzoyl Peroxide

FEATURE	PURPOSE	REFERENCE
8. Passive Mitigation		
8.1 Maintain separation distances from other buildings and equipment based on NFPA 43B spacing requirements that are a function of the concentration, form, and quantity of dibenzoyl peroxide stored, storage area sprinkler protection, and whether the storage area is detached or segregated	Proper separation distances reduce the likelihood of the storage building initiating a follow-on effect with adjacent equipment	NFPA 43B, 1993, 6
9. Detection, Warning, and Isolation		
No detection, warning, and isolation design considerations unique to dibenzoyl peroxide		
10. Fire Prevention/Protection/Mitigation		
10.1 Configure sprinkler systems in dibenzoyl peroxide storage areas according to NFPA 43B recommendations, which are a function of the form and concentration of the dibenzoyl peroxide	Sprinkler systems are effective for remotely putting out fires, and for keeping exposed dibenzoyl peroxide cool under fire conditions	NFPA 43B, 1993, 6
11. Post-Release Mitigation		
11.1 Absorb spills with water-wetted noncombustible material such as vermiculite and place in plastic container for disposal	Some absorbing agents such as cellulosic materials or spark-generating metals amplify the fire hazard and are incompatible with dibenzoyl peroxide	NFPA 49, 1991, 114
11.2 Do not sweep up spills of dry dibenzoyl peroxide; use a vacuum or wet method to reduce dust during clean up	Friction generated through sweeping could potentially ignite dibenzoyl peroxide dust	NJDOH, 1994
12. Codes and Standards		
ORGANIZATION	TITLE	REFERENCE
National Fire Protection Association	Storage of Organic Peroxide Formulations	NFPA 43B, 1993

7.5. Organic Peroxides: MEK Peroxide

Methyl ethyl ketone (MEK) peroxide is a mixture of peroxides produced by the reaction of methyl ethyl ketone with hydrogen peroxide. MEK peroxide is used in room-temperature curing of unsaturated polyester resins.

Commercial MEK peroxide is a colorless, oily, combustible, shock-sensitive liquid that has a flash point of 137°F (58°C) and a self-accelerating decomposition temperature (SADT) ranging from 140 to 160°F (60 to 71°C) in typical shipping containers for formulations ranging from 5.4 to 9.0% active oxygen (ATO, 1991). It is a suspected carcinogen with experimental tumorigenic data (Sax and

Lewis, 1989, 2319). Vapor inhalation should always be avoided by local exhaust ventilation and/or respiratory protection. MEK peroxide also has pronounced irritant effects on skin and eyes, in some cases leading to blindness (Varjavandi and Mageli, 1971).

MEK peroxide is not commercially available in concentrations above 60%, and must have an active oxygen level no greater than 9% to be transported. The diluted compound is less shock-sensitive than the pure form, and is much easier to handle. However, diluted MEK peroxide is still a hazardous chemical and has been involved in several industrial accidents. In Tokyo, 8000 pounds of MEK peroxide exploded during a fire, killing 19 and resulting in $1.25 million in damages. There are also several other recorded events of MEK peroxide undergoing a detonation as a result of a small fire.

In summary, the four most important objectives for the safe storage and handling of MEK peroxide are

(a) personnel protection,
(b) maintaining the proper diluent concentration,
(b) protection against contamination, and
(c) protection against fire.

Specific design considerations for the safe storage and handling of MEK peroxide that are *in addition to the precautions to be taken with organic peroxides that are listed in Table 7.3* are listed in the "Feature" column of Table 7.5. The reason or explanation for each feature is given in the "Purpose" column of the table, with references given in the last column if further information is needed. Note that the specific design considerations in Table 7.5 for MEK peroxide do not necessarily apply to other organic peroxides. In all cases, whether designing or evaluating a storage and handling arrangement for MEK peroxide or for any other organic peroxides, suppliers and manufacturers should be contacted for current MSDSs and design considerations.

TABLE 7.5
Specific Design Considerations for MEK Peroxide

FEATURE	PURPOSE	REFERENCE
1. Compatibility		
1.1 Acetone must never be used as a diluent for MEK peroxide	Acetone reacts with MEK peroxide to form acetone peroxide crystals, which are extremely shock-sensitive	Walrod and Gimbarski, 1979, 4
1.2 Separate MEK peroxide storage and handling from, and take particular precautions to avoid contact of MEK peroxide with: strong acids, strong alkali, strong oxidizers, transition metal salts, and reducing agents	Contamination by many common reagents may initiate explosively rapid decomposition	

TABLE 7.5
Specific Design Considerations for MEK Peroxide

FEATURE	PURPOSE	REFERENCE
2. Storage Time/Shelf Life		
2.1 Expect the shelf life of MEK peroxide to be approximately three to six months, depending on the storage temperature	The shelf life is 6 months at a storage temperature of 65-85 °F (18-29 °C) and 3 months at 85-100 °F (29-38 °C)	ATO, 1991
2.2 Dispose of too-old MEK peroxide by burning or hydrolysis; MEK peroxides can be hydrolyzed by treatment with ten times its weight of sodium hydroxide solution	Limited quantities of the material to be burned can be placed in a shallow trench and ignited with a 6 ft (1.8 m) torch, where allowed	Luperox, n.d., 5; Walrod and Gimbarski, 1979, 4
3. Storage Quantity and Configuration		
3.1 Storage containers of MEK peroxide should be protected from shock and/or physical damage	MEK peroxide is a shock-sensitive explosive	Sax and Lewis, 1989, 2319
4. Air and Moisture Exclusion		
No air and moisture exclusion design considerations unique to MEK peroxide		
5. Monitoring and Control		
5.1 Store MEK peroxide below 100 °F (38 °C)	At temperatures over 100 °F, there is a loss in activity of the peroxide; at higher storage temperatures the risk of autoignition is amplified	
5.2 Do not allow the temperature of diluted MEK peroxide to go low enough to allow the MEK peroxide to crystallize out of solution	Will depend on diluent and concentration; MEK crystals may be shock-sensitive	
6. Handling and Transfer		
No handling and transfer design considerations unique to MEK peroxide		
7. Last-Resort Safety Features		
No last-resort safety features unique to MEK peroxide		
8. Passive Mitigation		
8.1 Segregate or isolate MEK peroxide storage areas from other storage or operating areas by at least the distances specified in NFPA 43B	Proper spacing reduces the likelihood of a fire involving MEK peroxide initiating a follow-on effect with other buildings and equipment	NFPA 43B, 1993
9. Detection, Warning, and Isolation		
No detection, warning, and isolation design considerations unique to MEK peroxide		
10. Fire Prevention/Protection/Mitigation		
10.1 Use foam on MEK fires if the MEK has been diluted with ethyl acetate or similar flammable solvent	Water alone will be ineffective on this type of fire	Walrod and Gimbarski, 1979, 4
11. Post-Release Mitigation		
No post-release mitigation design considerations unique to MEK peroxide		
12. Codes and Standards		
National Fire Protection Association	Storage of Organic Peroxide Formulations	NFPA 43B, 1993

7.6. Temperature-Sensitive Materials: Ethylene Oxide

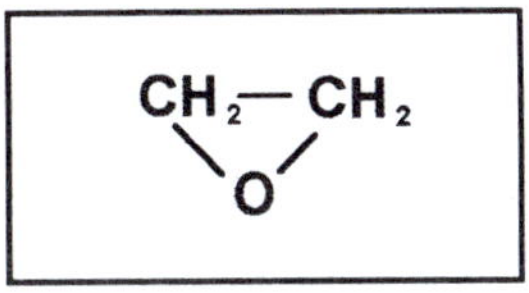

Ethylene oxide, produced by the vapor phase oxidation of ethylene, is used primarily in the production of ethylene glycol. It is also used as a raw material for the manufacture of a variety of commercially important materials, such as detergents and epoxide resins. With a boiling point of 50.7°F (10.4°C), it is a colorless, flammable gas at room temperature and pressure, and is easily liquefied at conditions of 7.3 psig (50 kPa) and 70°F (21°C) (Matheson, 1980, 322). Autodecomposition temperatures ranging from 840 to 1040°F (450 to 560°C) have been observed under experimental conditions.

The lower flammable limit of ethylene oxide in air is 3% by volume, and there is no upper flammable limit because at high concentrations (up to pure ethylene oxide) in air, explosive decomposition can occur (Lees, 1980, 263; MCA SD-38, 5; Matheson, 1980, 322). Ethylene oxide has an ether-like odor; however, since its odor threshold of 430 ppm is well above exposure limits, odor provides an inadequate warning of exposure. Continuous exposure to low concentrations of ethylene oxide may result in a loss of smell, and at high concentrations irritation of the respiratory system and eyes may occur. Ethylene oxide has been identified as an experimental mutagen and a carcinogen, so all contact should be reduced to the lowest possible level (MCA SD-38, 8; NJDOH, 1994; Matheson, 1980, 323).

Ethylene oxide is a highly reactive, flammable, and toxic chemical. For this reason, special precautions must be taken in its handling and storage. Processes involving ethylene oxide have been involved in several plant-scale explosions. In a 1991 incident at Seadrift, Texas, a redistillation column exploded, causing one fatality and extensive plant damage. The incident was initiated by an iron-oxide-catalyzed decomposition reaction near the top of a reboiler tube, then accelerated upward through the column (Viera et al., 1993). In another incident, a large ethylene oxide feed tank ruptured when unreacted ammonia backflowed into the tank. The ensuing cloud of ethylene oxide vapors ignited, generating blast overpressures and resulting in one fatality and nine injuries (Lees, 1980, 914).

Specific design considerations for the safe storage and handling of ethylene oxide that are *in addition to the usual precautions to be taken with flammable vapors and liquids* are listed in the "Feature" column of Table 7.6. The reason or explanation for each feature is given in the "Purpose" column of the table, with references given in the last column if further information is needed. Note that the design considerations in Table 7.6 for ethylene oxide do not necessarily apply to other temperature-sensitive materials. In all cases, whether designing or evaluating a storage and handling arrangement for ethylene oxide or for any other temperature-sensitive material, suppliers and manufacturers should be contacted for **current** MSDSs and design considerations.

In summary, the most important items to remember when storing and handling ethylene oxide are

(a) personal protection,
(b) protection against forming explosive mixtures of ethylene oxide vapor with air,
(c) protection against uncontrolled heating and/or fire, and
(d) avoidance of rust, contamination, or incompatible materials of construction that may initiate a runaway reaction.

TABLE 7.6
Specific Design Considerations for Ethylene Oxide

FEATURE	PURPOSE	REFERENCE
1. Compatibility		
1.1 Avoid contamination that could lead to runaway reaction of liquid ethylene oxide or of aqueous solutions. Avoid catalytic self-reaction of heated ethylene oxide vapor. Avoid possible cross-ties with reactive materials that could contaminate ethylene oxide, and ensure the use of dedicated hoses, etc. Refer to MSDS and literature to determine compatibility	Ethylene oxide reacts with numerous other materials such as acids and bases (including amines). Traces of some liquid contaminants may promote reaction with water. Some solid materials can act as catalysts for ethylene oxide self-reaction	Bretherick, 273-6; Hess and Tilton, 1950, 1254; Matheson, 1980, 324; MCA SD-38, 8; NFPA 49, 156; NFPA 491M, 88; Sax and Lewis, 1989, 1611
1.2 Consider the use of cellular glass insulation on heated tanks or lines containing ethylene oxide; sufficient thickness of cellular glass should be used to ensure an adequate degree of fire endurance	Ethylene oxide reacts with water in porous refractory insulations, forming polyethylene glycols that may accumulate and spontaneously ignite in hot insulation under certain conditions. Cellular glass may not give equivalent thermal protection as other refractory insulations under fire exposure conditions	Britton, 1991
1.3 Evaluate lubricants and other fluids contacting ethylene oxide for chemical reactivity	Reaction with ethylene oxide could lead to fluid degradation, undesired solids formation, or liquid phase runaway of ethylene oxide	Current industry practice
1.4 Evaluate gaskets and solid sealants for compatibility with ethylene oxide; reinforced PTFE or graphitic gaskets are commonly used after evaluation for chemical compatibility and mechanical suitability	Reaction or sealant degradation may occur after extended periods of exposing incompatible gaskets or solid sealants to ethylene oxide; emissions can be caused by seal leakage	Current industry practice

TABLE 7.6
Specific Design Considerations for Ethylene Oxide

FEATURE	PURPOSE	REFERENCE
2. Storage Time/Shelf Life		
2.1 Carefully analyze and monitor any storage of aqueous ethylene oxide solutions, particularly if greater than 1-2 wt% ethylene oxide; see reference for general relationships between process variables	While aqueous solutions are stable under certain conditions, ethylene oxide and water undergo a hydrolysis reaction that might cause a rise in temperature and pressure; key parameters are the ratio of liquid to vapor volume, initial temperature and pressure, concentration of ethylene oxide in water, the amount of the contained ethylene oxide that is hydrolyzed, and the availability of cooling and/or vapor space venting capacity	Curtis, 1990
2.2 Keep the ethylene oxide storage environment free of contaminants	Pure liquid ethylene oxide requires a temperature of about 390 °F (200 °C) for exothermic reaction to be readily observed in the laboratory using adiabatic calorimetry; exothermic reaction begins at much lower temperatures in the presence of common contaminants	Britton, 1990, 75
2.3 Ethylene oxide may have a limited shelf life as specified by the supplier for the type of container and storage conditions; refrigerated storage will extend shelf life	Ethylene oxide may slowly polymerize on prolonged storage, forming a non-volatile residue that may appear as a haze; precipitation from solution may occur at temperatures less than about 55-60 °F	Matheson, 1980, 324; Current industry practice
3. Storage Quantity and Configuration		
3.1 Where possible, minimize stored inventories of ethylene oxide	Reducing the quantity of ethylene oxide stored on-site makes the facility inherently safer in the event of loss of containment, loss of control, or an area fire	MCA SD-38, 20
3.2 Avoid rust formation, such as by the use of stainless steel, properly cleaned carbon steel, proper welding and inspection practices, and nitrogen inerting of equipment not in service	Iron oxides catalyze polymerization of ethylene oxide producing nonvolatile residue; certain high-surface-area iron oxides can produce fast polymerization and, in some cases, disproportionation reactions; these fast reactions may lead to ethylene oxide vapor decomposition under certain temperature and pressure combinations	Lees, 1980, 263; Viera et al., 1993; Current industry practice
3.3 Avoid stagnant volumes where iron oxides and other potentially catalytic solids may accumulate in the presence of non-inerted ethylene oxide vapor	Stagnant volumes may accumulate active catalyst under conditions where heat may accumulate and promote the fast reactions	Current industry practice

TABLE 7.6
Specific Design Considerations for Ethylene Oxide

FEATURE	PURPOSE	REFERENCE
4. Air and Moisture Exclusion		
4.1 Exclude air from equipment containing ethylene oxide. Inert storage tanks using an inert gas that is free of contaminants	Air can lead to ignition of ethylene oxide under milder conditions than for the pure vapor. Proper inerting of ethylene oxide will prevent flame propagation from an initiation site; the diluent requirement varies with storage conditions and the inert gas used. To prevent contamination of ethylene oxide, the inert gas should be of known purity such as from a dedicated source	Britton, 1990, 85; Lees, 1980, 264; MCA SD-38, 10
4.2 Take appropriate measures to minimize moisture in equipment where this may cause formation of rust	Iron oxides accelerate polymer formation; this is both a quality issue and a potential handling problem should polymers precipitate (see 5.1). Iron oxides may, under certain conditions, lead to fast reactions that can be hazardous (see 3.2)	Matheson, 1980, 324; Current industry practice
5. Monitoring and Control		
5.1 Store ethylene oxide at or below ambient temperature	Lower temperatures reduce the rate of polymer formation and provide a greater safety margin against uncontrolled exotherm in the event of contamination. However, internal cooling can potentially mask an exotherm in the tank caused by a contaminant. If polymer is already present in the liquid, it may precipitate from solution at temperatures below 55-60 °F and cause problems in liquid transfer	Lees, 1980, 263; MCA SD-38, 20; Current industry practice
5.2 Continuously monitor tank temperatures; install an alarm system that will automatically signal an increase in temperature	An increase in temperature can warn of the onset of a reaction due to contamination of the ethylene oxide	Hess and Tilton, 1950, 1252; MCA SD-38, 20
5.3 Do not permit DOT drums or cylinders of ethylene oxide to stand in the sun for more than a short period of time; drums can be cooled by water spray in hot weather or kept in a properly designed air-conditioned building	Contents can be heated by solar heating more rapidly than the contents of large tanks; drums and cylinders are usually not equipped with insulation and/or cooling necessary to maintain the proper storage handling temperature	MCA SD-38, 20
5.4 Keep storage tanks well-insulated against any sharp rises in temperature	Minimizing heat input to the tank will provide an additional safety factor against an uncontrolled temperature rise	NFPA 49, 1994
5.5 Low-pressure steam or hot water is recommended for heating ethylene oxide or mixtures in which ethylene oxide is used	The use of a lower-temperature heating fluid reduces the likelihood of overheating ethylene oxide	MCA SD-38, 10

TABLE 7.6
Specific Design Considerations for Ethylene Oxide

FEATURE	PURPOSE	REFERENCE
6. Handling and Transfer		
6.1 Design handling and transfer systems to avoid the presence of "hot spots," and avoid exposing pipelines to potential heat sources and fire hazards	Inadvertent heating of ethylene oxide can rapidly lead to vaporization and explosive decomposition	Bretherick, 273; Hess and Tilton, 1950, 1252; MCA SD-38, 10
6.2 Design and locate all piping, valves, gauges, vents, flame arresters, tank openings, emergency relief systems, and engineering controls such that they may be readily and periodically inspected and cleaned	Ethylene oxide tends to form polymers that can plug equipment	MCA SD-38, 12
6.3 Consider transfer of ethylene oxide by gravity or pressurizing with inert gas where feasible (however, pumps are commonly used in liquid ethylene oxide service)	Pressure or gravity transfer avoids potential hazards associated with pump installations; e.g., dead-head pumping	Hess and Tilton, 1950, 1256; Lees, 1980, 263; Matheson, 1980, 324
6.4 Vent lines from ethylene oxide tanks should extend to a safe area outside the process area; all manual vents containing flammable ethylene oxide mixtures should include flame arresters	Flashback from an external ignition source to a flammable atmosphere inside a tank must be prevented	MCA SD-38, 20; Matheson, 1980, 324
6.5 Pumps should be equipped with tandem mechanical seals with compatible seal fluid, or, instead, sealless pumps can be used	Seal fluid can be a source of contamination	MCA SD-38, 12; Current industry practice
6.6 Reliably protect against dead-head pumping of ethylene oxide by multiple levels of design and/or operating safeguards	Dead-head pumping, such as against a closed valve, quickly generates heat and may lead to ignition of vapor or polymerization initiation	MCA SD-38, 12
7. Last Resort Safety Features		
7.1 Provide an emergency relief/blowdown system for all ethylene oxide storage tanks; develop an emergency plan for the consumption, destruction, and/or removal of ethylene oxide when it contains a reactive contaminant	The evaporative cooling provided by the actuation of the depressurizing system might mitigate the consequences of an incipient runaway reaction, depending on the magnitude of the contamination or other initiating cause	MCA SD-38, 20; Hess and Tilton, 1950, 1252; Current industry practice
7.2 Consider providing a continuous inert gas purge through flame arresters, vents, and emergency relief systems and across rupture disk surfaces	The inert gas stream can prevent plugging by polymer formation	Current industry practice
8. Passive Mitigation		
8.1 Design dikes and drains so that ethylene oxide is kept out of confined spaces such as sewers	Ethylene oxide vapors can build up to explosive concentrations in confined spaces	NJDOH, 1994

TABLE 7.6
Specific Design Considerations for Ethylene Oxide

FEATURE	PURPOSE	REFERENCE
9. Detection, Warning, and Isolation		
9.1 Combustible gas detectors are commercially available for ethylene oxide determinations within the range of 0.01-3.5%	Detectors can give rapid warning of loss of containment of ethylene oxide	MCA SD-38, 9
10. Fire Prevention/Protection/Mitigation		
10.1 Ethylene oxide storage tanks are recommended to have adequate insulation and installed water spray systems; another option is to use earth-mounded ethylene oxide storage tanks. Cool all equipment handling ethylene oxide if exposed to fire	Vapor ignition or uncontrolled polymerization may occur under fire conditions; external heating can cause loss of inerting as ethylene oxide evaporates	Hess and Tilton, 1950, 1254; NFPA 49, 156; Current industry practice
10.2 Extinguish ethylene oxide liquid fires with an excess of water	Ethylene oxide liquid fires are difficult to extinguish; unconfined fires will continue to burn until diluted to 22 parts water to 1 part ethylene oxide; confined fires require even further dilution	Hess and Tilton, 1950, 1254; MCA SD-38, 15; NFPA 49, 1994
10.3 Handle waste from an extinguished liquid fire with care	The waste may continue to generate toxic ethylene oxide vapor; also, depending on the temperature and degree of confinement, the vapor from aqueous solutions can be flammable at only a few percent by weight of ethylene oxide	MCA SD-38, 15
10.4 Where containers of ethylene oxide are involved in a fire, fight fire from a protected location where possible	Liquid ethylene oxide containers can undergo internal reaction under fire conditions with ensuing blast, missile, and fireball effects	NFPA 49, 1994
11. Post-Release Mitigation		
11.1 Dilute a liquid spill of ethylene oxide with water to its nonflammable concentration and dispose of properly	The addition of water to ethylene oxide reduces the vapor concentration of the oxide (see 10.2 above)	Lees, 1980, 536
12. Codes and Standards		
ORGANIZATION	TITLE	REFERENCE
Factory Mutual Engineering & Research	Ethylene Oxide	FM 7-92, 1992
U.S. Occupational Safety and Health Administration	Ethylene oxide	29 CFR 1910.1047

7.7. Pyrophoric Materials: Aluminum Alkyls

$$R_3\text{-}Al$$

Aluminum alkyls, produced primarily by the reaction of aluminum with hydrogen and olefins, are a group of chemicals commonly used as catalyst components. The more commonly used aluminum alkyls, along with their abbreviations and chemical formulas, are listed in Table 7.7.

Aluminum alkyls are colorless, pyrophoric liquids with atmospheric boiling points that range from 261°F (127°C) to 682°F (361°C) depending on the size of the alkyl substituent groups (Ethyl, 1993; Heck and Johnson, 1962, 36). The *n*-alkyl derivatives decompose at temperatures greater than 350 to 450°F (177 to 232°C), to form the corresponding flammable, unsaturated hydrocarbon gases (Ethyl, 1993; NFPA 49, 1994). Aluminum alkyls are highly reactive with air and water, with the lower-molecular-weight compounds tending to have greater reactivity. Depending on the alkyl chain length, aluminum alkyls can spontaneously ignite or give off a dense smoke with air (Governale et al., 1965, 105). Aluminum alkyls in solution at concentrations less than 15% are considered nonpyrophoric. However, these low-concentration solutions still evolve heat and should be treated as if pyrophoric. Aluminum alkyls can cause severe skin burns due to their reaction with air and body moisture. Inhalation of large quantities of aluminum alkyls can cause the symptoms of "metal fume fever," which include throat irritation, headache, fever, chills, and nausea (Ethyl, 1993; NFPA 49, 39).

Due to the high reactivity of aluminum alkyls, several precautions must be taken to ensure safe handling. In 1973, a process containing aluminum alkyls exploded while being cleaned when caustic material in the cleaning solution reacted with aluminum alkyls remaining in the lines to form hydrogen (Lees, 1980, 923).

TABLE 7.7
Commonly Used Aluminum Alkyls

CHEMICAL NAME	ABBREVIATION	CHEMICAL FORMULA
Trimethylaluminum	TMA	$(CH_3)_3Al$
Triethylaluminum	TEA	$(C_2H_5)_3Al$
Tripropylaluminum	TnPA	$(C_3H_7)_3Al$
Triisobutylaluminum	TIBA	$(i\text{-}C_4H_9)_3Al$
Diethylaluminum chloride	DEAC	$(C_2H_5)_2AlCl$
Ethylaluminum dichloride	EADC	$C_2H_5AlCl_2$

The main objectives for the safe storage and handling of aluminum alkyls (Heck and Johnson, 1962, 35) are

(a) prevention of uncontrolled escape of alkyls and contact with reactive materials
(b) protection of operators and equipment from potential leaks, sprays, or splashes.

Specific design considerations for the safe storage and handling of aluminum alkyls that are *in addition to the usual precautions to be taken with flammable liquids* are listed in the "Feature" column of Table 7.8. The reason or explanation for each feature is given in the "Purpose" column of the table, with references given in the last column if further information is needed. Note that the design considerations in Table 7.8 for aluminum alkyls do not necessarily apply to other pyrophoric materials. In **all** cases, whether designing or evaluating a storage and handling arrangement for aluminum alkyls or for any other pyrophoric material, suppliers and manufacturers should be contacted for **current** MSDSs and design considerations.

TABLE 7.8
Specific Design Considerations for Aluminum Alkyls

FEATURE	PURPOSE	REFERENCE
1. Compatibility		
1.1 Separate aluminum alkyl handling from, and take particular precautions to avoid contact of the aluminum alkyls with: acids, alcohols, amines, carbon tetrachloride, chloroform, n,n-dimethylformamide, halogenated hydrocarbons, strong oxidizers, and water	Aluminum alkyls will react violently and/or explosively with many common solvents. Be sure to keep all incompatible materials away from areas containing aluminum alkyls	MSDS 62711; NFPA 49, 1994; NFPA 491M, 1991, 215; Sax and Lewis, 1989, 3368
1.2 Design and operate to avoid contact with air	Aluminum alkyls are spontaneously flammable in air	Sax and Lewis, 1989, 3368
2. Storage Time/Shelf Life		
2.1 The shelf life of aluminum alkyls can vary from one year to as much as ten years, depending on the type of alkyl	Most aluminum alkyls decompose over time to form hydrides; TIBA has shown a significant amount of decomposition after one year, while TEA has a shelf life of up to ten years	Ethyl, 1993
3. Storage Quantity and Configuration		
3.1 Store aluminum alkyls in an isolated area, away from any combustible or flammable materials; an outside or detached facility is preferred	Storing hazardous materials in a remote area makes the facility inherently safer in the event of loss of containment or control or an area fire	Governale et al., 1965, 106; MSDS 62711; NFPA 49, 1994

TABLE 7.8
Specific Design Considerations for Aluminum Alkyls

FEATURE	PURPOSE	REFERENCE
3.2 Place pipelines containing aluminum alkyls as low as possible in a storage facility	Low-lying pipelines lessen any hazards from overhead leaks from the pipelines	Heck and Johnson, 1962, 38
4. Air and Moisture Exclusion		
4.1 Store aluminum alkyls under a dry nitrogen blanket at all times (dry nitrogen requirements are typically less than 30 ppm O_2 and less than 0.4 ppm water)	Aluminum alkyls are pyrophoric and form explosive mixtures with water; air and water contamination should be avoided at all times	Ethyl, 1993; Heck and Johnson, 1962, 38; Governale et al., 1965,106
4.2 Do not route water lines inside an enclosure containing aluminum alkyls	Aluminum alkyls form explosive mixtures with water	Governale et al., 1965, 109
5. Monitoring and Control		
5.1 Measurement by weight is preferred for tank inventory measurement	By keeping the tank closed, there is less chance for the aluminum alkyls to be exposed to air or water	Ethyl, 1993
5.2 For tank level measurement, use displaced-float-type measurement devices	This prevents the use of sight glasses and other fragile materials of construction that can easily fail and discharge aluminum alkyls into the atmosphere	Ethyl, 1993
5.3 When thawing frozen aluminum alkyls, do not allow the temperature to rise above 180 °F (82 °C); for the more easily decomposable alkyls such as TIBA, the temperature should not exceed 150 °F (66 °C)	Aluminum alkyls decompose at high temperatures to form flammable gases	Ethyl, 1993; NFPA 49, 1994
6. Handling and Transfer		
6.1 Avoid bottom outlets on aluminum alkyl storage tanks; transfer aluminum alkyls through dip pipes in the top of the tank	By placing all fittings above the liquid level in aluminum alkyl tanks, the hazard of a severe release is generally reduced: a bottom outlet avoids the need for fittings on the bottom or side of the tank, where leaks in bottom or side fittings would be fed by gravity	Ethyl, 1993; Governale, 106; Heck and Johnson, 1962, 37
6.2 Valves should be operated from a remote location, whenever possible	Remote operation reduces the risk of accidental exposure to aluminum alkyls	Ethyl, 1993; Governale et al., 1965, 106; Heck and Johnson, 1962, 38
6.3 Transfers are usually made by differential pressure; If pumps are used, both diaphragm and centrifugal pumps work well	With pressure transfer, seal problems with pumps are eliminated and hazardous maintenance is reduced; however, pressure transfer has pressurized leak hazards	Ethyl, 1993; Governale et al., 1965, 108
6.4 Design all lines and equipment to be completely drained and purged; there should be no low spots in the lines	There should be no possibility of any aluminum alkyls remaining in a line when lines are broken for maintenance	Ethyl, 1993; Governale et al., 1965, 110; Heck and Johnson, 1962, 38;

TABLE 7.8
Specific Design Considerations for Aluminum Alkyls

FEATURE	PURPOSE	REFERENCE
6.5 Cleaning of aluminum alkyl storage tanks is not recommended	Significant quantities of hazardous waste are generated during the cleaning process; it is often less expensive to obtain a new vessel; however, proper disposal of used vessel must also be considered	Ethyl, 1993
7. Last-Resort Safety Features		
7.1 Rupture disks should be used under relief valves; a pressure gauge should be placed in between the valves	Disk protects relief valve from fouling from aluminum oxides; pressure gauge indicates any leaks present in the disk	Ethyl, 1993; Governale et al., 1965, 108
7.2 Relief valves should be sized considering both the heat from an external fire vaporizing the aluminum alkyl (which usually calls for a high relief temperature due to the low vapor pressures of aluminum alkyls) and the potential for exothermic decomposition	Water entry is usually excluded as a sizing basis because extreme precautions are taken in design of aluminum alkyl facilities to prevent water entry; test data indicates the potential for exothermic decomposition below the boiling point, such as ARC data indicating an exotherm at 180 °C for TEA as compared to its boiling point of 194 °C	Ethyl, 1993; Current industry practice
7.3 Ten percent of the tank volume should remain unfilled	Allows for thermal expansion and keeps the liquid level below all lines and fittings in the storage vessel	Ethyl, 1993
8. Passive Mitigation		
8.1 Isolate aluminum alkyl storage areas from other equipment by 60 to 130 ft (18 to 40 m) or by a barrier wall	Adequate spacing reduces the chance of an incident with one tank initiating a chain reaction with adjacent equipment	Ethyl, 1993
8.2 Install storage tanks on a concrete pad sloped to drain spills to a safe, dry area	Major spills can be confined to the area around a single tank, and reducing the dike size will reduce the surface area for evaporation of aluminum alkyls from a spill and for mitigating pool fire consequences	Ethyl, 1993
9. Detection, Warning, and Isolation		
9.1 Install emergency shutdown stations, 60 to 130 ft (18 to 40 m) from the process area, which close valves in transfer lines, depressurize vessels and tanks, and stop flow	Remote isolation will mitigate the consequences of a leak or failure	Ethyl, 1993
10. Fire Prevention/Protection/Mitigation		
10.1 Specify all valves to be of a fire-safe design	Prevent valve failure in the event of a fire	Heck and Johnson, 1962, 38
10.2 Control small aluminum alkyl fires by blanketing and absorbing with sand, diatomaceous earth, or vermiculite	Enough material must be used to completely absorb the aluminum alkyl (usually 10 times the weight of the alkyl); for larger fires, it may be best to let the fire burn out	Ethyl 1993; Heck and Johnson, 1962, 36; NFPA 49, 1994; Sax and Lewis, 1989, 3368

TABLE 7.8
Specific Design Considerations for Aluminum Alkyls

FEATURE	PURPOSE	REFERENCE
10.3 The following materials should NOT be used in aluminum alkyl fires: water, water-based foams, carbon tetrachloride, and carbon dioxide	These materials react violently and/or form noxious gases with aluminum alkyls	Ethyl, 1993; Heck and Johnson, 1962, 36; NFPA 49, 1994; Sax and Lewis, 1989, 3368
10.4 Dispose of blanketed aluminum alkyls carefully after burning has subsided	Active alkyls may remain under the blanket after burning has subsided; carefully bring any unreacted alkyls to the surface for controlled burning	Ethyl, 1993; Heck and Johnson, 1962, 36
11. Post-Release Mitigation		
11.1 Water fog systems are beneficial for non-flaming aluminum alkyl spills; remote activation of the water fog is recommended	Water fogs effectively control aluminum alkyl spills; they dilute the hydrocarbon vapor cloud, impede the flame front, lower the temperature of the alkyl pool, destroy the spilled alkyl, and cool adjacent equipment	Ethyl, 1993
12. Codes and Standards		
No codes and standards unique to storage and handling of aluminum alkyls		

7.8. Peroxide Formers: 1,3-Butadiene

$$CH_2{=}CH{-}CH{=}CH_2$$

1,3-Butadiene is produced primarily as a coproduct of the thermal cracking of petroleum for the manufacture of olefins. It is commonly used in the production of rubber and plastics.

Butadiene is a colorless gas with a mildly aromatic, pleasant odor, or a colorless liquid below 24°F (NJDOH, 1994). It has an atmospheric boiling point of 24°F (–4°C), and a flash point of –105°F (–76°C) (Sax and Lewis, 1989, 572). Its lower flammable limit is reported as 2% by volume, and its reported upper flammable limit is 12% (MCA SD-55, 6). Butadiene, with an odor threshold of 1.6 ppm, has good warning properties for avoidance of exposure to dangerous vapor concentrations (NJDOH, 1994). Blurring of vision and nausea are also characteristic symptoms of exposure to butadiene (MCA SD-55, 6).

Butadiene is a mildly toxic chemical and is a suspected carcinogen, so care should be taken in handling at all times (NJDOH, 1994). In high concentrations, butadiene vapor is an anesthetic that can cause respiratory paralysis and death. At lower concentrations, there may be a slight irritation of the eyes, nose, and throat. Skin contact with liquid butadiene can cause frostbite (MCA SD-55, 6).

Butadiene is extremely flammable, and when mixed with air, forms unstable organic peroxides. In the presence of air or peroxides, "popcorn" polymer, a crystalline-like white polymer, can be formed. In confined spaces, the growth of "popcorn" polymer is capable of causing the development of several thousand psig of pressure, sufficient to rupture condenser shells, valves, tubes, etc. (MCA SD-55, 7). When stored and handled properly, there is little chance of polymers or peroxides forming.

Specific design considerations for the safe storage and handling of butadiene that are *in addition to the usual precautions to be taken with flammable gases or liquefied flammable gases* are listed in the "Feature" column of Table 7.9. The reason or explanation for each feature is given in the "Purpose" column of the table, with references given in the last column if further information is needed. The design considerations in Table 7.9 for butadiene do not necessarily apply to other peroxide formers. In all cases, whether designing or evaluating a storage and handling arrangement for butadiene or for any other peroxide former, suppliers and manufacturers should be contacted for current MSDSs and design considerations.

In summary, the most important objectives for the safe storage and handling of butadiene are

(a) personnel protection,
(b) exclusion of oxygen,
(c) protection of the quality and monomeric status of the material under normal circumstances, and
(d) protection from fire.

Maintaining the integrity of the butadiene containment can be considered an integral part of both (a) and (b).

It should be noted that storage arrangements are generally different for butadiene producers as for users. Producers store large quantities of butadiene in large, low-pressure, refrigerated tanks, whereas users are more likely to store butadiene in horizontal storage ("bullet") tanks similar to the handling of liquefied petroleum gases.

TABLE 7.9
Specific Design Considerations for 1,3-Butadiene

FEATURE	PURPOSE	REFERENCE
1. Compatibility		
1.1 Separate butadiene handling from, and take particular precautions to avoid contact of butadiene with: strong Lewis acids such as aluminum chloride; strong mineral acids; acidic and large-pore molecular sieves; buten-3-yne (with heat and pressure); cobalt; crotonaldehyde (above 180 °C); oxidizers including chlorine dioxide, halogens, and nitrogen oxides (NO_x)	The rate of butadiene polymerization increases with many common reagents, with this hazard being reduced if the butadiene is properly inhibited; also, butadiene can form explosive compounds by reaction with nitrogen oxides (NO_x)	Bretherick, 1990, 435; NFPA 491M, 1991, 40; Sax and Lewis, 1989, 589
1.2 Design and operate butadiene storage and handling facilities to avoid contact with oxygen	Polymerization is accelerated and peroxides are formed in the presence of oxygen	Bretherick, 1990, 435; MCA SD-55, 7; NFPA 49, 1994; Current industry practice
1.3 Do not use copper, silver, and their alloys in any equipment or part that may come into contact with butadiene	Trace amounts of acetylene present in the butadiene may react with these materials, forming spontaneously explosive compounds	MCA SD-55, 14; Current industry practice
1.4 Use PTFE for seals and gasket materials	PTFE avoids compatibility problems	Current industry practice
2. Storage Time/Shelf Life		
2.1 Maintain a concentration of *t*-butylcatechol (TBC) inhibitor of at least 50-100 ppm during storage and handling of butadiene (Note: other inhibitors are di-n-butylamine, phenyl-beta-naphthylamine, and phenyl-alpha-naphthylamine; however, TBC is the most commonly used inhibitor)	The function of the inhibitor is to react with oxygen before the oxygen has a chance to react with the butadiene; use of an inhibitor such as TBC prevents the formation of peroxides and "popcorn" polymer in the liquid phase	Matheson, 1980, 80; MCA SD-55, 7; NFPA 49, 1994; Current industry practice
2.2 Determinations of inhibitor content, polymer content, and oxygen levels should be made on a monthly basis; more frequent monitoring should be done if tank turnover is low or weather conditions are extreme	TBC concentration can decrease over time as it performs its inhibiting function	Current industry practice
3. Storage Quantity and Configuration		
3.1 Storage tanks of butadiene can be passivated, for example, with a 3% solution of sodium nitrite is circulated in the tank for 16-24 hours and drained by displacement with nitrogen; passivation frequency will depend on the presence of polymers or peroxides or if the system has been exposed to oxygen	Passivation is a systematic technique for desensitizing metal surfaces and eliminating oxygen; it reduces the formation of peroxides and "popoorn" polymers	Current industry practice; Enjay, n.d., 3; Current industry practice

TABLE 7.9
Specific Design Considerations for 1,3-Butadiene

FEATURE	PURPOSE	REFERENCE
3.2 Tanks containing liquid butadiene should be inerted with the appropriate inert gas; the vapor space should not contain more than 0.3% by volume oxygen before filling the tank; more importantly, the oxygen concentration should be less than 150 ppm (0.015%) after filling	TBC inhibitor has no effect on butadiene vapor in contact with oxygen, so an inert gas is necessary to prevent peroxide or polymer formation	MCA SD-55, 7; Current industry practice
4. Air and Moisture Exclusion		
4.1 Air should be totally excluded from the storage containment; storage of butadiene at elevated pressures or under an inert atmosphere is recommended	Butadiene forms explosive peroxides on contact with air; the recommended maximum oxygen level is 150 ppm to prevent peroxide formation	MCA SD-55, 7; NFPA 49, 1994; Current industry practice
4.2 Rust should be excluded from butadiene storage and handling facilities, and moisture should not be allowed to accumulate in the storage tank. Provide facilities to drain condensed moisture from the inert gas	Polymerization of butadiene may be catalyzed by rusty steel; the presence of rust and moisture in storage tanks rapidly consumes the inhibitor. Moisture can condense from the gas if butadiene is being stored cold; e.g., with refrigeration	Bretherick, 1990, 435; Current industry practice
5. Monitoring and Control		
5.1 Use high-temperature indicators and alarms on storage tanks; consider multiple instruments, alarms, and interlocks to protect against high-temperature deviations	Detection of a temperature rise will give warning if a reaction occurs	Current industry practice
5.2 The temperature for butadiene in storage should be kept as low as possible, preferably below 95 °F (35 °C)	At temperatures above 95 °F (35 °C), the rate of dimerization increases rapidly, lowering the quality of the butadiene product	Matheson, 1980, 80; MCA SD-55, 15
5.3 Paint the outside of outdoor storage tanks with a reflective paint, usually white	Minimizing heat input to the tank will provide an additional safety factor against an uncontrolled temperature rise	Current industry practice
5.4 Maintain the oxygen concentration in the vapor space of butadiene tanks and tank cars at 150 ppm by volume oxygen or less	Inhibitors have no influence on butadiene in contact with oxygen in the vapor phase; detection of oxygen in the vapor space of a butadiene storage tank can forewarn the presence of peroxide and polymer formation that will accumulate in the liquid	Enjay, n.d., 4
6. Handling and Transfer		
6.1 Design and locate all piping, valves, gauges, vents, flame arresters, tank openings, emergency relief systems, and engineering controls such that they may be readily and periodically inspected and cleaned	Butadiene tends to form polymers that can plug equipment	NFPA 49, 1994; Current industry practice

TABLE 7.9
Specific Design Considerations for 1,3-Butadiene

FEATURE	PURPOSE	REFERENCE
6.2 Install low point drains in lines	An insoluble water layer may accumulate in low points; peroxides also tend to accumulate in low points	Current industry practice
7. Last-Resort Safety Features		
7.1 Install relief valves in pairs to facilitate testing and inspection. Consider installing rupture disks before relief valves with a pressure indicator between the rupture disk and the relief valve	Butadiene vapor tends to form "popcorn" polymer buildup on relief valves	Current industry practice
7.2 Add TBC if TBC has been depleted	TBC prevents polymer formation by destroying peroxides; peroxides may also be destroyed by heating the butadiene with dilute alkalies, water or ferrous salts	Exxon, 1974, 11
8. Passive Mitigation		
No passive mitigation design considerations unique to butadiene		
9. Detection, Warning, and Isolation		
No detection, warning, and isolation design considerations unique to butadiene		
10. Fire Prevention/Protection/Mitigation		
10.1 Maintain maximum possible distance when fighting fires involving butadiene, or fight fire from a protected location	Initiation of butadiene polymerization by heat, particularly in closed drums, storage tanks, and blocked-in lines can result in rupture of the containment with ensuing explosion blast, missiles, and fireball effects	NFPA 49, 1994; Rohm and Haas, 1992, 2
11. Post-Release Mitigation		
No post-release mitigation design considerations unique to butadiene		
12. Codes and Standards		
No codes and standards unique to storage and handling of butadiene		

7.9. Water-Reactive Materials: Sodium

Na

Sodium, produced primarily through the reduction of sodium salts such as sodium chloride, is a metal that has been used in the production of antiknock agents for gasoline. Sodium is used as a catalyst for polymerization and other organic reactions, as a reduction agent in manufacturing and metals production, in metal descaling, and in the production of Na_2O_2, dyes, pharmaceuticals, herbicides, and many organometallic compounds (Austin, 1984, 254; Lemke, 1983). It also has specialized uses such as in nuclear applications as a heat transfer medium in molten form.

Sodium is a silvery-white/grey, soft, waxy metal. At temperatures above 239°F (115°C), the metal will autoignite in the presence of air (MCA SD-47, 5). Cold sodium metal is odorless, and produces no toxic vapors. Burning sodium, however, emits clouds of sodium oxide fumes that absorb moisture to give toxic, irritating sodium hydroxide "dust" (MCA SD-47, 6). In the absence of moisture, cold sodium metal causes no damage to the skin. Molten sodium and sodium exposed to moisture can cause severe thermal and alkali burns to the skin and severe eye damage (MCA SD-47, 6). For these reasons, goggles, face shields, gloves, and flame-retardant protective clothing are recommended when working with sodium (Lemke, 1983).

When kept dry and free from contaminants, sodium is thermally stable under both normal and fire exposure conditions. However, when exposed to water, sodium reacts rapidly at room temperature to give sodium hydroxide and hydrogen. Sufficient heat is given off during the reaction to melt the sodium and to ignite the hydrogen present (Sittig, 187).

Specific design considerations for the safe storage and handling of sodium metal are listed in the "Feature" column of Table 7.10. The reason or explanation for each feature is given in the "Purpose" column of the table, with references given in the last column if further information is needed. Note that the design considerations in Table 7.10 for sodium do not necessarily apply to other water-reactive materials. In all cases, whether designing or evaluating a storage and handling arrangement for sodium or for any other water-reactive material, suppliers and manufacturers should be contacted for current MSDSs and design considerations.

The key factor in the safe storage and handling of sodium is the total exclusion of air and, particularly, all moisture. Sodium combined with water and air will generally result in an explosion. When an explosion occurs, the unreacted metal is usually scattered over a wide area, where it can cause further damage. For this reason, storage is generally in hermetically sealed metal shipping containers or in molten form in enclosed bulk storage if used in large quantities. (Bulk shipments of sodium are made in specially designed tank trucks or tank cars, with molten sodium solidified in the tank prior to shipment, then melted with hot oil circulating through channels in the tank jacket upon arrival, and vacuum-transferred out of the tank truck or tank car.) Proper safety precautions for the storage and handling of sodium metal minimizes the possibility of fire and resultant injury to personnel, damage to equipment, and loss of the valuable liquid metal (Griffin and Piccot, 1978, 194).

TABLE 7.10
Specific Design Considerations for Sodium

FEATURE	PURPOSE	REFERENCE
1. Compatibility		
1.1 Design and operate to avoid contact with water	Sodium **reacts violently with water** to produce sodium hydroxide, and at 104 °F (40 °C), the heat of reaction is adequate to ignite the hydrogen liberated	Bretherick, 1990, 1370; MCA SD-47, 6; NFPA 491M, 1991, 190; Sax and Lewis, 1989, 3057
1.2 Separate sodium handling from, and take particular precautions to avoid contact of sodium with: bromine, chloroform, iodine monobromide, nitrogen-containing explosives, and many inorganic halogenated compounds such as aluminum chloride, ferric chloride, ferrous bromide, mercuric chloride, silicon tetrachloride, silver fluoride, vanadium pentachloride, and zinc bromide	A mixture of sodium and these compounds **produces a shock-sensitive material** that explodes on impact	Bretherick, 1990, 1370; NFPA 491M, 1991, 187; Sax and Lewis, 1989, 3057
1.3 Sodium **reacts violently or forms explosive and/or flammable mixtures** when combined with many common chemicals. Separate sodium handling from, and take particular precautions to avoid contact of sodium with: water, any oxidizing materials, aqueous acids, ammonia + aromatic hydrocarbons, ammonium nitrate, bromoazide, calcium + mixed oxides, carbon monoxide + ammonia, carbon tetrachloride, chlorinated hydrocarbons, chlorine, chlorine trifluoride, diazomethane, diethyl ether, dimethylformamide, fluorine, hydrazine hydrate, hydrogen peroxide, hydrogen sulfide, hydroxylamine, iodine, iodine monochloride, iodine pentafluoride, lead oxide, maleic anhydride, mercurous oxide, mercury, monoammonium phosphate, phosphorus, phosphorus pentoxide, phosphorus trichloride, potassium oxides, potassium ozonide, potassium superoxide, sodium peroxide, sulfur, sulfur dioxide, tellurium, thiophosphoryl fluoride, vanadyl chloride		Bretherick, 1990, 1370; NFPA 491M, 1991, 190; Sax and Lewis, 1989, 3057
1.4 Take care in handling sodium with or near any of the following compounds: chromium trioxide, cupric oxide, molybdenum trioxide, nitrogen peroxide, nitrosyl fluoride, nitrous oxide, phosgene, phosphoryl chloride, selenium, and stannic oxide	When mixed with certain chemicals, sodium reacts with luminescence	Bretherick, 1990, 1370; NFPA 491M, 1991, 190; Sax and Lewis, 1989, 3057
1.5 Do not use concrete in places where a possible sodium leak could occur	Sodium spilled onto concrete may react with water or chemicals in the concrete and damage or destroy the surface, particularly at high temperatures	Griffin and Piccot, 1978, 149
2. Storage Time/Shelf Life		
2.1 Reactive materials like alkali-metal dispersions in volatile solvents should not be stored indefinitely, but be clearly labeled after receipt to show the disposal date	Sodium metal dispersions in volatile solvents may form unstable peroxides that could be concentrated over an extended time	Bretherick, 1990, 1370

TABLE 7.10
Specific Design Considerations for Sodium

FEATURE	PURPOSE	REFERENCE
2.2 Sodium wastes should be disposed of by incineration, with scrubbing of the off-gases	Due to the water reactivity of sodium, it cannot be disposed of directly in biological waste treatment plants, chemical landfills, or by flushing with water into process sewers	Bretherick, 1990, 1379; MCA SD-47, 14
3. Storage Quantity and Configuration		
3.1 Store and handle in a detached, fire resistant building; avoid storage in locations where water is used or may be needed to fight fires	Storage areas that are fire resistant and do not house water lines or water handling facilities minimize the danger of fire, explosion, and health problems	MCA SD-47, 6; NFPA 49, 1994
3.2 A waterproof enclosure should be installed over the tank manhole of outdoor tanks	Allows maintenance to be performed under all weather conditions	Sittig, 1956, 147
3.3 Tanks should be designed to permit both pressure and vacuum operations	Operations involving sodium commonly use both pressured and vacuum systems	Sittig, 1956, 147
3.4 Molten sodium should be stored under a nitrogen blanket at all times	The following specifications should be met on the blanketing nitrogen: $<$ 35 mg/m^3 moisture $<$ 0.5% by volume oxygen $<$ 1.5% by volume hydrogen	Sittig, 1956, 147
4. Air and Moisture Exclusion		
4.1 Air should be totally excluded from the storage containment; solid sodium should be stored in a non-volatile solvent such as kerosene or other liquid hydrocarbons	Dispersions of sodium can ignite spontaneously in air or oxygen; oxygen reacts with sodium to form sodium oxide, reducing the product quality	NFPA 49, 1994; NFPA 491M, 1991, 187; Sax and Lewis, 1989, 3057
4.2 Moisture and water should be totally excluded from all sodium storage and handling facilities	When exposed to water or moisture in any form, sodium reacts violently	Bretherick, 1990, 1378; NFPA 49, 1994; NFPA 491M, 1991, 190; Sax and Lewis, 1989, 3057
5. Monitoring and Control		
5.1 Areas containing sodium should be heated to a constant temperature (but not with steam or hot water)	A heated room prevents condensation of moisture in the room due to extreme changes in temperature	Sittig, 1956, 145
5.2 Electric heater coils or circulating hot oil jackets are recommended for use in the storage of molten sodium	Water or steam-based heating should not be used on equipment containing sodium	Sittig, 1956, 146
6. Handling and Transfer		
6.1 Use only ***dry*** nitrogen or other dry, inert gases in sodium handling	Small amounts of moisture present in various gases could ignite the sodium and cause a significant fire hazard	Sittig, 1956, 144

TABLE 7.10
Specific Design Considerations for Sodium

FEATURE	PURPOSE	REFERENCE
6.2 Keep automatic sprinkler, water, and steam lines out of sodium handling areas	Failure of the water or steam lines could cause water to come into contact with sodium	Sittig, 1956, 144
6.3 Once sodium has been removed from its original container, a special metal container with a tight-fitting lid should be used for transporting sodium bricks to other plant areas	Prevents sodium from being contaminated by air, moisture, or other non-compatible compounds	Sittig, 1956, 146
6.4 Unload tank cars of sodium by circulating hot oil to liquefy the sodium and withdrawing by sealless pump	Even though transferring by vacuum reduces the hazards in the event of an equipment leak or break, the use of, e.g., magnetic-drive pumps is faster and also reduces potential leak points over other pump designs	Sittig, 1956, 103; Current industry practice
6.5 Completely welded construction is preferred in sodium handling facilities	Welded construction reduces the possibility of contamination by sodium oxide	Lemke, 1983, 196
7. Last-Resort Safety Features		
7.1 The transfer lines and receiving vessel for a molten sodium dump system, if employed, must be maintained at a temperature above the melting point of sodium at all times	In order to be effective, the dump system should ensure fluidity of the molten sodium	Lee, 1955, 387
8. Passive Mitigation		
8.1 Equipment and vessels containing sodium should be isolated from other facilities by walls and floors of fire-resistant construction	Fire walls prevent a severe fire from spreading to other areas or equipment	MCA SD-47, 6
9. Detection, Warning, and Isolation		
9.1 Equipment for indicating the presence of hydrogen during emergency spill-control situations is desirable	Hydrogen-indicating devices can warn of the build-up of explosive concentrations of hydrogen in the event of a sodium release	MCA SD-47, 7
9.2 Conduct moisture-determination tests for vessels prior to receiving sodium	Moisture tests ensure vessels are completely moisture free before the introduction of sodium	Current industry practice
10. Fire Prevention/Protection/Mitigation		
10.1 Provide natural ventilation by installing vents at the highest points of possible hydrogen accumulation	Adequate ventilation can prevent explosion hazards from hydrogen accumulation	MCA SD-47, 13
10.2 Fight fires involving sodium with soda ash, dry sodium chloride, graphite, or other <u>compatible</u> dry powder; do not use water, carbon dioxide, halogenated extinguishing agents, or water-based foams	Common fire extinguishing agents aggravate the situation by introducing additional reaction or explosion hazards	NFPA 49, 1994; Sax and Lewis, 1989, 3057
10.3 Post signs in areas containing sodium that warn against the use of water for extinguishing fires	Water should not be used on sodium fires due to the water-reactivity of sodium	Sittig, 1956, 146

TABLE 7.10
Specific Design Considerations for Sodium

FEATURE	PURPOSE	REFERENCE
10.4 Design equipment to confine any possible sodium fires	Sodium fires are easier to extinguish if it is possible to exclude air and flood the equipment with nitrogen or other inert gas	Lemke, 1983, 196
10.5 Soda ash for extinguishing sodium fires should be kept in moisture-proof containers in a dry area	Soda ash absorbs water from the air, and loses its effectiveness as a fire fighting agent; soda ash used in sodium fire-fighting should contain less than 3% moisture	MCA SD-47, 10
10.6 Enforce no smoking rule around all handling areas	There should be no potential ignition sources in the area due to the possibility of hydrogen being present	Griffin and Piccot, 1978, 134
10.7 Full protective clothing, including self-contained breathing apparatus, must be worn when combatting a major fire involving sodium	Sodium fires evolve large quantities of caustic smoke	MCA SD-47, 10
11. Post-Release Mitigation		
11.1 Shallow pans of welded sheet steel having sufficient capacity to hold contents of the storage/handling system should be used under all the equipment in which sodium is used; keep the pans dry such as by periodically wiping with soda ash	If sodium leaks into the pans, contact with water is avoided and sodium releases can be confined and treated with dry soda ash	MCA SD-47, 7
11.2 Provide emergency shower facilities where there is a possibility of skin exposure; the shower should be located near an exit, and so that water cannot come into contact with the sodium	Even though water reacts with sodium to liberate hydrogen, a safety shower is the best method of combatting burns	MCA SD-47, 11
12. Codes and Standards		
No codes and standards unique to storage and handling of sodium		

7.10. Water-Reactive Materials: Chlorosulfonic Acid

$ClSO_2OH$

Chlorosulfonic acid (chlorosulfuric acid, sulfuric chlorohydrin) is a multi-purpose acid. Some of the major uses of chlorosulfonic acid are in detergents, pharmaceuticals, dyes and pigments, ion-exchange resins, catalysis, and the preparation of many common materials (Burrus, 1983, 878).

Chlorosulfonic acid is a colorless to light yellow, fuming liquid with an atmospheric boiling point of 306°F (152°C). It is a nonflammable compound, but is a strong oxidizing agent. Chlorosulfonic acid has a pungent odor, which gives it good warning properties for avoidance of exposure to dangerous vapor concentrations. Chlorosulfonic acid is capable of producing severe acid burns. Through the action of the moisture in air or water, it is broken down into hydrochloric and sulfuric acids, both of which are extremely irritating to the eyes, lungs, and mucous membranes (NFPA 49, 1994). Chronic overexposure can cause lung damage, so the proper ventilation and/or respiratory protection should be used when the possibility of vapor exposure exists.

Chlorosulfonic acid is considered relatively safe to handle compared to many other water-reactive materials. Chlorosulfonic acid reacts violently with water, but not with the explosive force of many other water-reactive materials such as sodium. However, extreme care should be taken in handling and storage to exclude water and air from the chlorosulfonic acid containment. Chlorosulfonic acid is also incompatible with many other common reagents, so it is recommended that operations involving chlorosulfonic acid take place in a remote area.

Specific design considerations for the safe storage and handling of chlorosulfonic acid are listed in the "Feature" column of Table 7.11. The reason or explanation for each feature is given in the "Purpose" column of the table, with references given in the last column if further information is needed. Note that the design considerations in Table 7.11 for chlorosulfonic acid do not necessarily apply to other water-reactive materials. In all cases, whether designing or evaluating a storage and handling arrangement for chlorosulfonic acid or for any other water-reactive material, suppliers and manufacturers should be contacted for current MSDSs and design considerations.

In summary, the most important aspects for the safe storage and handling of chlorosulfonic acid are

(a) exclusion of moisture,
(b) exclusion of incompatible materials, and
(c) protection against a possible release.

Maintaining the integrity of the chlorosulfonic acid containment can be considered an integral part of all aspects.

TABLE 7.11
Specific Design Considerations for Chlorosulfonic Acid

FEATURE	PURPOSE	REFERENCE
1. Compatibility		
1.1 Separate chlorosulfonic acid handling from, and take particular precautions to avoid contact of chlorosulfonic acid with: aqueous solutions (even, e.g., the small amount of water in 98% sulfuric acid under some conditions), combustibles, organic matter, acetic acid, acetic anhydride, acetonitrile, acrolein, acrylic acid, acrylonitrile, allyl alcohol, allyl chloride, 2-aminoethanol, ammonium hydroxide, aniline, n-butyraldehyde, creosote oil, cresol, cumene, dichloroethyl ether diethylene glycol monomethyl ether, diisobutylene, diisopropylether, diphenyl ether, epichlorohydrin, ethyl acetate, ethyl acrylate, ethylene chlorohydrin, ethylene cyanohydrin, ethylene diamine, ethylene glycol, ethylene glycol monoethyl ether acetate, ethylene imine, glyoxal, hydrochloric acid, hydrofluoric acid, hydrogen peroxide, isoprene, mesityl oxide, metal powders, methyl ethyl ketone, nitric acid, 2-nitropropane, phosphorus, β-propiolactone, propylene oxide, pyridene, silver nitrate, sodium hydroxide, sulfolane, styrene, vinyl acetate, vinylidene chloride		Bretherick, 1990, 951; NFPA 491M, 1991, 63; Sax and Lewis, 1989, 877
2. Storage Time/Shelf Life		
No storage time/shelf life considerations unique to chlorosulfonic acid		
3. Storage Quantity and Configuration		
3.1 Slowly and carefully vent stored drums at least two times per month; more often in warmer weather	Hydrogen is produced through corrosion of the drum by chlorosulfonic acid; pressure should be relieved as necessary in order to avoid explosive conditions	NFPA 49, 1994; Sax and Lewis, 1989, 877
3.2 Protect open steelwork by an acid-resistant paint	Chlorosulfonic acid is very corrosive to metals, especially when in the presence of atmospheric moisture	NFPA 49, 1994
3.3 Store drums with plugs up	Chlorosulfonic acid could corrode an upside-down plug, leading to a release of acid	NFPA 49, 1994
3.4 Storage tanks containing chlorosulfonic acid should not have bottom outlet valves	Uncontrolled leakage through the bottom outlet, such as due to bottom valve leak-through, can lead to a significant spill	Current industry practice
3.5 Vent storage tanks to a scrubber having a high-pressure alarm	Chlorosulfonic acid fumes should not be released to the atmosphere, where they can react with moisture to form noxious gases	Current industry practice
3.6 Positively prevent scrubber solution backup into the vessel	Backup of scrubber solution into chlorosulfonic acid storage tanks could cause an uncontrolled reaction with scrubber solution	Rohm & Haas (1992), Current industry practice
3.7 Allow no direct water connections to any equipment containing chlorosulfonic acid	This prevents water from being accidentally introduced into the chlorosulfonic acid containment, causing violent reactions	Current industry practice
4. Air and Moisture Exclusion		
4.1 Totally exclude moist air from chlorosulfonic acid storage and handling facilities; a silica gel drier should be installed on the air inlet to chlorosulfonic acid storage tanks	When exposed to the atmosphere, chlorosulfuric acid fumes react with the moisture of the air and release highly irritating and corrosive hydrochloric acid fumes and sulfuric acid mist	Burrus, 1983, 877; Current industry practice

TABLE 7.11
Specific Design Considerations for Chlorosulfonic Acid

FEATURE	PURPOSE	REFERENCE
4.2 Totally exclude moisture and water from all chlorosulfonic acid storage and handling facilities	Chlorosulfonic acid is highly reactive with water or moisture	NFPA 49, 1994; NFPA 491M, 1991, 63; Sax and Lewis, 1989, 877
5. Monitoring and Control		
No monitoring and control design considerations unique to chlorosulfonic acid		
6. Handling and Transfer		
6.1 Electrical fixtures, equipment, and wiring should be corrosion-protected	Corrosive action of the chlorosulfonic acid could lead to electrical shorts	NFPA 49, 1994
6.2 Chlorosulfonic acid should never stand in a line completely sealed between two closed valves or check valves	Excessive pressures caused by liquid thermal expansion can create hazardous acid sprays or pipe rupture	Burrus, 1983, 878
6.3 Use of flange guards on all flanged joints is recommended	Flange guards offer additional protection against a major release in the case of flange failure	Burrus, 1983, 878
6.4 Install a parallel railroad track around the area where the chlorosulfonic acid rail car is loaded/unloaded so that other rail cars may be routed around it	A separate rail track for the chlorosulfonic acid rail car reduces the possibility of the rail car being hit, causing a large release of chlorosulfonic acid	Current industry practice
6.5 Cover transfer and storage areas and dikes with plastic or other suitable material	Covered transfer areas prevent the collection of rainwater in dikes	Rohm & Haas (1992), Current industry practice
6.6 Incoming rail cars are verified with supplier data and sampled if supplier is not certified	Verification and analysis of tank car contents reduces the likelihood of incompatible material being introduced to the chlorosulfonic acid containment	Current industry practice
7. Last-Resort Safety Features		
No last-resort safety features unique to chlorosulfonic acid		
8. Passive Mitigation		
No passive mitigation systems unique to chlorosulfonic acid		
9. Detection, Warning, and Isolation		
9.1 Install liquid level indicating systems, high level alarm systems, and switches on all tanks containing chlorosulfonic acid	Alarms are a good method of reducing the possibility of a chlorosulfonic acid release	Current industry practice
9.2 All push buttons for isolation valves and all alarms should be placed in both field and control room locations	Remote operated valves allow for a fast response if the area containing the local alarms is unsafe for personnel to enter	Current industry practice

TABLE 7.11
Specific Design Considerations for Chlorosulfonic Acid

FEATURE	PURPOSE	REFERENCE
10. Fire Prevention/Protection/Mitigation		
10.1 Include consideration of hydrogen gas presence in hot work permit systems pertaining to chlorosulfonic acid processes	Although chlorosulfonic acid itself is not flammable, hydrogen gas formed by that action of chlorosulfonic acid on metal forms explosive mixtures with air	Burrus, 1983, 878
10.2 Take measures to ensure fire-fighting water does not come into contact with chlorosulfonic acid	Chlorosulfonic acid is reactive with water, and highly acidic runoff water will be formed	NFPA 49, 1994
11. Post-Release Mitigation		
11.1 Install a system containing mineral oil near the dike to provide coverage for large spills of chlorosulfonic acid within the dike area	A layer of mineral oil on chlorosulfonic acid reduces fuming and allows for safer cleanup	Current industry practice
12. Codes and Standards		
No codes and standards unique to storage and handling of chlorosulfonic acid		

Bibliography

Center for Chemical Process Safety, *Guidelines for Engineering Design for Process Safety,* American Institute of Chemical Engineers, New York, 1993.

American Institute of Chemical Engineers-Center for Chemical Process Safety, *Guidelines for Storage and Handling of High Toxic Hazard Materials,* American Institute of Chemical Engineers, New York, 1988.

American Institute of Chemical Engineers-Center for Chemical Process Safety, *Guidelines for Vapor Release Mitigation,* American Institute of Chemical Engineers, New York, 1988.

References

AIChE-CCPS (1988). *Guidelines for Vapor Release Mitigation.* American Institute of Chemical Engineers-Center for Chemical Process Safety, New York.

AIChE-CCPS (1993). *Guidelines for Engineering Design for Process Safety.* American Institute of Chemical Engineers-Center for Chemical Process Safety, New York.

ATO (1993). "Safe Handling, Storage and Transportation of Peroxides Requiring Refrigeration." Elf Atochem North America, Inc., Philadelphia, Pennsylvania 19102.

ATO (1992). "Diacyl Peroxides," Product Bulletin No. Q631-5C. Elf Atochem North America, Inc., Organic Peroxides Division, Buffalo, New York 14240.

ATO (1991). "Lupersol Ketone Peroxides," Product Bulletin No. Q590-1M. Elf Atochem North America, Inc., Organic Peroxides Division, Buffalo, New York 14240.

Austin, G. T. (1984). *Shreve's Chemical Processing Industries,* 5th Ed. McGraw-Hill, New York.

Badische et al. (n.d.). "Acrylic Acid, A Summary of Safety and Handling." Compiled by Badische Corporation; Celanese Chemical Company, Inc.; Rohm and Haas Company; and Union Carbide Corporation.

BASF et al. (1991). "Acrylic Acid, A Summary of Safety and Handling," 2nd Ed. Compiled by BASF Corporation; Hoechst Celanese Chemical Group, Inc.; Rohm and Haas Company; and Union Carbide Chemicals and Plastics Company Inc. January.

BASF et al. (1993). "Prevention of Acrylic Acid Polymerization in Storage." Compiled by BASF Corporation; Hoechst Celanese Chemical Group, Inc.; Rohm and Haas Company; and Union Carbide Chemicals and Plastics Company Inc. December.

Bretherick, L. (1990). *Bretherick's Handbook of Reactive Chemical Hazards,* 4th Ed. Butterworths, London.

Britton, L. (1990). "Thermal Stability and Deflagration of Ethylene Oxide." *Plant/Oper. Prog.* 9(2, April), 75-85.

Britton, L. (1991). "Spontaneous Fires in Insulation." *Plant/Oper. Prog. 10*(1, January), 27-44.

Bond, J. (1985). *Loss Prev. Bull. 065,* Institution of Chemical Engineers, Rugby, Warwickshire, U.K.

Burrus, H. O. (1983). "Chlorosulfonic Acid." In M. Grayson, ed., *Kirk-Othmer Encyclopedia of Chemical Technology 5,* 3rd Ed. John Wiley & Sons, New York.

Catalyst Resources (1983). "Benzoyl Peroxide-77," Product Bulletin No. 31.1. Catalyst Resources, Inc., Houston, Texas.

Curtis, J. S. (1990). "Safe Storage of Dilute Ethylene Oxide Mixtures in Water." *Plant/Oper. Prog.* 9(2, April), 91-97.

DIERS (1992). *Emergency Relief System Design Using DIERS Technology, DIERS Project Manual,* Design Institute for Emergency Relief Systems, American Institute of Chemical Engineers, New York.

DOT (1984). *Chemical Hazard Response Information System (CHRIS) Hazardous Chemical Data.* Commandant Instruction M.16465.12A. U.S. Department of Transportation, U.S. Coast Guard, 14 November. U.S. Government Printing Office, Washington, D.C.

DOT (1993). *Emergency Response Guidebook,* U.S. Department of Transportation, Washington, D.C.

Dow (1966). "Styrene-Type Monomers," Product Bulletin No. 170-151B-3M-366. The Dow Chemical Company, Midland, Michigan 48674.

Dow (1967). "Storage and Handling of Styrene-Type Monomers," Product Bulletin No. 170-280-6M-1067. The Dow Chemical Company, Midland, Michigan 48674.

Dow (1989). "The Dow Family of Styrenic Monomers," Product Bulletin No. 115-00608-1289X-SAI. The Dow Chemical Company, Midland, Michigan 48674.

Enjay (n.d.). "Storage and Handling of Gaseous Olefins and Diolefins," Technical Bulletin No. 2. Enjay Company, Inc., New York.

Ethyl (Albemarle) Corp. (1993). "Handling Procedures for Aluminum Alkyl Compounds." Ethyl (Albemarle) Corporation, Baton Rouge, Louisiana 70820.

Exxon (1974). "Storing and Handling Liquefied Olefins and Diolefins." Exxon Chemical Company, USA.

FM 7-92 (1992). "Ethylene Oxide." *Loss Prevention Data Sheet* 7-92 (periodically updated). Factory Mutual Engineering and Research, Norwood, Massachusetts.

Governale, L. J., Ruhlin, W. B., and Silvus, R. F. (1965). "How to Handle Aluminum Alkyl Compounds." *Chem. Eng. Prog. 61*(4, April).

Griffin, C. W. and Piccot, A. R. (1978). "Safety." In O. J. Foust, ed., *Sodium NaK Engineering Handbook,* Vol. 3. Gordon and Breach, Science Publishers, New York.

Hagopian, C., Koch, P., and Lewis, P. J. (1983). "Styrene." In M. Grayson, ed., *Kirk-Othmer Encyclopedia of Chemical Technology 21,* 3rd Ed. John Wiley & Sons, New York.

Hatayama, H. K., Chen, J. J., DeVera, E. R., Stephens, R. D., and Storm, D. L. (1980). "A Method for Determining the Compatibility of Hazardous Wastes." EPA-600/2-80-076, U.S. Environmental Protection Agency, Washington, D.C., April. Available through National Technical Information Service, Springfield, Virginia.

Heck, W. B. and Johnson, R. L. (1962). "Aluminum Alkyls Safe Handling." *Ind. Eng. Chem. 54*(12, December).

Hess, L. G. and Tilton, V. V. (1950). "Ethylene Oxide Hazards and Methods of Handling." *Ind. Eng. Chem. 42*(6, June).

Johnson, D. P., Critchfield, F. E., and Hess, L. G. (1961). "Low Temperature Storage of Acrylic Acid." *Ind. Eng. Chem. 53*(4, April).

Kirch, L. S., Kargol, J. A., Magee, J. W., and Stuper, W. S. (1988). "Stability of Acrylic Monomers." *Plant/Oper. Prog. 7*(4, October).

Lee, R. E. (1955). "Safety and Fire Protection, Non-Radioactive Sodium." In C. B. Jackson, ed., *Liquid Metals Handbook Sodium (NaK) Supplement,* 3rd Ed. U. S. Government Printing Office, Washington, D.C.

Lees, F. P. (1980). *Loss Prevention in the Process Industries* (2 volumes). Butterworth, London.

Lemke, Charles H. (1983). "Sodium and Sodium Alloys." In M. Grayson, ed., *Kirk-Othmer Encyclopedia of Chemical Technology 21,* 3rd Ed. John Wiley & Sons, New York.

Levy L. B. and Penrod, J. D. (1989). "The Anatomy of an Acrylic Acid Runaway Polymerization." *Plant/Oper. Prog. 8*(2, April).

Luperox (n.d.). "Safe Handling of Organic Peroxides." Pennwalt/Luperox (now Elf Atochem North America, Inc.), Philadelphia, Pennsylvania.

Matheson (1980). Braker, W. and Mossman, A. L., Eds., *Gas Data Book,* 6th Ed. Matheson Gas Products, Inc., Seacaucus, New Jersey.

MCA SD-37 (1971). "Properties and Essential Information for Safe Handling and Use of Styrene Monomer." Chemical Safety Data Sheet SD-37, Manufacturing Chemists Association (now Chemical Manufacturers Association, CMA), Washington, D.C. Note: this information is no longer published or kept up to date by CMA.

MCA SD-38 (1971). "Properties and Essential Information for Safe Handling and Use of Ethylene Oxide." Chemical Safety Data Sheet SD-38, Manufacturing Chemists Association (now Chemical Manufacturers Association, CMA), Washington, D.C. Note: this information is no longer published or kept up to date by CMA.

MCA SD-47 (1974). "Properties and Essential Information for Safe Handling and Use of Sodium." Chemical Safety Data Sheet SD-47, Manufacturing Chemists Association (now Chemical Manufacturers Association, CMA), Washington, D.C. Note: this information is no longer published or kept up to date by CMA.

MCA SD-55 (1974). "Properties and Essential Information for Safe Handling and Use of Butadiene." Chemical Safety Data Sheet SD-55, Manufacturing Chemists Association (now Chemical Manufacturers Association, CMA), Washington, D.C. Note: this information is no longer published or kept up to date by CMA.

Nemec, J. W. and Bauer, W. (1983). "Acrylic Acid and Derivatives." In M. Grayson, ed., *Kirk-Othmer Encyclopedia of Chemical Technology 1,* 3rd Ed. John Wiley & Sons, New York.

Nicolson, A. (1991). "The Effect of O_2 Concentration on Methacrylic Acid Stability." *Plant/Oper. Prog. 10*(3, July).

Niehaus, W. R. (1971). "Safe Handling Practice for Acrylate Monomers." *Paint and Varnish Production*, May.

NFPA 30 (1993). *Flammable and Combustible Liquids Code.* National Fire Protection Association, Quincy, Massachusetts.

NFPA 43B (1993). *Storage of Organic Peroxide Formulations.* National Fire Protection Association, Quincy, Massachusetts.

NFPA 49 (1994). *Hazardous Chemicals Data.* National Fire Protection Association, Quincy, Massachusetts.

NFPA 325M (1994). *Fire Hazard Properties of Flammable Liquids, Gases, and Volatile Solids.* National Fire Protection Association, Quincy, Massachusetts.

NFPA 491M (1991). *Manual of Hazardous Chemical Reactions.* National Fire Protection Association, Quincy, Massachusetts.

NJDOH (1994). New Jersey Department of Health, Right to Know Program, Trenton, New Jersey.

Noury (1985). "Organic Peroxides Safety and Handling." Noury Chemicals/Akzo Chemie America product bulletin.

Rohm and Haas (1992). "Material Safety Data Sheet for 1,3-Butadiene." Rohm and Haas Company, Philadelphia, Pennsylvania 19105.

@Ref = Sabaa, M. W. (1990). "Thiocarbohydrazide as Inhibitor for the Liquid and Gas-Phase Polymerisation of Acrylic Acid." *Chemistry and Industry.* 5 March.

Sax, N.I., and Lewis, Sr., R. J. (1989). *Dangerous Properties of Industrial Materials,* 7th Ed. Van Nostrand Reinhold, New York.

Shelly, P. G. and Sills, E. J. (1969). "Monomer Storage and Protection." *Chem. Eng. Prog. 65*(4, April).

Sittig, Marshall. (1956). *Sodium Its Manufacture, Properties and Uses.* Reinhold Publishing Corporation, New York.

Slater, D. H. (1978). "Vapour Clouds." *Chemistry and Industry.* May 6, 295.

SPI (1978). "The Storage and Handling of Organic Peroxides in the Reinforced Polyester Fabricating Plant." *Safety and Loss Prevention Bulletin No. 19.* Society of the Plastics Industry, 335 Lexington Ave., New York, New York 10017.

TCE, 1994. *The Chemical Engineer.* December 15.

UN (1990). *United Nations Recommendation on Transport of Goods, Tests and Criteria,* 2nd Ed. United Nations, New York.

Varjavandi, J. and Mageli, O. L. (1971). "Safe Handling of Organic Peroxides." *J. Chem. Ed. 48*(7, July).

Viera, G. A., Simpson, L. L., and Ream, B. C. (1993) "Lessons Learned from the Ethylene Oxide Explosion at Seadrift, Texas." *Chem. Eng. Prog. 89*(8, August), 66-75.

Walrod, D. H. and Gimbarski, R. G. (1979). "Safe Handling and Use of Organic Peroxides." *Plastics Compounding.* January/February 1979.

Wampler, F. M. (1988). "Formation of Diacrylic Acid During Acrylic Acid Storage." *Plant/Oper. Prog.* 7(3, July).

APPENDIX A

Reactive Chemicals Literature Sources

The references in this appendix are selected for their relation to Section 3.1 on the identification of reactivity hazards from literature sources. These references are grouped into six subject areas:

Procedures for Hazard Evaluation and Testing

Industrial Dust Explosions

Edited by K. L. Cashdollar and M. Herzberg
1986   372 pages   ISBN-0-8031-0957-1

This ASTM publication is a collection of papers covering the latest research and data on dust explosions. It contains papers on the effects of temperature and pressure on explosive characteristics, preventive and design measures for protection, and sizing of dust explosion vents. This compilation of papers is an excellent reference tool for industrial safety and chemical engineers. The papers were presented at a symposium on Industrial Dust Explosions held in Pittsburgh, Pennsylvania during June, 1986.

Available from ASTM,
1916 Race Street, Philadelphia, PA 19103
(215) 299-5585

"Chemical Reaction Hazards: An Integrated Approach"

N. Gibson, R. L. Rogers, and T. K. Wright

1987 Published in *Inst. Chem. Eng. Symp. Ser., 102*, 61

This review article describes how to avoid reaction hazards by integrating hazard evaluation into process design.

Guidelines for Hazard Evaluation Procedures, 2nd Edition with Worked Examples

1992 461 pages ISBN 0-8169-0491-X

This book provides a summary of hazardous materials evaluation techniques that are useful to engineers and plant operators. The book covers methods of identifying, assessing, and reducing hazards in chemical plants.

Available from American Institute of Chemical Engineers
Publication Sales, Dept. CAT-81,
345 East 47th Street, New York, NY 10017
(215) 705-7338

Hazard Assessment of Chemicals

Edited by Jitendra Saxena

436 pages ISBN 0-89116-450-2

This book (Volume 5) covers pesticide rating systems, potential hazards occurring in the areas of biotechnology and oil-well drilling fluids, and measurement of environmental pollutants by electrochemical techniques. Provides an anatomy of the methyl isocyanate incident in Bhopal.

Available from Taylor & Francis,
Distribution Center
242 Cherry Street, Philadelphia, PA 19106-1906
(215) 238-0940

Hazards in the Process Industries: Hazards IX

331 pages ISBN 0-85295-198-1

These conference (April, 1986, Manchester) proceedings review areas of risk assessment and process safety. Topics covered include fires and dust explosions, safe practices and equipment, dispersion and ventilation. The book is useful in identifying hazards and assessing their risks. It also covers expert systems for risk and safety analysis.

Available for The Institution of Chemical Engineers,
165-171 Railway Terrace,
Rugby CV21 3HQ, United Kingdom

Major Chemical Hazards

V. C. Marshall
1987 600 pages ISBN 470-20813-9

This book provides case histories on the prevention of fires and explosions. It examines chemical plant hazards and their risk potentials. Detailed accounts on the formation and dispersal of vapor clouds are presented.

Available from John Wiley & Sons, Inc.,
1 Wiley Drive, Somerset, NJ 08873-1272
(201) 469-4400

Hazards from Pressure

1987 400 pages ISBN 0-08-034820-3

This The Institution of Chemical Engineers Symp. Ser. reviews pressure hazards in the chemical industry. The latest developments in accident prevention through pressure control are presented and discussed in detail.

Available from The Institution of Chemical Engineers,
165-171 Railway Terrace, Rugby, Warks CV21 3 BR, UK

Chemical Process Hazard Review

Edited by John H. Hoffman and Daniel C. Maser
1985 121 pages ISBN 0-8412-0902-2

ACS Symp. Ser. 274 (1985). Accounts are presented of hazard reviews performed by industrial chemists and chemical engineers on various chemicals processes. This collection of papers also covers the estimation of thermochemical data and the application of reactor modeling to hazard evaluation.

Available from American Chemical Society Publications,
1155 16th Street NW, Washington, DC 20036
(202) 872-4600

Safety of Reactive Chemicals

Tadao Yoshida
1987 404 pages ISBN 0-444-42748-1

Industrial Safety Series, Volume 1. This book describes case histories of industrial accidents involving reactive chemicals. It also covers modern techniques that are used for evaluating the dangers of hazardous materials. Other topics covered include fire and explosion hazards, thermochemical calculations, and emergency response systems.

Available from Elsevier Science Publishing Company, Inc.,
52 Vanderbilt Avenue, New York, NY 10017
(212) 370-5520

Fundamentals of Fire and Explosion

Daniel R. Stull 1977

This compendium of findings on control of chemical hazards has been written in a short course format and reviews the fundamentals of fire and explosion technology in a simplified, easy-to-understand manner.

"Safety"

Richard W. Prugh

1988 Published in *Encyclopedia of Polymer Science Engineering.*, *14*, 804

This review article covers many facets of safety in the polymer industry. Guidelines are presented on how to design polymerization reactors and associated equipment and on how to operate them safely. Recommendations are presented on how to protect workers and the public from the hazards of polymers and the equipment used to produce them. This review provides 70 references that will lead the reader to additional information on toxicity, reactivity, hazards of fires and explosions, and legal matters dealing with polymers.

Available from John Wiley & Sons, Inc.,
1 Wiley Drive, Somerset, NJ 08873-1272
(201) 469-4400

Self-Heating: Evaluating and Controlling the Hazards

P. C. Bowes

1984 500 pages ISBN 0-444-99624-9

This monograph considers the theoretical aspects of thermal explosions resulting from the self-heating of materials mainly solids. Steady-state and transient models of thermal explosions are developed and validated by laboratory scale studies. The latter part of the book is concerned with the application of these models to practical problems. The practical examples considered include the oxidative self-heating of charcoals, wood products, hay, grains, and the exothermic decomposition of unstable materials such as peroxides and calcium hypochlorite. Emphasis is placed on how these theoretical treatments can lead to predictions of the conditions necessary for ignition and explosion and how to design countermeasures.

Available from Elsevier Science Publishing Company, Inc.,
52 Vanderbilt Avenue, New York, NY 10017
(212) 633-3650

Runaway Reactions, Unstable Products and Combustible Powders

ISBN 0-85295-136-1

The Institution of Chemical Engineers Symp. Ser. No. 68 (1981). These papers cover the theoretical aspects of runaway reactions and thermal explosions, hazards of combustible powders, the assessment of reaction hazards by calorimetry, and hazard evaluation programs. New test data are provided on the thermal properties and explosion hazards of organic peroxides.

The Institution of Chemical Engineers (IChemE),
George E. Davis Building, 165-171 Railway Terrace,
Rugby, Warks CV21 3HQ UK

Note: Out of print: Available at Virginia Polytechnic Institute and State University Library QD 516 .R86

The Recognition and Reduction of Ignition Hazards in the Chemical Industry

These conference (London, 1981) proceedings cover the thermal hazards of chemical reactions, the ignition of fires and explosions, and hazards associated with organic peroxides.

Available from Scientific and Technical Studies
Bath House, 56 Halborn Viaduct, London EC1A 2EZ, UK

Managing Hazardous Substances Accidents

A. J. Smith

This book describes steps that can be taken to prevent accidents during the handling of hazardous materials.

Available from McGraw Hill Book Company,
1221 Avenue of the Americas, New York, NY 10020
(212) 512-2000

A Review of Hazard Identification Techniques and Their Application to Major Accidents Hazards

S. T. Parry
ISBN 0-85-356200-8

This book reviews methods that can be used to identify potential hazards associated with storing, handling, and processing of dangerous materials. The subject matter is presented at a level that will be most useful to those who have limited experience in identifying potential chemical hazards.

Available from Safety & Reliability Directorate,
UKAEA, Culcheth, Warrington, WA3 4NE, UK

Chemical Process Hazards with Special Reference to Plant Design

The Institution of Chemical Engineers Symp. Ser. No. 58;
EFCE Event No. 230 (1980)

These papers cover topics on dust explosion hazard assessment, flammability and explosions of gases, and autoignition temperature determinations. Emphasis is placed on process hazards and plant design.

Available from The Institution of Chemical Engineers,
George E. Davis Building, 165-171 Railway Terrace,
Rugby, Warks CV21 3 HQ, England

Combustion, Flames and Explosions of Gases

Bernard Lewis and Guenther VonElbe, 1951

The first edition of this book provides a review of the subject matter comprising the theory and chemical reactions between fuel gases and oxygen, the hydrodynamics of combustion waves, detonation waves and jet flames, and the thermodynamics of combustion gases. The second edition appeared in 1961 and included much newer material. It emphasizes wave propagation in combustion, flame stretch and diffusional stratification. It is shown how these concepts led to an understanding of flammability limits. Tabulated in the appendices are heats of combustion/decomposition, flammability limits for gases under various conditions, and flame temperatures.

Available from Academic Press, Inc.,
465 South Lincoln Drive, Troy, MO 63379
1-800-321-5068

Unconfined Vapour Cloud Explosions

Keith Gugan

Case histories are presented for 12 major accidents that have involved unconfined vapor cloud explosions. A general discussion is included on the fundamentals of combustion in gaseous mixtures and ways to predict hazards, prevent occurrences, and protect against these disasters.

Available from Gulf Publishing Company,
P. O. Box 2608, Houston, TX 77001
(713) 529-4301

Explosion Hazards and Evaluation

W. E. Baker, P. A. Cox, P. S. Westine, J. J. Kulesz, and R. A. Strehlow

Covers the fundamentals and theoretical concepts of combustion and explosion phenomena. In-depth treatments are included for blast waves, fragments and missiles caused by explosions, and thermal radiation effects. The sections on explosion evaluation procedures and hazard assessment for large industrial complexes will be of particular interest to those involved with process safety and plant design.

Available from Elsevier Science Publishing Company, Inc.
52 Vanderbilt Avenue, New York, NY 10017

(212) 989-5800

Explosions Course Prevention Protection

W. Bartknecht; Translated by H. Burg and T. Almond

Part one of this book covers the fundamentals of explosions and detonations in vessels and pipes, including coverage on explosion limits, pressure effects, and ignition energies. Part two describes steps to take to prevent explosions and describes safety measures that can be used to guard against the effects of explosions in confined areas. Protective measures are also described in part three, which covers the use of specific production equipment such as grinders, bag slitting machines, fluid bed driers, and granulators.

Available from Springer-Verlag,
44 Hartzway, Secaucus, NJ 70794
(201) 348-4033

Dust Explosions and Fires

K. N. Palmer

This book is a comprehensive technical assessment of the field of dust explosions. It includes information on testing dust explosibility, flame propagation, ignition sources, and preventive measures. The book provides guidance on how to design and protect various types of industrial explosion-proof equipment often used to process dusts.

Available from Routledge, Chapman and Hall,
29 West 35th Street, New York, NY 10001
(212) 244-3336

Dust Explosions

Peter Field

This book is a comprehensive review on the hazards and prevention of dust explosions. It covers the explosibility tests used to study dust explosions in the U.S., UK, and other European countries. The reader will learn how to interpret explosibility data and how to protect against explosions by proper sizing of relief vents and the application of suppression techniques.

Available from Elsevier Science Publishing Company, Inc.,
52 Vanderbilt Avenue, New York, NY 10017
(212) 989-5800

International Symposium on Runaway Reactions

This international symposium (Cambridge, Massachusetts, March, 1989) was sponsored by the Center for Chemical Process Safety of AIChE and the Institution of Chemical Engineers. Thirty-five papers were presented in seven sessions on reactor stability, relief and safety. The reader of these papers will gain the latest insights into the prediction and assessment of thermal runaway reactions, ARC data modeling, the use of various types of calorimetry to study runaway reactions, vent sizing, and much more. These excellent contributions provide current knowledge and cover the most recent developments in chemical reaction hazards assessment and DIERS technology.

Available from American Institute of Chemical Engineers,
Publication Sales, Dept. CAT-81,
345 East 47th Street, New York, NY 10017
(212) 705-7657

Accident and Loss Prevention

What Went Wrong? Case Histories of Process Plant Disasters

Trevor A. Kletz
1988 238 pages ISBN 0-87201-919-5

This excellent reference book provides case histories on a wide variety of incidents that have occurred in the process industries. Each incident is described in terms of what happened, why it happened, and how it might be prevented from reoccurring in the future. Advice is presented on how to eliminate accidents during maintenance and process modifications, on how to prevent explosions in storage tanks, stacks, and tank trucks and cars, and how to control and detect leaks. This second edition has been vastly expanded to cover accidents in computer-controlled plants, and to offer new insights into the hazards of static electricity and materials of construction. All chemists, safety engineers, and plant operators should benefit from reading and applying the information presented in this book.

Available from Gulf Publishing Company,
P. O. Box 2608,
Houston, TX 77252-2608
(713) 529-4301

International Symposium on Preventing Major Chemical Accidents

Edited by John L. Woodward, 1987
ISBN 0-8169-0411-1

These proceedings were jointly sponsored by the Center for Chemical Process Safety of the American Institute of Chemical Engineers, The United States Environmental Protection Agency, and The World Bank. The session held on reactive chemicals covered topics such as process hazards, calorimetric methods, pressure relief, hazard evaluation, and research needs in reactive chemicals.

Available from American Institute of Chemical Engineers,
Publication Sales, Dept. CAT-81,
345 East 47th Street, New York, NY 10017

Loss Prevention

Loss Prevention is a 19-volume set of papers presented at symposia sponsored by the AIChE. These symposia have been held from 1967 to 1995, with objectives aimed at the management of safety problems and prevention of accidents through the free exchange of information. In general, a new volume is issued annually. These papers cover a variety of facets of plant safety and loss prevention, including hazard analysis programs, case histories and means of prevention, safe design and operation of chemical plants, the fundamentals of fires and explosions, transportation, and much more. Some of the earlier volumes may not be available except in reference libraries.

Available from American Institute of Chemical Engineers,
Publication Sales, 345 East 47th Street,
New York, NY 10017
(212) 705-7657

Learning from Accidents in Industry

Trevor A. Kletz
1988 158 pages ISBN 0-408-02696-0

This book outlines principals useful in identifying the multiple, underlying causes of industrial accidents and discusses how to prevent their reoccurrences. Fifteen major industrial accidents are examined as to their causes and what corrective measures could have lessened or eliminated the resulting damage. The basic principals of safety are presented as well as suggestions on what questions should be asked during the investigation of an accident. All chemists and chemical engineers in need of information on the safe use of chemicals will benefit from what is presented in this book.

Available from Butterworths,
80 Montvale Avenue, Stoneham, MA 02180
(617) 438-8464

Major Industrial Hazards: Their Appraisal and Control

John Withers
262 pages ISBN 0-470-21067-2

This book addresses the risk assessment of major industrial hazards. The early part of the book covers the mathematics of risk analysis and considers case histories in order to develop a classification system for major industrial risks. Next, the author covers how to estimate what will happen in an accident and how to assess the amount of damage that might be caused. The final parts of the book cover transport risks, mitigation of hazards, and the costs and benefits of risk prevention. The book will serve as useful reading to those who want to understand risks and how to minimize them in plant operations.

Available from John Wiley & Sons, Inc.,
1 Wiley Drive, Somerset, NJ 08873-1272 (201) 469-4400

Preventing Major Chemical and Related Process Accidents

ISBN 0-85295-225-2

The Institution of Chemical Engineers Symp. Ser. No. 110. These papers cover research on risk assessment, safety management and control, use of test data in designing plants, consequence analysis, regulatory issues, and accident investigation. The case histories presented cover incidents at Bhopal, Sandoz, and Mossmorran.

Available from The Institution of Chemical Engineers,
165-171 Railway Terrace, Rugby CV21 3HQ UK

Loss Prevention Bulletin

Published bimonthly by the Institution of Chemical Engineers, this bulletin contains case histories of industrial disasters along with several other topics of interest in the reactive chemicals area. In particular, Loss Prevention Bulletin 066 covers dust explosion testing and prevention of and protection from dust explosions.

Available from The Institution of Chemical Engineers,
165-171 Railway Terrace, Rugby CV21 3HQ UK

Process Safety Progress

Published quarterly by the AIChE, these papers cover plant safety, loss prevention, fires and explosions, reactors, descriptions of industrial accidents, and other topics concerning plant operations. This is an excellent resource for announcements on new books, future symposia, and conferences related to process safety. Formerly *Plant/Operations Progress.*

Available from Process Safety Progress, AIChE
345 East 47th Street, New York, NY 10017
(212) 705-7338

Safety and Accident Prevention in Chemical Operations

H. W. Fawcett and W. S. Wood
1982 910 pages

This book, in its second edition, covers topics on the hazards involved in the operation of chemical and petroleum plants.

John Wiley & Sons, Inc.,
1 Wiley Drive, Somerset, NJ 08873-1272
(201) 469-4400

Note: Out of print, but still available in reference libraries.

Loss Prevention and Safety Promotion in the Process Industries,
Volume 3, Chemical Process Hazards VIII

259 pages The Institution of Chemical Engineers Symp. Ser. No. 82, 1983

These conference (September, 1983, Harrogate) proceedings cover runaway reactions, fires, and explosions.

Available from The Institution of Chemical Engineers,
165-171 Railway Terrace, Rugby CV21 3HQ UK

CHRIS Manual II. Hazardous Chemical Data Manual

U.S. Coast Guard
1985 2200 pages Order No. 050-012-00215-1

This reference manual provides data needed for the hazard assessment of chemicals that are frequently transported by water. This resource is part of the Coast Guard's Chemical Hazard Response Information System (CHRIS). The manual describes the physical and chemical properties for over 1200 industrial chemicals.

Available from Superintendent of Documents,
Government Printing Office, Washington, DC 20402
(202) 275-3648

Hazardous Substances, A Reference

Melvin Berger
1986 128 pages ISBN 0-89490-116-8

This is a quick reference guide to information on more than 230 materials identified as presenting potential environmental problems. The information provided includes common names, acronyms, less common names, formula, descriptive characteristics, exposure limits, uses, and health effects. Additional sections include references, definitions of terms, description of federal laws, and government agencies that provide information on hazardous materials. Accounts are given for several recent disasters that have resulted from hazardous substances.

Available from Enslow Publishers, Inc.,
Bloy Street & Ramsey Avenue, Box 777, Hillside, NJ 07205
(201) 964-4116

Toxic and Hazardous Industrial Chemicals Safety Manual

Topics covered include handling and disposal along with toxicity and hazard data. This manual presents data on 702 chemicals that are used in the Japanese chemical industry. The manual was compiled from a comprehensive study of the literature and expertise in Japan. The eleventh edition was published in 1986. It contains a new section listing dangerous interchemical reactions for 285 of the 702 chemicals.

Available from the international Technical Information Institute,
Toranomon-Tachikawa Building, 605, 1 Chome, Nishi Shim-bashi,
Minato-ku, Tokyo, Japan

Data Sources and Compilations

Hawley's Condensed Chemical Dictionary

N. I. Sax and R. J. Lewis
1288 pages

Provides information on the flammability and explosive risks of chemicals, trade mark products, and chemical reactions. This revised and updated book is useful as a source of technical data and descriptive information. It contains over 20,000 entries.

Available from Van Nostrand Reinhold,
P. O. Box 668, Florence, KY 41042-9979
(606) 525-6600 or 1-800-323-4958

Solvents Safety Handbook

Edited by D. J. De Renzo
1986 696 pages

This book presents safety data on 335 frequently-used solvents. The information presented includes fire hazards, reactivity, and physical and chemical properties.

Available from Noyes Publishers,
120 Mill Road, Park Ridge, NJ
(201) 391-8484

Hazardous Chemicals Chart

This 1985 wall chart provides hazard information on over 190 industrial chemicals. For each chemical, NFPA ratings on flammability and reactivity, flash points, ignition temperatures, and firefighting information are summarized.

Available from Laboratory Safety Supply Company,
P. O. Box 1368, Janesville, WI 53547
(608) 754-2345

The Sigma-Aldrich Library of Chemical Safety Data

Edited by R. E. Lenga
1985 1999 pages Order No. 214,000-7

This excellent reference work is highly recommended for anyone concerned with the hazards of chemicals. This comprehensive compilation presents safety information on over 7,000 materials. The reactive chemicals information summarized for each chemical include physical properties, incompatibility data, handling and storage requirements, and firefighting procedures.

Available from Aldrich Chemical Company, Inc.
1001 West St. Paul Avenue, Milwaukee, WI 53233
(414) 273-3850

Fire Protection Guide on Hazardous Materials

This guide is an excellent source of information on hazardous chemicals in commercial use today. The guide is comprised of a collection of four standards and manuals published by National Fire Protection Association (NFPA). The four sections have been divided and arranged so that one can locate information quickly in case of an emergency. The first section on fire hazard properties provides flash points and flammable limits for more than 1300 flammable substances. Hazardous chemical data covering fire, explosion, and toxicological properties are listed for approximately 415 chemicals in the second section. This section also includes recommendations on storage and firefighting of these chemicals, and each chemical is identified by the NFPA system as to the degree of health, flammability, and reactivity hazards that it poses. The manual (third section) on hazardous chemical reactions will be particularly useful to research and development personnel. This section summarizes data on 3550 mixtures deemed potentially dangerous from a review of incidents reported in the literature. References to descriptions of the fires or explosions caused by these reactions are presented. The last section of the guide describes the identification system developed by NFPA to determine the degree of health, flammability, and reactivity hazards that a chemical poses. This system also covers reactivity hazards with water. The ninth edition was published in 1986.

Available from National Fire Protection Association,
470 Atlantic Avenue, Boston, MA 02210

Handbook of Reactive Chemical Hazards

L. Bretherick
1985 1852 pages ISBN 0-408-01388-5

This updated edition (third) of this authoritative reference work contains hazard data on nearly 9000 chemicals. This excellent book presents data on runaway reactions, potential explosions, ignition, and fire hazards. It is highly recommended for anyone concerned with the safety of reactive chemicals.

Available from Butterworths,
80 Montvale Avenue, Stoneham, MA 02180
(617) 438-8464

Hazardous Chemicals Information and Disposal Guide

Margaret-Ann Armour, Lois M. Browne and Gordon L. Weir
1987

The third edition of this guide presents data on 462 chemicals. The information has been compiled from published materials and the results of spillage and waste disposal research conducted at the University of Alberta. Presented for each chemical are NFPA hazard ratings on health, fire and reactivity/stability, chemical and physical properties, fire hazards, hazardous reactions, physiological properties and health hazards, spillage disposal, and waste disposal. The spillage and disposal methods are explained in detail, and these sections include a listing of the chemical reactions that take place to form the innocuous products that are involved with each disposal procedure. References to the original sources of information are provided. Also included is an incompatibility table providing guidance on conditions to avoid for specific chemicals or classes of compounds. This book was printed at the University of Alberta. Parts of the book are available as the HAZINF file on CIS (see pp. 42).

Available from The Lab Store,
3888 North Fratney Street, Milwaukee, WI 53212

"Flammability Characteristics of Combustible Gases and Vapors"
Michael G. Zabetakis

This bulletin contains a review of the field of vapor flammability as it existed in 1964 and presents data on the flammability limits, autoignition temperatures and burning rates for more than 200 combustible gases measured under a variety of conditions. The experimental data are grouped under various chemical families that allow readers to detect similarities and trends. The empirical rules and graphs presented allow readers to quickly estimate data for literally thousands of combustible gases in various oxidants at various temperatures and pressures. A section is included on preventive measures that covers inerting, flame arresters and relief diaphragms. This bulletin (No. 627, Bureau of Mines) supplements work (Bulletin 503) by H. F. Coward and G. W. Jones who summarized available flammability data through 1952.

Available from United States Government Printing Office
Superintendent of Documents, Washington, DC 20402
(202) 275-3648

Flash Points of Organic and Organometallic Compounds
Richard M. Stephenson

This compilation tabulates flash points for organometallics and a wide variety of different types of organic compounds.

Available from Elsevier Science Publishing Company, Inc.
52 Vanderbilt Avenue, New York, NY 10017
(212) 989-5800

Material Safety Data Sheets

Extremely Hazardous Substances Superfund Chemical Profiles
U. S. Environmental Protection Agency
1988 1807 pages ISBN 0-8155-1166-3

This two-volume reference guide contains detailed chemical profiles for each of the 366 chemicals listed as extremely hazardous substances by the U. S. Environmental Protection Agency in the 1987 *Federal Register*. For approximately 65 percent of the chemicals, the profiles contain information taken from the *National Library of Medicine's Toxicology Data Base* (TDB) or *Hazardous Substance Data Bank* (HSDB). For those chemicals not included in TDB/HSDB, the profiles were developed from information reported in the literature. The data have been reviewed for accuracy and completeness. Chemicals are listed alphabetically based on the common names used in section 302 of SARA Title III. An index of Chemical Abstracts Service (CAS) registry numbers has been included to assist in quickly finding information on the chemicals. Each profile contains chemical identification information and data on toxicity and health hazards, physical and chemical properties, fire and explosion hazards, reactivity, uses, precautions for safe handling and use, protective equipment, and emergency treatment. For approximately 120 chemicals, guidance for emergency first aid treatment is presented in non-medical language. Revisions are underway to provide first aid information for the remaining 246 chemicals. In the introductory material, a discussion on personal protective equipment including respirators and clothing, a list of abbreviations, and a glossary of medical terms are included.

Available from Noyes Publications,
Mill Road at Grand Avenue, Park Ridge, NJ 07656
(201) 391-8484

Chemical Safety Data Sheets Volume 1: Solvents

The Royal Society of Chemistry
1989 300 pages ISBN 0-85186-903-3

This compilation of safety data sheets provides detailed and up-to-date information on over 100 solvents, including chlorinated and aromatic hydrocarbons, alcohols, ketones, amines, and esters. The chemical, physical, and health hazards of each solvent are covered as well as information on safe handling, disposal, and legislation governing their uses. The information will be useful to anyone who comes into contact with these solvents.

Available from CRC Press, Inc.,
2000 Corporate Boulevard, NW, Boca Raton, FL 33431
1-800-272-7737

Trade Names Data Base

The Trade Names (TN) database is an electronic information resource containing nearly 50,000 Material Safety Data Sheets (MSDSs) supplied by over 200 chemical producers and distributors. This collection is compiled by Canadian Centre for Occupational Health and Safety (CCOHS), which provides full-text MSDSs in English or French as they are received from the contributors. TN is easy-to-use and readily accessible on the Centre's computerized on-line system, CCINFOLINE, and is also available on a CD-ROM, CCINFODISC, supplied in the latest version (ISO-9660) of the High Sierra Standard. TN provides information on chemical and physical properties, storage and handling precautions, cleanup and disposal, and health hazards. The annual subscription to the CD-ROM component is reasonably priced at $114 Canadian dollars.

Available from The Canadian Centre for Occupational Health and Safety,
250 Main Street East, Hamilton, Ontario, Canada L8N 1H6
(416) 572-2981

Material Safety Data Sheet Service

The material safety data sheets compiled by manufacturers provide an important source of safety data. This subscription service is a microfiche-based system comprising over 55,000 MSDSs produced by several hundred manufacturers. It is the world's largest single collection, with updates provided every 60 days. The system is indexed by CAS registry number, brand or trade names, and supplier or source of data. These printed indices give easy access to the filmed MSDSs. It is priced at $5000 per year.

Available from Information Handling Services,
15 Inverness Way East, P. O. Box 1154, Englewood, CO 80150
(303) 790-0600 or 1-800-241-7824

Materials Safety Data—The Genium Collection

The Genium Collection is a three-volume set of material safety data sheets on over 900 hazardous materials found in the industrial workplace. These MSDSs are a comprehensive source of safety information, and the data have been compiled and reviewed by industrial hygienists and occupational physicians. The sources of data for each sheet are given. The collection is updated three times yearly adding over 120 new or revised sheets. Genium also offers personal computer software, known as MSDS Engine, to manage MSDS information. This software package can be purchased preloaded with the Genium Collection of MSDSs.

Available from Genium Publishing Corporation,
Department DS8A, 1145 Catalyn Street, Schenectady, NY 12303-1836
(518) 377-8854

Compendium of Safety Data Sheets for Research and Industrial Chemicals Parts I–VI

Edited by Lawrence H. Keith and Douglas B. Walters

This six-volume set is a compilation of safety data for 1607 chemicals used in industry and research. The types of reactive chemicals information provided include upper and lower flammability limits, flash points, and procedures for safe handling and storage. The compilation contains a useful section on gloves that can be used to select appropriate materials for personal protection. These six volumes list for $525.

Available from VCH Publishers, Inc.,
303 N. W. 12th Avenue, Deerfield Beach, FL 33441
(212) 683-8333

Chemtox

Chemtox is a microcomputer-based MSDS collection covering over 3,200 chemicals. In addition to providing toxicological data, it covers physical and chemical properties, emergency response data, and personal protection equipment. The database management software allows users to add and modify their own information.

Available from VNR Information Services,
115 Fifth Avenue, New York, NY 10003
(212) 254-3232

Material Safety Data Sheets on CD-ROM

OHS MSDS on Disc is a CD-ROM product that provides material safety data sheets on about 10,000 individual chemical substances or mixtures, including all chemicals available from Fisher Scientific. Developed by Occupational Health Services, Inc., a subscription to this product includes regular updates and supporting software used to retrieve information from the disk.

Available from Occupational Health Services, Inc.,
Suite 2407, 450 Seventh Avenue, New York, NY 10123
(212) 967-1100

The **Sigma-Aldrich Material Safety Data Sheets** is a collection on CD-ROM of over 24,000 MSDSs that include the chemical structures for the Aldrich products. The annual subscription includes updated disks that are issued every 3 months with new safety information and data on new products from Aldrich Chemical. Information is retrieved from the disk by catalog number, chemical name, CAS registry number, or molecular formula. The collection is reasonably priced at $1300 per year.

Available from Aldrich Chemical Company, Inc.,
1001 West St. Paul Avenue, Milwaukee, WI 53233
(414) 273-3850

Computerized On-line Databases

The Chemical Hazard Response Information System (CHRIS)

CHRIS was developed by the U.S. Coast Guard to provide emergency response information for those handling spills and dealing with disposal problems involving hazardous materials. The system contains reactive chemicals information on fire hazards, chemical reactivity, physical and chemical properties, National Fire Protection Association Hazard classifications, and the National Academy of Sciences hazard ratings. A section has been included that provides procedural advice on actions that should be taken in case of emergencies involving over 1,150 transported chemicals. This source of safety information is available on-line through Chemical Information Systems, Inc., Baltimore, MD 21212. Phone 800-CIS-USER or (301) 321-8440.

MICROCHRIS is a personal computer version of CHRIS. It is available from AIA Corporation, 134 Middle Neck Road, Suite 210, Great Neck, NY 11021 for $995. The CHRIS file is also part of CHEM-BANK DATABASE that is available on a CD-ROM disc from SilverPlatter Information, Inc., 37 Walnut Street, Wellesley Hills, MA 02181. Phone (617) 239-0306. The annual subscription is priced at $1350, which includes quarterly updates.

Hazardous Substances Data Bank (HSDB)

HSDB is an excellent on-line source of reliable safety information on over 4200 chemical substances. This factual file covers chemical substances that are known or suspected of being toxic, are produced in high volume, or are regulated by government agencies. The areas of reactive chemicals information covered include physical and chemical properties, physical hazards including fire, explosion and reactivity, safe handling and use, and emergency procedures. All data added to the file are completely referenced and fully peer reviewed. The file is searchable on the National Library of Medicine's (NLM) Toxicology Data Network (TOXNET), which offers users flexible, user-friendly searching.

CHEMPRO is a microcomputer-based version of HSBD. It covers all of the chemicals in HSDB, but the technical information available is not as extensive. All reactive chemicals information in HSDB has been included in the subset of data loaded into ChemPro. A quarterly update service can be purchased, and a communications package has been built into the software that allows users to readily access HSDB at NLM.

For information on HSDB, contact Specialized Information Services Division, National Library of Medicine, 8600 Rockville Pike, Bethesda, MD 20894 (301) 496-6531 or 496-1131

For information on CHEMPRO, contact Fisher Scientific at (412) 562-8383.

Chemical Information System (CIS)

The Chemical Information System is a computerized retrieval service that provides access to five files containing reactive chemicals information. The CHRIS component developed by the U.S. Coast Guard is described on page 40. Baker is an MSDS file developed by J. T. Baker Company. It contains OSHA-formatted MSDSs on 1600 chemical substances or mixtures. This file is the simplest to use on CIS and is searchable by CAS registry number, chemical name, or fragmentary name. The Hazardous Chemical Information and Disposal (HAZINF) file developed at the University of Alberta provides hazard ratings (fire, health, and reactivity), chemical properties, and hazardous chemical reactions for 220 substances. This database provides concise but detailed information on how to handle spills, including procedures for cleaning them up and getting rid of what is left over. The most extensive source of safety data on CIS is the Oil and Hazardous Materials/Technical Assistance Data (OHM/TADS) files covering 1,402 chemicals deemed potentially hazardous by the U.S. Environmental Protection Agency. The information in this file has been compiled from published sources by contractors to EPA. This file is an excellent place to start in compiling information for MSDSs. Special print formats are available that allow users to extract MSDS data in OSHA formats. The chemicals covered are selected from those likely to be involved in spills, have large production volumes, or are highly toxic. The reactive chemicals information provided includes reactivity, fire protection and explosiveness, chemical and physical properties, and storage and handling procedures. Very recently, CIS announced the addition of another new component that contains reactive chemicals information. MALLIN is a file containing material safety data sheets for 1,415 laboratory or electronic chemicals supplied by Mallinckrodt, Inc. This file is searchable by chemical name or CAS registry number. In addition, some of the numeric fields and all text are searchable.

A microcomputer version (MICRO OHM/TADS) is available from AIA Corporation, 134 Middle Neck Road, Suite 210, Great Neck, NY 11021 for $995. This software will run on portable computers that can be carried to an emergency scene and used to locate safety information on an unknown material simply from its description. The OHM/TADS file is also part of CHEM-BANK DATABASE, which is available on a CD-ROM disc from SilverPlatter Information, Inc. 37 Walnut Street, Wellesley Hills, MA 02181. The annual subscription is priced at $1350, which includes quarterly updates.

For information on CIS, contact Chemical Information Systems, Inc.
7215 York Road, Baltimore, MD 21212
(301) 321-8440 or 1-800-CIS-USER

Hazardline

Hazardline is an on-line resource available from BRS that contains safety data on over 3000 chemicals. The database developer is Occupational Health Services who input information in areas such as chemical and physical properties, recommendations on firefighting, and emergency procedures for handling leaks and spills.

Available from BRS Information Technologies
1200 Route 7, Latham, NY 12110
(518) 783-1161 or 1-800-833-4707

Chemical Hazards in Industry (CHI) and Laboratory Hazards Bulletin (LHB)

The Royal Society of Chemistry

CHI and LHB are computerized, on-line databases covering various aspects of hazardous materials that are likely to be encountered in the industrial workplace. Subjects primarily covered in CHI include fires and explosions and their prevention, safe practices in storage and transportation of chemicals, legislative matters, industrial hygiene, and waste management. Items included in LHB cover primarily reactivity hazards, waste disposal, biological effects of chemicals, first aid, and legislative matters relating to use. The extent of coverage in these databases is very broad with over 200 international journals, conference proceedings, books, news articles, and audio-visual materials scanned each month for relevant information. The two databases combined contain over 18,000 records with updates of nearly 300 new records added monthly. Many of the citations contain abstracts. Printed copies of the same material are available from the producer.

Until recently, these databases were separately loaded on STN International, the Scientific & Technical Information Network, as supplied by FIOZ-Chemie GmbH of Berlin. The Chemical Safety NewsBase (file CSNB) has now replaced files LHB and CHI on STN International. The Chemical Safety newsBase is also searchable on Orbit. The advantages of searching the merged file on Orbit or STN is that duplicate records are eliminated and that CAS Registry numbers are searchable in the Registry Number (RN) field or basic index. The citations in CHI and LHB will be of interest to anyone involved with chemical safety management.

Available from The Royal Society of Chemistry,
The University of Nottingham, NG7 2RD, UK
0602 507411
Available on STN International, c/o Chemical Abstracts Service,
A Division of the American Chemical Society, 2540 Olentangy River Road,
P. O. Box 02228, Columbus, OH 43202
(614) 447-3600 (FAX-614/447-3713)
Available on Pergamon Orbit Infoline,
8000 Westpark Drive, McLean, VA 22102
(703) 442-0900

Chemical Abstracts

CA is an on-line database containing comprehensive coverage of chemical information taken from the worldwide literature in all areas of chemistry and chemical engineering. The file contains over eight million records. The file includes safety-related information from about 1976 on the hazardous properties of chemical substances and chemical reactions, chemical laboratories and industry concerns, flammability, accidents, fires and explosions, and test methods. This file is a useful resource when looking for generic safety information on materials or when one file includes recent safety information on hazardous chemical reactions or laboratory procedures that are routinely published as letters to the editor in a few chemical journals and magazines.

Available on STN International, c/o Chemical Abstracts Service,
2540 Olentangy River Road,
P. O. Box 02228, Columbus, OH 43202
(614) 447-3600 (FAX-614/447-3713)

Educational and Training Materials

Hazardous Materials in the Workplace: Meaning, Measurement and Control

S. S. Chissick, K. Briggs, C. Dunbar and T. Dooher
1984 ISBN 0471-90382-5

An educational program designed for use by safety training coordinators. The package contains a 27-minute video, a manual, 17 slides, and chemical and health questionnaires. This program emphasizes the storage and handling of hazardous substances.

Available from John Wiley & Sons, Inc.,
1 Wiley Drive, Somerset, NJ 08873-1272
(201) 469-4400

Chemsafe

This complete training program on chemical health and physical hazards consists of eight modules, handouts, and a trainer's manual. It can be purchased as a slide-tape program or as a video. Topics covered include corrosives, solvents, oxidizers, poisons, gases, explosives and carcinogens.

Available from BNA Communications Europe,
Tennyson House, 159 Great Portland Street, London W1N 5FD, UK

Chemistry of Hazardous Materials

Eugene Meyer
1977 370 pages ISBN 0-13-129239-0

This intermediate level textbook has been particularly useful in teaching fire-fighters of the hazards associated with chemicals. Basic chemical concepts are included as well as detailed information on thousands of commercially available hazardous materials. The chemical and physical properties of a wide variety of hazardous materials are highlighted. Excellent discussions are included on the principles of chemical reactions, corrosive materials particularly acids, water-reactive materials, toxic materials, explosives, and radioactive materials. The appendices contain information on first aid measures and emergency treatment of poisons and a tabulation of safety data on 52 common solvents.

Available from Prentice-Hall,
200 Old Tappan Road, Old Tappan, NJ 07675
(201) 592-2000

Safety in the Chemistry Laboratory

This comprehensive 42-minute program is available as a videotape or in slide/audio tape format and comes with a training log, self-test, and full script. It was developed primarily for professionals working in chemical laboratories. Topics covered include Material Safety Data Sheets, labels, hazard classes, safe handling of hazardous chemicals, spills, and disposal. The classes of materials covered include corrosives, flammables, health toxins, and reactive chemicals.

The contents of the "Safety in the Chemistry Laboratory" program is available as three shorter videotapes covering MSDSs, spills and disposal, and the handling of toxic chemicals. Each tape is approximately 15 minutes in length. Titles and ordering information are presented below:

SAF-201-57 "Hazard Communication: The Chemical Worker's Right to Know"

SAF-301-57 "Chemical Safety Measures, Spills and Disposal"

SAF-401-57 "Handling Hazardous Chemicals"

Available from Savant,
801 East Chapman, Suite 101, P. O. Box 3670, Fullerton, CA 92634
(714) 870-7880 or 1-800-472-8268, ext. 57 (FAX 714/526-7400)

Preventing Emergencies in the Process Industries

This training program includes a videotape, an instruction manual, and 60 color slides. The videotape presents case studies on seven industrial disasters that emphasize the need and importance of pre-emergency planning. Presented in the format of a short course, group discussion and participation of the trainees is encouraged.

Available from Gulf Publishing Company Video,
P. O. Box 2608, Houston, TX 77252-2608
(713) 529-4301

The Chemical Safety Factor System

Provides a basic training program covering the essentials of safe handling and storing of hazardous chemicals. Outlines a method that can be used to identify and to remember potential hazards associated with the use of chemicals. Factor is an acronym standing for Flammability and Corrosive, Toxic or Reactive. Materials supplied include audio-visual aids, books, and posters.

Available from Comprehensive Loss Management, Inc.
7671 Old Central Avenue N.E., Minneapolis, MN 55432
(612) 944-1959

Handling Hazardous Chemicals Safety

This 16 minute film teaches workers how to handle and store chemicals safety, and how to minimize the dangers of spills. The types of chemicals covered include oxidizers, flammables, and those that are water sensitive.

Available from BNA Communications,
9417 Decoverly Hall Road, Rockville, MD 20850
(301) 948-0540

Chemical Hazard Communication Guidebook

This book provides step-by-step guidance on how to comply with the OSHA Hazard Communication Standard and SARA Title III Emergency Planning and Community Right-to-Know Requirements. The Hazardous Materials Transportation Act is also covered. Topics discussed include compliance planning, incident reporting of spills, and toxic chemical release reporting. Comprehensive reference guides are included on how to prepare Material Safety Data Sheets, performing hazard determinations, labeling, training, determining proper packaging, and preparing shipping papers. This guidebook is reasonably priced at $75.

Available from Executive Enterprises Publications Co., Inc.,
22 West 21st Street, New York, NY 10010-6904
1-800-332-1105

Lowry's Handbook of Right-to-Know and Emergency Planning—SARA Title III

George G. Lowry and Robert C. Lowry
1988 421 pages ISBN 0-87371-112-2

Handbook of Compliance for Worker and Community, OSHA, EA, and the States.

This comprehensive volume is divided into three parts that provide extensive guidance on how to comply with the Hazard Communication Standard and the 1986 Superfund Amendments and Reauthorization Act (SARA). In part A, the book covers the basic and major elements of these two laws, providing detailed overviews of what is required. Part B covers identification and characterization of hazardous chemicals, labeling, preparing MSDSs, recordkeeping, and a written Hazard Communication program. All required aspects of the laws are addressed, and sources of more detailed guidance are provided. Part C looks at policy issues such as trade secrets and the legal question of preemption raised when other federal, state, or local laws may apply.

Copies of the text of the two federal statutes are presented in full in the appendix. Other useful information compiled in the appendix includes sources of information covering books, computer data banks, training programs, lists of regulated materials indexed by CAS registry number, and federal forms including the EPA-MSDS form 174.

Available from Lewis Publishers, Inc.,
121 South Main Street, Chelsea, MI 48118
(313) 475-8619

The Facts about the Hazard Communication Standard

This training program, designed to inform workers about the Hazard Communication Standard, consists of a slide show and a 44-page booklet. The program covers the physical hazards of chemicals, including reactivity and their potential health hazards. Material safety data sheets are explained as well as proper spill procedures and correct handling and storage of hazardous chemicals.

Available from National Safety Council,
444 North Michigan Avenue, Chicago, IL 60611
(312) 527-4800

Chemical Safety in the Health Products Industry

This training program consists of a 22-minute video providing background information on the Hazard Communication Standard. It was designed to help employers comply with the standard concerning employee training. You will learn what you need to know about a hazards communication program and how to use material safety data sheets. Other subjects covered include labeling, storage, handling spills, and controlling exposures.

Available from Health Industry Manufacturers Association,
1030 15th Street NW, Suite 1100, Washington, DC 20005
(202) 452-8240

Teach. Telling Employees about Chemical Hazards

This training program was developed to prepare workers for compliance to the Hazard Communication Standard. It describes the use of material safety data sheets, chemical labeling, chemical storage and handling, and health hazard recognition. The program, which includes an instruction manual, is available as slides or as a video.

Available from Natlsco,
K-3, Long Grove, IL 60049
(312) 540-2400

Hazard Communication Compliance Manual: The Complete Guide for Worker Right-to-Know

This new manual is a step-by-step guide designed to help ensure compliance with the requirements of the OSHA Hazard Communication Standard. It provides comprehensive information on material safety data sheets, labels, employee training, and a written program plan. Supplemental materials include a directory of information sources, full text of the OSHA standard, and ready-to-copy forms and memos.

Available from The Bureau of National Affairs, Inc.,
Circulation Department, P. O. Box 40947, Washington, DC 20077-4928
1-800-372-1033

APPENDIX B

Industry Practice Survey Results

Summarized in this appendix are the results of a survey of current industry practice related to the storage and handling of reactive materials. The survey form, reproduced on pages 336–350 at the end of this appendix, was sent to all 83 companies that were Center for Chemical Process Safety sponsor companies at the time of the survey. A total of 28 responses were received and used in this work. It should be noted that most of the 28 responses were from larger chemical companies, many of which have in-house expertise and other factors that should be kept in mind when interpreting the results of the survey.

It must also be kept in mind that the responses to a survey of this type report the practices that have been adopted for specific situations; also that there may be other, equally important practices that were not reported. It may also be that the practices reported may not be appropriate for other situations. In addition, there is no representation that the practices reported do not violate any law, regulation, or other legal requirement. Neither AIChE, CCPS, the Subcommittee, the writers/editors/reviewers, nor any other party participating in this project can be held responsible for the application of the reported practices or for any omissions or errors in this report.

The table on the next two pages lists the survey questions grouped by topics, and gives the Summary Number in this appendix that tabulates a summary of responses to each question. The most frequent responses are listed first in the summary table; however, it should be noted that this is only a general arrangement and is affected by how the answers are grouped together.

It should also be noted that the survey questions and responses are intended to only apply to the **on-site storage and handling of reactive chemicals** and not to process operations, flammable materials in general, or materials with high toxic hazard properties only. Each summary topic is discussed in Chapter 5 of this book; the corresponding Chapter 5 section numbers are given in the left column of the index on the following two pages.

For comparison purposes, the maximum number of responses in each reactive material category was as follows:

• polymerizing	21
• shock-sensitive	8
• thermally decomposing	17
• isomerizing	6
• disproportionating	5
• pyrophoric	14
• peroxide-forming	10
• water-reactive	17

Thus, for example, for the first entry in Summary 1, 7 out of the 21 companies or facilities handling polymerizing materials use a chemical interaction matrix to identify potential incompatibility problems. These numbers do not add up to 28, since most companies that responded to the survey handle more than one type of reactive material.

Index of Survey Practice Questions

Topic	Reactive Materials Storage/Handling Question	Summary #
Compatibility	• How are potential incompatibility problems identified?	1
[discussed in Section 5.2 of this book]	• What design/operating practices are used for preventing incompatibility incidents related to...	
	- process materials/reagents?	2
	- impurities?	3
	- heat transfer fluids?	4
	- materials of construction and corrosion products?	5
	- insulation?	6
	• What potential hazards have been identified with respect to compatibility of fire extinguishing agents with on-site reactive materials?	7
	• What design/operating practices are used for preventing incompatibility incidents related to any other materials?	8
	• Any other practices related to compatibility issues?	9
Storage Time/ Shelf Life	• What limitations are there on the shelf life of reactive materials stored at your facilities?	10
[Section 5.3]	• What design and/or operating practices are employed for increasing shelf life?	11
	• How is too-old material handled and disposed of?	12
Storage Quantity and Configuration	• What determines maximum inventories for reactive materials stored at your facilities?	13
[Section 5.4]	• In what configuration are the reactive materials stored (tanks, cylinders, cartons, etc.)?	14
	• If tank storage is used, are outlet nozzle(s) on the top or the bottom of the tank(s)?	15
	• What are your siting practices for reactive materials storage, such as tank spacing or distance to plant boundaries?	16
	• Are there any restrictions on container shape or configuration?	17
	• Are the storage container contents circulated or mixed? If so, how and why?	18

Topic	Reactive Materials Storage/Handling Question	Summary #
Air and Moisture Exclusion [Section 5.5]	• What design/operating practices are used to exclude air from the reactive materials containment?	19
	• What design/operating practices are used to keep moisture out of the reactive materials containment?	20
Monitoring and Control [Section 5.6]	• How are the following storage condition parameters monitored?	
	- oxygen concentration	21
	- humidity/moisture content	22
	- pressure	23
	- temperature	24
	• What kind of systems are used for controlling the temperature of the stored material?	25
Handling and Transfer [Section 5.7]	• What practices are employed for the manual handling of reactive materials on-site?	26
	• What design specifications are used for on-site transfer (piping) systems?	27
	• What operating practices related to on-site transfer (piping) systems are employed?	28
Last-Resort Safety Features [Section 5.8]	• What last-resort safety systems are employed?	29
	• What emergency relief system configurations are employed?	30
	• How are emergency relief devices sized?	31
	• What is your current practice for use of header systems?	32
	• What is your current practice related to treatment systems for handling reactive materials released through relief devices?	33
	• Are explosion suppression systems used inside any storage or handling equipment?	34
Passive Mitigation [Section 5.9]	• What passive mitigation systems are used to reduce the severity of consequences of a reactive materials storage/handling incident?	35
Detection, Warning, and Isolation [Section 5.10]	• What systems are in place to detect loss of containment of reactive materials?	36
	• What systems are in place to warn of their release?	37
	• What is in place to reduce the duration of a release by isolating the source?	38
Fire Prevention/ Fire Protection [Section 5.11]	• What is used for ignition source control in storage/handling areas?	39
	• What is used for mitigating fires in storage/handling areas?	40
Post-Release Mitigation [Section 5.12]	• What is in place to be used to counter a loss-of-containment accident?	41
	• What personal protective equipment is used to protect against a spill or release?	42
	• What emergency response plan features are specifically associated with the storage and handling of reactive chemicals on site?	43
Hazard Reviews [Section 5.13]	• If different categories of hazard severity are used, how are they defined?	44
	• What reviews are conducted for each hazard severity category?	45
Codes and Standards [Section 5.14]	• What publicly available consensus standards do you utilize in the design of reactive chemicals storage and handling facilities?	46

SUMMARY 1

Compatibility—Hazards Identification

Polymerizing	Shock-Sensitive	Thermally Decomposing	Isomerizing	Disproportionating	Pyrophoric	Peroxide-Forming	Water-Reactive	***How are Potential Incompatibility Problems Identified?***
7	2	4	1	1	5	3	6	Chemical interaction matrix
5	1	4			4	3	6	MSDSs; information from supplier
6	1	4			3	2	3	Hazards reviews (process hazard analyses): What-If, HAZOP studies
3	2	2	1	1	2	2	3	Knowledge and expertise of research chemists, safety personnel, and operating personnel; internal expert group; input from chemists and technical experts; identified by developmental chemist using existing information
2	1	2			2	2	4	Literature search/review; hazardous material references (Brethericks...)
1	2	2	1	1	2	1	2	Lab-scale experiments; test mixing results
2	1	2	1	1	1	1	2	Team-based review system or Reactive Chemical Review before introducing or changing any chemical or facility
2		2			2	1	2	Results from development work
2		1			2	2	2	Key chemical groups; structure search for reactive groups
2		1			1	1	2	Experience
2		1			1	1	2	Test if necessary
1		3			1	1	1	DSC, ARC, DTA, other calorimetry techniques
2		1			1		1	Manufacturer's recommended practices
1		1			1	1	1	Known or suspected chemistry and kinetics
	1	1			1	1	1	Hazard identification techniques
2								List of incompatible materials; in-house database of incompatibles
1					1			No formal system; examined at Corporate Development, Plant Technical, Plant Engineering, and Plant Process Engineering levels
1								Safety, Health, Environmental review
	1							All materials screened; those that meet certain guidelines are tested by drop weight
		1						ASTM CHETAH code
							1	Hazards are well-known

Five responses provided no information pertinent to this portion of the survey.

SUMMARY 2

Compatibility—Process Materials/Reagents

What Design and/or Operating Practices Are Used for Preventing Incompatibility Incidents Related to Process Materials/Reagents?

Polymerizing	Shock-Sensitive	Thermally Decomposing	Isomerizing	Disproportionating	Pyrophoric	Peroxide-Forming	Water-Reactive	
10	1	3			4	3	8	Dedicated equipment - dedicated unloading spots - dedicated loading spots and lines for products - product classification in dedicated process/storage areas - storage tanks in segregated area away from major process units - dedicated storage tanks - dedicated piping; unique fittings/connections - dedicated equipment with no cross-connections
5	1	4			3	1	2	Labeled, dedicated lines, piping, pumps, tanks, equipment and unloading/areas facilities for incoming materials; lines and vessels labeled where incompatibility potential exists
2	1	1	1	1	1	1	2	Labeling/"Right-to-Know" labels
1		3			3	1	2	Isolated and/or segregated storage; drums separated by categories and by walls, dikes, or spaces; storage and handling in separate buildings
2	1	1	1	1	1	1	1	Analysis of all incoming raw materials prior to use, including catalyst, seal fluids, heating fluids; each shipping container sampled to verify contents or impurities before unloading
1	1	1			2	1	1	Standard operating procedures from material safety data sheets
2					2	2		Inert gas blanketing protection; air excluded by inerting
1	1	1			1	1	1	Stored in drums; depends on operator training to properly store and transport materials to batch reactors where they are loaded per detailed procedures; operators supported by computerized hazard information systems; special ID lab
2							2	Training and operating/identification instructions/systems
1	1	1					1	In some cases, piping is physically disconnected/reconnected to avoid inadvertent mixing of incompatible chemicals
1	1	1					1	In some cases, double blocks and bleeds are used to isolate incompatible chemicals
1		1				1	1	Laboratory identification tests
		1			1		2	Limited storage quantity/inventory control
1		1				1		Manufacturers' loading/analysis sheets
		1			1		1	Computer and operator checks and balances
		2						Tanks rinsed with inert liquid and purged with inert gas/railcars or trailers cleaned and sealed prior to loading
1								Monitor stability of contents of storage tanks
1								Either use dedicated equipment or use a cleaning/inspection process
1								Avoid combustible, corrosive, and incompatible materials
1								Stored in original containers (usually plastic)
1								Avoid materials that may catalyze violent polymerization
1								Boil-out/steam-out/water rinse of multi-use equipment
1								Various techniques used; if hazard is severe, two devices required; sometimes two sign-offs required
		1						Certification of all process materials, analyses of trailer, railcar prior to transfer; cleaning with fresh fluid and sealing of dedicated railcars, trailers prior to loading
		1						Stored in labeled drums in dedicated, refrigerated drum shelters
		1						Administrative controls
				1				Only fresh product stored; all returns and off-specs are reprocessed to remove impurities (no cross-ties between storage and processing)
						1		Lab sampling to identify incoming material if no certificate of analysis with material
							1	Interlocks on water hookups to permit water line attachment while process is operating

Four responses provided no information pertinent to this portion of the survey.

SUMMARY 3

Compatibility—Impurities

What Design and/or Operating Practices Are Used for Preventing Incompatibility Incidents Related to Impurities?

Polymerizing	Shock-Sensitive	Thermally Decomposing	Isomerizing	Disproportionating	Pyrophoric	Peroxide-Forming	Water-Reactive	
18	2	10	2	2	9	7	10	Analysis; sampling - analyze for purity and/or presence of inhibitors - analysis made on material after major bulk transfers or at specific time intervals - analysis by vendor of incoming raw materials and spot-checked by plant operators - analysis before unloading; analysis of incoming materials; product analysis before transfer to storage; lab "certifies" prior to use - Manufacturers'/shipper's loading and/or analysis sheets and/or certification or "certificate-of-analysis" - analytical analysis - identify/sample materials or lab analysis prior to unloading for storage - analysis of trailer or railcar prior to transfer - quality control/identification; letters of analysis - drums sampled (square root method) - oxygen content of vinyl chloride checked prior to receipt (must be below 1,000 ppm by volume)
3	1	5			4	2	5	Dedicated, labeled lines/systems; dedicated/segregated storage and/or piping, handling/unloading
1	1	1			1	1	1	Material used per detailed procedures
	1	1			1	1	1	Impurities reviewed from initial analysis and through process change protocols
1		1			1		1	Inspection of shipping containers
1							2	Administrative controls; inventory controls; training and identification
1								Inert gas blanketing protection
1								Delivery lines operated every 2 weeks by standard operating procedure
1								Inhibitor added in recycle product prior to storage
		1						Tanks rinsed with inert liquid and purged with inert gas
		1						Refrigerated storage
		1						Pilot plant stage testing
		1						ARC testing with suspected material
		1						Cleaning with fresh fluid and sealing of dedicated railcars, trailers prior to loading
				1				Only fresh product stored; all returns and off-specs reprocessed to remove impurities, no cross-ties between storage and processing
				1				Stainless steel cone tanks for excess quantities/regular inspections done
					1			Empty containers discarded

Six responses provided no information pertinent to this portion of the survey.

SUMMARY 4

Compatibility—Heat Transfer Fluids

What Design and/or Operating Practices Are Used for Preventing Incompatibility Incidents Related to Heat Transfer Fluids?

Polymerizing	Shock-Sensitive	Thermally Decomposing	Isomerizing	Disproportionating	Pyrophoric	Peroxide-Forming	Water-Reactive	
1	1	3			2	2	2	Identification through HAZOP/hazards reviews/review by experienced personnel
1	1	1	1	1	1	1	1	Incoming materials analyzed
1		2			1	1	1	Safety committee review/Experience
1		1			1		2	Heat transfer fluids are compatible with product
1		1			1	1	1	Literature (manufacturer's, general)
1		1			1	1	1	Material safety data sheets
1		1			1	1	1	DTA or ARC testing
	1	1			1	1	1	Evaluation through DIERS testing
	1	1			1		1	Incompatibility matrix
	1	1			1		1	Small-quantity test mixing results
1		1						Limit temperature and heat transfer areas
1								Water used as cooling medium and steam for heating
1								Molten salt, hot oil, refrigerants, limited usage
							1	Air coolers used
							1	Use of non-aqueous heat transfer fluids eliminates reaction in case of coil leak
							1	Fluids used in double tubesheet shell and tube exchangers to prevent addition of water into cold tanks

Seventeen responses provided no information pertinent to this portion of the survey.

SUMMARY 5

Compatibility—MOC and Corrosion Products

What Design and/or Operating Practices Are Used for Preventing Incompatibility Incidents Related to Materials of Construction and Corrosion Products?

Polymerizing	Shock-Sensitive	Thermally Decomposing	Isomerizing	Disproportionating	Pyrophoric	Peroxide-Forming	Water-Reactive	
1	3	3	1	1	3	2	3	Corrosion/compatibility charts; incompatibility matrix developed
2	2	3			3	2	4	Materials evaluation testing; corrosion tests; corrosion coupons used extensively; corrosion coupons installed in actual operating system; corrosion coupon testing conducted by research corrosion lab under wide range of operating conditions (temperature, pH, velocity, etc.)
4		3		1	2	1	2	Safety reviews (for "not same kind" equipment replacement); hazard reviews; HAZOP considers materials of construction/corrosion products and autocatalysts
3	1	3			2	1	2	Literature, recommendations from manufacturers/suppliers and general
2		3		1	2		3	Engineering piping standards; corporate/internal design standards; piping and vessel specifications set at design stage
1	1	2			2	1	3	Operating experience
1	1	2	1	1	1	1	1	Material specified for all process streams
1		2			1	1	1	ARC testing in pilot plant stage with suspected material; DTA or ARC testing "spiked" with suspected corrosion products to test effect
1	1	1			1	1	1	Detailed procedures, subjected to process hazard analysis
1		1			1	1	1	Material safety data sheets
1		1			1	1	1	Materials engineer's recommendation based on literature
1		1					2	Stainless steel used, based on supplier's recommendations, experience, and corrosion lab data
	1	1			1		1	Research-quantity test mixing results used
2		1						Consult corrosion group/experts; in-house corrosion experts
1					1			Dedicated unloading and piping each with standard operating procedures
1							1	Team-based review system before any chemical is introduced or changed; use all available information from suppliers, literature, and laboratory to perform these reviews; have access to thermal stability apparatus if needed (ARC, DSC, etc.)
1								Detailed listing of recommended and not-recommended materials pertaining to given monomer; part of SHE review
1								Each product is reviewed by materials of construction experts to ensure materials are compatible; prior tested products listed in manual
1								For styrene, elastomers are difficult to specify; literature usually calls for VITON®, but acrylics attack it; EPDM, fluoroelastomers best overall
1								Design for low flow rates
1								For vinyl chloride monomer, follow recommended use of carbon steel or stainless steel; aluminum, copper, and brass not used due to potential for forming chloroalkyls or acetylides
1								Avoid metals and metal alloys that may initiate polymerization
				1				Peroxide storage for materials stored in excess of drum quantities are stored in stainless steel cone tanks that are periodically emptied, cleaned, and internally inspected for deposits, corrosion, etc.

Four responses provided no information pertinent to this portion of the survey.

SUMMARY 6

Compatibility—Insulation

What Design and/or Operating Practices Are Used for Preventing Incompatibility Incidents Related to Insulation?

Polymerizing	Shock-Sensitive	Thermally Decomposing	Isomerizing	Disproportionating	Pyrophoric	Peroxide-Forming	Water-Reactive	
3	1	3	1	2	2	1	1	Engineering standards; piping standards; internal design standards; piping and vessel specifications
2	1	2	1	1	2	2	1	Experience/judgment
2		2		1	2		2	Hazard reviews; process and operability hazard analyses when modifications/changes made
1	1	1	1	1	1	1	1	Capability testing with all materials and insulation or adsorbents
1		1			1	1	1	Literature (manufacturers' and general)
1		1			1	1	1	Material safety data sheets
	1	1			1		1	Incompatibility matrix developed
	1	1			1		1	Research quantity test mixing results
1								Use foam glass and calcium silicate with aluminum and stainless steel coverings and banding along with mastic sealed joints *(Caution: calcium silicate has a very high surface area, is very reactive with some materials, and has a high water content)*
1								Products reviewed with materials of construction experts to ensure compatibility
1								Oil on insulation is greater problem than materials in survey

Nineteen responses provided no information pertinent to this portion of the survey.

SUMMARY 7

Compatibility—Fire Extinguishing Agents

What Potential Hazards Have Been Identified with Respect to Compatibility of Fire Extinguishing Agents with On-Site Reactive Materials?

Polymerizing	Shock-Sensitive	Thermally Decomposing	Isomerizing	Disproportionating	Pyrophoric	Peroxide-Forming	Water-Reactive	
1					3		6	Incompatibility of fire water and water-based fire extinguishing agents with pyrophorics and water-reactive materials - Unless deluging is a solution to the problem where minor leaks are a problem (i.e., P_2S_5), water sprinklers usually not used with water-reactives - Water-based foams incompatible with water-reactive materials - Pyrophorics and water-reactive materials react with most extinguishing agents; use soil or sand for containment if practical; allow spilled material to burn out - Foam ineffective because of reactivity with water-reactive material - Pyrophorics incompatible with fire water - Use of sprinklers on pyrophoric/water reactive materials - Water is not used on anhydride due to compatibility problems; water is used on diketene in selected applications
1	1	1	1	1	1	1	1	Compatibility of foams and system have been tested with reactive chemicals
1	1	1	1	1	1	1	1	Research and testing used to identify incompatibilities
1	1	1			1	1	1	Case-by-case determination of where to store materials (i.e., drums of water-reactive material not stored in a sprinkled area)
	1	1			1	1	1	MSDSs are used as a guide to identify incompatibilities with extinguishing media
1		1			1		1	Fire protection, containment system tailored for compatibility with stored chemical based on MSDS, vendor, and reasonable chemical practice
1								Correct type of foam must be used
1								After an incident in which AFFF (aqueous film-forming foam) had been used, experienced foaming problems in equipment
				1				Deprive oxygen from release area (CO_2 or water)

Eleven responses provided no information pertinent to this portion of the survey.

SUMMARY 8

Compatibility—Other Materials

What Design and/or Operating Practices Are Used for Preventing Incompatibility Incidents Related to Any Other Materials?

Polymerizing	Shock-Sensitive	Thermally Decomposing	Isomerizing	Disproportionating	Pyrophoric	Peroxide-Forming	Water-Reactive	
1	1	1	1	1	1	1	1	Capability testing with all materials and adsorbents
1		1			1	1	1	Decontamination and cleaning fluids Procedures, blinds, double block valves, material safety data sheets, literature (manufacturers' and general), hazards reviews, experience
1								Teflon®, metal gaskets; exotic metals; double mechanical seals with flush fluids; inert pump, valve and agitator packing where needed; Kalrez® O-rings; ceramic and carbon seal faces
1								Avoid other chemicals that will initiate polymerization reactions
		1						Instrument fluids, oils, etc.: lab testing for catalysis
						1		Sodium nitrite added to liquid ring seal water to prevent peroxide formation at vacuum Nash equipment (liquid-ring pump or compressor)
							1	Gasketing, pipe linings, valve linings (experience applied to other devices with similar materials of construction)

Twenty responses provided no information pertinent to this portion of the survey.

SUMMARY 9

Compatibility—Other Practices

Any Other Practices Related to Compatibility Issues?

Polymerizing	Shock-Sensitive	Thermally Decomposing	Isomerizing	Disproportionating	Pyrophoric	Peroxide-Forming	Water-Reactive	
1	1	1			1	1	1	PHA team classifies all materials
1						1		Butadiene usually stored in mild steel; tank and lines pickled with sodium nitrate before using
					1		1	Personnel must be trained to handle materials
1								Construction materials/corrosion products/auto catalyst considered in HAZOP
1								Product kept out of direct sunlight
1								Set order of production so any cross-contamination is acceptable between products and product grades
1								Dedicated storage, remote and temperature controlled
1								Test inhibitor level and stability of all shipments
		1						ARC testing
				1				Only fresh product stored (all returns and off-specs are reprocessed to remove impurities), no cross-ties between storage and processing
							1	Storage tanks and unloading areas including containment dikes are covered and enclosed and kept away from rain water
							1	No permanent water connections allowed on anhydride or diketene storage vessels

Eighteen responses provided no information pertinent to this portion of the survey.

SUMMARY 10

Storage Time/Shelf Life—Limitations

What Limitations Are There on the Shelf Life of Reactive Materials Stored at Your Facility?

Polymerizing	Shock-Sensitive	Thermally Decomposing	Isomerizing	Disproportionating	Pyrophoric	Peroxide-Forming	Water-Reactive	
1	2	2	1	1	2	1	2	Database to manage shelf life of materials (Materials Management Department); computerized inventory system controls and monitors assigned chemical shelf-life
2	1	1	1	1	1	1	3	Storage time limits (set by quality control/R&D); quality of product issue
1	1	1	1	1	1	1	1	Storage and retention times monitored on raw materials and products (tests have been run to determine onset of potential problems due to extended storage)
2		1			2		1	Turnover sufficient to avoid problems; inventory and storage practices avoid shelf-life concerns
1	1	1			1		1	Compounds, storage drums, and packages examined on case-by-case basis and shelf-life limitations established as necessary
3						1		Company standards / guidelines set by monomer handling team / administrative restrictions imposed on most polymerizing chemicals
1		2	1					Suppliers'/manufacturers' recommendations followed
1		1	2					Inventory monitored/minimized/controlled
1		2						Products have high temperature alarms or have temperature checked and logged maintained; storage temperature specified (emphasize eliminating freezing as well as overheating, hazards introduced if monomer needs thawing)
1		1						Materials tested using adiabatic calorimetry and a time to maximum rate calculated
1						1		Inhibitor must be added in recycle product prior to storage
	1					1		Formation of shock-sensitive materials from peroxide-forming materials avoided
					1		1	None known for process material used
1								Styrene: check of inhibitor levels (inhibitor levels in other monomers are only checked by exception); manage inventories with respect to storage life
1								Maximum storage time specified for each monomer, approximately 1 year
1								Static inventory requires treatment every 4 weeks
1								Shelf-life studies conducted under extreme circumstances (elevated temperature simulates hot warehouse to determine shelf-life as needed)
1								Depletion of inhibitors or O_2 to activate inhibitors used
1								If no water or oxygen present, vinyl chloride monomer may be stored indefinitely
				1				None known for type of storage used (refrigerated storage or boiling or subcooled liquid)
						1		"Date opened" label applied to dispensing containers of peroxide formers that must be open to air; contents discarded as specified
						1		Storage of peroxide formers: 1 month opened, 1 year unopened

Five responses provided no information pertinent to this portion of the survey.

SUMMARY 11

Storage Time/Shelf Life—Increasing

What Design and/or Operating Practices Are Employed for Increasing Shelf Life?

Polymerizing	Shock-Sensitive	Thermally Decomposing	Isomerizing	Disproportionating	Pyrophoric	Peroxide-Forming	Water-Reactive	
								Inhibitors
7	1	1				6		Inhibitor(s) used; no details given
4	1	1	1	1	1	1	1	Inhibitor levels monitored to ensure proper concentration (concentration is key specification); inhibitors and stability analysis run periodically; inhibitor levels checked for styrene, inhibitor levels in other monomers checked by exception
3	1	1	1	1	1	1	1	Storage and retention times monitored/regularly scheduled checks performed
1								Inhibitor levels retested based on type of inhibitor and type of monomer
1								Periodic level checks and at times over-inhibition if it is known product will not be moved or will be held at abnormal conditions (high temperatures); care to prevent freezing that will separate inhibitors from the material
1								Inhibitors added by vendors. In no case is material stored longer than estimated viability of inhibitors. This includes acrylonitrile that is stored both with and without oxygen contact (to ensure activity of methoxyphenol inhibitor)
1								*t*-butyl catechol scavenges oxygen
1								Inhibitor added and certified at production area
1								Inhibitors added as early as possible in manufacturing process
1								Inhibitor levels determined by material specific stability testing
1								Suppliers'/manufacturers' recommendations followed
1								Static inventory requires treatment every 4 weeks
		1						Most products in this category have high temperature alarms or are checked and logged; freezing prevented
								Sampling systems
3	2	2	1	1	2	3	1	Sampling system(s) used; no details given
1	1	1	1	1	1	1	1	Designed to minimize personnel exposure
1	1	1	1	1	1	1	1	Designed to prevent contamination
1						1		Monitor O_2 purity on stored materials
1								System circulation to obtain a good representative sampling and also to generate even inhibitor mixing
1								Requires special lab set-up/equipment
1								Sampling with dedicated system
1								Tanks monitored weekly for inhibitor
1								Specific requirements for sampling outgoing materials for inhibitor and stability; also applies to incoming materials at receiving locations; also stability sampling required for material kept more than an assigned number of days
1								Daily HCN color check (using bulls eye)
1								Continuous analyses for inhibitor

SUMMARY 11

Storage Time/Shelf Life—Increasing

What Design and/or Operating Practices Are Employed for Increasing Shelf Life?

Polymerizing	Shock-Sensitive	Thermally Decomposing	Isomerizing	Disproportionating	Pyrophoric	Peroxide-Forming	Water-Reactive	
								Inventory records
6	5	8	4	4	7	8	7	Inventory records used; no details given
4		1				1		First-in/first-out policy
1						1		Records kept on receipt of incoming raw materials, tank levels
		1					1	Warehouse records kept
1								Accurate level measurements to ensure correct inhibitor addition for minimum concentrations
1								Severe restrictions on most polymerizing chemicals imposed administratively; for diketene in particular, inventories are minimized and production tailored to immediate usage needs
1								Styrene inventories are managed to prevent excessive storage life
					1			Lowest possible on-site inventory maintained
								Other practices
1	1	2	1	1	1			Temperature control/refrigerated storage
	1	1			1	1	1	Dry shed storage
2								Some inhibitors need oxygen; will maintain air or 8% O_2 on atmosphere; will avoid N_2 connections in the area
1			1					Rely on high turnover to keep material fresh; sized to maintain high turnover rate
			1				1	Rotate stock; inventory control
1								Styrene cooled; also small quantity of air injected to maintain the inhibitor active, otherwise styrene is N_2 blanketed
1								Storage tanks cooled with water spray/deluge
1								Bulk storage facilities must have provision for adding inhibitor in case material fails inhibitor concentration or stability tests
1								Dilution
1								Use of dished-bottom tanks to reduce possible holdup and in turn aged material
			1					Minimize inventory
			1					Work with vendor for low dimer specifications and quick shipments
			1					Quick-unloading railcars
			1					Cool cars before leaving vendor
				1				Continuous filtering to remove impurities
						1		Clean tanks on scheduled basis to minimize peroxide formation
							1	Container quality (material double-packed/sealed)

One response provided no information pertinent to this portion of the survey.

SUMMARY 12

Storage Time/Shelf Life—Outdated Material

Polymerizing	Shock-Sensitive	Thermally Decomposing	Isomerizing	Disproportionating	Pyrophoric	Peroxide-Forming	Water-Reactive	***How is Too-Old Material Handled and Disposed Of?***
2	2	2	1	1	2	2	2	On a case-by-case basis
3	1	1	1	2	1			Reclaimed (if possible) by recycling through production facility; disproportionating and off-specification polymerizing material is reprocessed
1	1	2	2	1	1	1	1	Reacted, recycled, burned, etc.
1	1	1	1	1	1	1	1	In accordance with good environmental practices, preferably thermal oxidation
3					1	1	2	Thermal oxidation
1		1			2		1	Inventory control; do not hold materials long enough to be "old"; all material used in process; materials are used or resold
1		2					1	Applicable hazardous waste regulations followed for material that cannot be reworked; approved hazardous waste disposal
						2	1	Sump and then waste treatment; dosing to waste water treatment plant
						1	2	Quench; neutralize by reacting slowly with water
2								Inhibit, then send back to manufacturing process for rework or incinerate
2								Analyzed or stability tested prior to use if beyond shelf life
					1		1	Returned to manufacturer in original container
1								Thermal oxidation or controlled polymerization, dewatering, and disposal in approved landfill
1								Material polymerized and disposed of as a polymer
		1						Depending on type of material, handling/disposal is based on reaction products, solubility, treatability

Six responses provided no information pertinent to this portion of the survey.

SUMMARY 13

Storage Quantity and Configuration—Maximum Inventories

What Determines Maximum Inventories for Reactive Materials Stored at Your Facilities?

Polymerizing	Shock-Sensitive	Thermally Decomposing	Isomerizing	Disproportionating	Pyrophoric	Peroxide-Forming	Water-Reactive	
5	4	4	2	2	5	5	6	Production requirements/usage rates
5	1	2	1		3	3	2	Delivery time limitations; shipment limitations and/or sizes
2	2	2		1	2	2	3	Storage capacity limitations
1	2	2	1	1	2	1	2	Minimum inventories maintained for smooth operating processes. F&EI and CEI calculations are used to justify inventories; "just-in-time" inventory control
4		1			1	1	1	Risk analysis and usages; consequence analysis
1		1			1		2	Inventory control
	1	1			1	1	1	Regulatory guidelines
1		1				1		Use rate and shipment size for incoming materials
1					1			Shelf life
					1		1	Economic factors
1								Program for highly hazardous materials management requires inventory control to the minimum consistent with stable operations
1								Maximum inventory set during design by tank size, based on need versus risk evaluation; tanks containing toxic or reactive materials are kept at lowest inventories by procedure and interlock
1								Diketene is stored only for process control reasons, usually less than 3 days and is not shipped over the road
		1						Fire regulations limit inventories both in units and in plant
		1						Evaluated by individuals and committees during design, process considerations, consequences, and increased frequency with larger number of parallel processing units
			1					One-month maximum

Four responses provided no information pertinent to this portion of the survey.

SUMMARY 14

Storage Quantity and Configuration— Storage Configuration

In What Configuration Are Reactive Materials Stored (Tanks, Cylinders, Cartons, etc.)?

Polymerizing	Shock-Sensitive	Thermally Decomposing	Isomerizing	Disproportionating	Pyrophoric	Peroxide-Forming	Water-Reactive	
12	2	8	1		6	3	6	Tanks
6	1	2			2	2	4	Drums; fiber drums
1	1	2	1	1	6	2	3	Cylinders; pressurized cylinders
1	2	1			1	2	1	Cartons
1	1	1	1	1	1	1	1	Sound engineering practices coupled with safety reviews results in a wide variety of storage container configurations
1	1	2			1	1	1	Paper sacks; bags
4		1						Railcars
1		1			1	1	1	Separated by class and stored in tanks, drums, cylinders or cartons
		2			1		1	DOT-approved portable tanks or containers
3								Tank trucks
1			1			1		Sphere, limited by earthquake structural integrity
	1					1	1	Cans; gallon container cans
							2	Tote tanks; pressurized tote bins for oxygen- and moisture-reactive materials
1								Horizontal cylindrical storage tanks
1								Pressurized and atmospheric storage tanks
1								Chemical barges
1								Special tanks—no rough surfaces, all welds and nozzle welds ground smooth; no tank internal bracing (styrene will vaporize, condense, and readily polymerize on all vapor-space surfaces); inorganic zinc used to coat tank and nozzle internals
			1					Per NFPA Codes
				1				Refrigerated, stationary tank; occasionally kept on-site in refrigerated railcars (cryogenic) for a short period of time
					1			Sealed pails or drums
							1	Super sacks
							1	Double plastic bags packed in sealed steel drums
							1	Nitrogen-inerted tanks
							1	Plastic sealed buckets

Two responses provided no information pertinent to this portion of the survey.

SUMMARY 15

Storage Quantity and Configuration—Tank Outlet Location

If Tank Storage is Used, Are Outlet Nozzle(s) on the Top or the Bottom of the Tank(s)?

Polymerizing	Shock-Sensitive	Thermally Decomposing	Isomerizing	Disproportionating	Pyrophoric	Peroxide-Forming	Water-Reactive	
2	1	2			2		3	**Top of the tank**
9		5		1	3	2	6	**Bottom of the tank**
1	1	1	1	1	1	1		Fill from top and empty from bottom
1							1	Storage is in typical API 650 storage vessel; nozzles are on the bottom of tanks, with special fireproof gaskets, fusible link hand valves on pump suction lines, and restricted access
1								Outlet on bottom nozzles of tanks; have remote-operated, fire-safe valves
1	1	1	1	1	1	1	1	Nozzle location is based on good engineering practices coupled with safety reviews that result in a wide variety of storage container configurations
2		1			2			Both top and bottom of the tank
1		1			1	1	1	Only toxics stored in top nozzle only
1								Most on bottom; one tank containing toxic/reactive material has outlet on top

Six responses provided no information pertinent to this portion of the survey.

SUMMARY 16

Storage Quantity and Configuration—Siting

What Are Your Siting Practices for Reactive Materials Storage, such as Tank Spacing or Distance to Plant Boundaries?

Polymerizing	Shock-Sensitive	Thermally Decomposing	Isomerizing	Disproportionating	Pyrophoric	Peroxide-Forming	Water-Reactive	
4	2	3	1	1	3	2	3	NFPA guidelines
4	3	4	1	1	3	3	3	Partially or wholly based on insurance recommendations/guidelines/ requirements; insurance siting/spacing practices used as minimum standard
1	1	1	1	1	1	1	1	Good engineering practices coupled with safety reviews
1	1	1	1	1	1	1	1	Standard building codes
1	1	1	1	1	1	1	1	Local ordinances
1	1	1	1	1	1	1	1	Special storage cubicles; e.g., drums of water-reactive materials are not stored with flammables and are stored in cubicles that do not have fire sprinklers
2		1			1	1	1	Risk assessment is used, relying on consequence analysis for distance
1	1	1			1	1	1	Process hazard analysis recommendations; siting and inventory determined during review process
1		1			1	1	1	Company standard based on quantity
1		1		1	1		1	Company standard separation for storage of flammable liquids
1		1		1	1		1	Storage located away from handling area
2		1					1	Storage tank spacing based upon external fire considerations/ determined by fire protection recommendations
1		1			1		1	Quantitative risk analysis (per CCPS) to ensure meeting individual and societal risk standards
1		1			1		1	Partly, no special considerations; presently revamping based on potential consequences
1			1			1		As remote as possible
1					1			Siting decisions made per applicable code and known hazards during storage facility design
1								Company-based guidelines for diking standards and/or release consequence analysis
1								Small tank/dike/unloading operations fairly close to process building
1								Consequence analyses used to assist in siting new tanks to minimize on-site and off-site impacts, with separate dikes for each new monomer tank
		1						Storage is separated for large quantities and reactors from general processing with free space; however, storage tanks, reactors, and parallel trains are not necessarily separated from each other
		1						Blast walls are used for lowest 20-30 ft between tanks, reactors
						1		Quantity-distance
							1	Separate storage for reactive material
							1	Computer dispersion model used on new installations to help determine optimum site
							1	Anhydride tanks are diked separately from all but organic acid tanks

Seven responses provided no information pertinent to this portion of the survey.

SUMMARY 17

Storage Quantity and Configuration—Container Shape/Configuration

Any Restrictions on Container Shape or Configuration, such as Free Space, Length-to-Diameter Ratio, Other?

Polymerizing	Shock-Sensitive	Thermally Decomposing	Isomerizing	Disproportionating	Pyrophoric	Peroxide-Forming	Water-Reactive	
								Free space
2	2	3	1	1	3	1	3	No restrictions
2	1	2			2	1	2	Sufficient outage maintained to prevent overflow/overpressure from thermal expansion
1	1	1	1	1	1	1	1	As recommended by suppliers
1			1		1	1	1	15 to 20 percent free space maintained; use high-level alarms and/or trips
3								Minimum headspace required to allow sufficient oxygen for inhibitor to function; free space in drums must be adequate for oxygen to last a year
1					1		1	Minimum free space is 5 percent
1					1			Maximum tank inventories
1								Controlled to a minimum of 16 percent by volume; all tanks have liquid-level indicators
1								Full capacity not used (not greater than 85 percent), leave 6 percent outage on tank cars and trucks
					1			Contents used completely so no partially open containers are left
					1			Phosphorus stored under water
							1	Determined by combination of storage tank sizes (2 tanks half-filled so inventory can be shifted to either tank if necessary)
								Length-to-diameter ratio
4	2	4	1	1	5	2	5	No restrictions
1	1	1	1	1	1	1	1	As recommended by suppliers
1								Tanks set 2-to-1 height-to-diameter ratio minimum
								Other
1								Company practice is moving completely away from bottom unloading to top unloading on tank trucks and railroad tank cars with closed circuit (vapor return) systems
1								Storage vessel complies with ASME code, Section 8

Twelve responses provided no information pertinent to this portion of the survey.

SUMMARY 18

Storage Quantity and Configuration—Mixing

Are the Storage Container Contents Circulated or Mixed? If So, How and Why?

Polymerizing	Shock-Sensitive	Thermally Decomposing	Isomerizing	Disproportionating	Pyrophoric	Peroxide-Forming	Water-Reactive	
4	2	6	1	1	6	5	8	Contents are <u>not</u> circulated or mixed
13	1	3	1	1	2	1	1	Circulation/recirculation - Pumping; pump recycle - Tank circulation via pumps generally maintained for temperature control - Monomeric acid is stored in circulated and temperature-controlled tanks - Circulated by pump through heat exchanger to warm in winter, cool in summer; pressure must be positive to prevent outside air from entering storage tank
3								Mixing eductor - New tanks, in some cases, use mixing eductors, which allow rapid mixing of short-stop agents; eductors also make temperature measurements more accurate, allow mixing of normal inhibitor, and allow representative samples to be taken to test for stability and inhibitor - For tanks with short-stopping, mixed by eductor to mix in short-stopping agent - Recirculation through mixing eductor to even out temperature fluctuations, prevent freezing, and ensure inhibitors are mixed
1		1						Circulated by internal stirrer or external pump and recirculation piping to maintain uniform temperature and/or composition
1					1			Depending on application: mix, dilute, agitate, circulate, and/or refrigerate; use inert gas padding
1								Static storage recirculated every 4 weeks to redistribute inhibitor
1								HCN and styrene cooled by circulating through an external cooler
1								Usually running in and out of same tank; provides some mixing
				1				Vapor recovery system in refrigerated tank; liquid product continuously filtered to remove products of disproportionation

Eight responses provided no information pertinent to this portion of the survey.

SUMMARY 19

Air Exclusion

What Design/Operating Practices Are Used to Exclude Air from the Reactive Materials Containment (Container Filling Procedures, Inert Gas Padding, Other)?

Polymerizing	Shock-Sensitive	Thermally Decomposing	Isomerizing	Disproportionating	Pyrophoric	Peroxide-Forming	Water-Reactive	
								Container filling procedures
3	1	2	1	1	2	1	4	Container filling procedures used; no details given
1	1	1	1	1	1	1	1	Based on nature of material, practices include dry disconnects, dedicated lines, etc.
1				1	2		1	Closed, pressurized system, purging with inert gas; storage in pressurized bulk tanks only under inert pad
1		1						Pressure vessel with vapor pressure control system letting vapor pressure down into lower pressure system
1		1						Vapor balancing with supply vessel
1								Nitrogen purging and/or evacuation of some reactors and tanks
1								Vapor or displacement gas sent to water scrubber
		1						Company procedures
		1						Purging and padding with inert gas
			1					Closed-loop pressure unloading
							1	Drums filled with nitrogen before filling
								Inert gas padding
6	3	6	1	1	5	3	6	Inert gas padding used; no details given
1	1	1	1	1	1	1	1	Based on nature of material, practices include dry disconnects, dedicated lines, etc.
1		1			1	1	1	Inert gas supply regulator or pressure controller
				1	1		1	Purging and padding with inert gas
1					1			Stored in pressurized bulk tanks only under inert pad
1							1	Dry (-40°C dew point) nitrogen from a dedicated air plant is used to inert all process and storage vessels where flash point is below 100°F or material is normally stored above its flash point. Nitrogen is used for all reactive materials regardless of flash point to keep out moisture
1								Inert gas padding is typically used at Asian and European company sites, but not used at U.S. sites
1								Inert gas padding is used where needed and where not contraindicated
1								Where flammability is a concern, special "mixed gas" (~8% oxygen) systems are provided
		1						Use dedicated nitrogen supply; alarms on loss of pressure
		1						Standard inerting system
			1					Tank pressure must be kept above atmospheric pressure; nitrogen system as backup to heating system; avoid nitrogen because of inerts pressure buildup
					1			Low pressure inert gas sweep when transferring material
						1		Inert gas padding with conservation vent vented to atmosphere
						1		Nitrogen padding is used

SUMMARY 19

Air Exclusion

What Design/Operating Practices Are Used to Exclude Air from the Reactive Materials Containment (Container Filling Procedures, Inert Gas Padding, Other)?

Polymerizing	Shock-Sensitive	Thermally Decomposing	Isomerizing	Disproportionating	Pyrophoric	Peroxide-Forming	Water-Reactive	
								Other
1		1			1		1	Start-up, shutdown, clean-out, purging, and maintenance procedures
1								Vessels usually reactors are evacuated with vacuum system
1								Closed looped unloading to reduce emissions
1								Use of nitrogen to flowlines, etc., is not allowed; lines must be designed to drain
				1				Venting to closed headers and return to processing
							1	Nitrogen make-up in combination with light vacuum on all process vessels, storage tanks
								Air not excluded
1								Air must not be excluded from acrylic monomer containers; oxygen is necessary to activate inhibitors. To make sure a certain amount of air is present in each drum, we use fill-level requirements
1								Do not attempt to exclude air for monomer
1								Inert gas cannot be used with monomers
1								Air is usually needed for inhibition
1								Air make-up/vent systems
1								Inert gas padding not used for storage but okay for reactors (can not use oxygen-free gas, oxygen needed to activate inhibitor; "poor air" $<8\%$ oxygen is acceptable)
							1	Dry air used; desiccant dryer installed on tank vent

Five responses provided no information pertinent to this portion of the survey.

SUMMARY 20

Moisture Exclusion

What Design and/or Operating Practices Are Used to Keep Moisture Out Of the Reactive Materials Containment?

Polymerizing	Shock-Sensitive	Thermally Decomposing	Isomerizing	Disproportionating	Pyrophoric	Peroxide-Forming	Water-Reactive	
								Start-up procedures
5	3	4	2	2	4	3	6	Start-up procedures used; no details given
1	1	1	1	1	1	1	1	Purging with dry inert gas
					1		2	Tanks/vessels dried with hot nitrogen before use; storage vessels dried with hot inert gas
					1		1	Moisture determination tests for vessels receiving water-reactive materials
					1		1	Purge gas checked for moisture content ($\leq$150 ppm)
1								Before commissioning, written procedures developed for cleaning tanks, lines, etc.
1								Nitrogen purging, evacuation
1								Purging and padding with inert gas
							1	Piping and tankage is flushed with solvents that start with water-compatible materials and finish with hydrophilic substances to remove residual moisture
							1	Dry nitrogen used to purge equipment; sometimes flush with suitable solvent
								Air drying systems
2	1	2	1	1	2	1	3	Air drying systems used; no details given
1	1	1			1	1	1	Air is not used in system
1							1	Dry air used; desiccant dryer installed on tank vent; dryers generally provided on tank normal vent lines
	1						1	Dry (-40°C dew point) nitrogen from a dedicated air plant is used to inert all process and storage vessels where flash point is below 100°F or material is normally stored above its flash point. Nitrogen is used for all reactive materials regardless of flash point to keep out moisture
1								Air drying not considered the best way for polymerizing material—purging with dry nitrogen or "poor air" (nitrogen with a small percentage of added oxygen) is better
							1	Nitrogen used
								Package waterproofing
2	2	3	2	2	3	3	4	Package waterproofing used; no details given
							1	Sealed storage bins
							1	Sodium bricks stored in waterproof, sealed containers
								Building waterproofing
1	1	2	1	1	2	1	3	Building waterproofing used; no details given
							1	Sumps used to contain accidental spills are protected from ambient rain, flooding, etc. and periodically checked for "dryness"
							1	Sodium bricks stored in shelter

SUMMARY 20

Moisture Exclusion

What Design and/or Operating Practices Are Used to Keep Moisture Out Of the Reactive Materials Containment?

Polymerizing	Shock-Sensitive	Thermally Decomposing	Isomerizing	Disproportionating	Pyrophoric	Peroxide-Forming	Water-Reactive	
								Building drainage
2	2	2	2	2	2	2	4	Building drainage used; no details given
					1		1	Floors sloped to outside; materials stored on pallets
								Flood control
2	2	2	2	2	2	2	2	Flood control used; no details given
	1	1			1	1	1	Site storm sewer system
					1			Drums stacked above floor level
								Other
1	1	1	1	1	1	1	1	Refrigerated warehouses to lower humidity
1					1			Stored in pressurized bulk tanks only, under inert pad
		1					1	Moisture control of other components that are mixed with this material
					1		1	Shutdown, clean-out, and maintenance procedures
1								Color check (rust from steel oxidation)
						1		Equipment sealed from atmosphere, vinyl chloride vapor pressure greater thank atmospheric at storage temperature
							1	Full package contents used at a time
							1	Storage in sealed packages

Seven responses provided no information pertinent to this portion of the survey.

SUMMARY 21

Oxygen Monitoring

How is the Storage Condition Parameter of Oxygen Concentration Monitored?

Polymerizing	Shock-Sensitive	Thermally Decomposing	Isomerizing	Disproportionating	Pyrophoric	Peroxide-Forming	Water-Reactive	
5								Oxygen not monitored (polymerizing) - If blanketed, then not monitored - Atmospheric exchange caused by tank breathing seems to tak care of most problems - Normally stored with air atmosphere; not usually monitored except in some cases where 8% oxygen is used - Nitrogen padding, monitoring not needed - No vinyl chloride with oxygen content greater than 1,000 ppm by volume accepted
					3			Oxygen not monitored (pyrophoric) - None in storage, low pressure inert gas sweep when handling the material - Pyrophoric materials are produced, stored, and consumed on site, design and operation precludes oxygen in system - Nitrogen padding; monitoring not needed
						2		Oxygen not monitored (peroxide forming) - Designed to exclude air/moisture but no monitoring - No vinyl chloride with oxygen content greater than 1,000 ppm by volume accepted
2						1		Scheduled sampling of inhibitor; sampled at minimum twice daily
1		1			1			Low pressure alarm to warn of loss of inert gas padding
1					1		1	Analyzers or MSA meter
1							1	Oxygen and humidity are controlled using nitrogen purges, with oxygen content monitored periodically with hand-held gauge
1								Oxygen is maintained at a low level by use of a nitrogen pad; the nitrogen system is typically monitored by local pressure gauges, low pressure alarms, and/or analysis

Fifteen responses provided no information pertinent to this portion of the survey.

SUMMARY 22

Humidity/Moisture Monitoring

How is the Storage Condition Parameter of Humidity/Moisture Content Monitored?

Polymerizing	Shock-Sensitive	Thermally Decomposing	Isomerizing	Disproportionating	Pyrophoric	Peroxide-Forming	Water-Reactive	
							4	Humidity/moisture not monitored (water-reactive) - Design and operation preclude moisture ingress - Oxygen and humidity are controlled using nitrogen purges - All drums are sealed to prevent moisture intrusion
2								Humidity/moisture not monitored (polymerizing) - Nitrogen padding; monitoring not needed - Oxygen and humidity are controlled using nitrogen purges
					2			Humidity/moisture not monitored (pyrophoric) - None in storage; low pressure inert gas sweep when handling the material - Nitrogen padding; monitoring not needed
1								Periodic testing of specifications (e.g., monthly)

Twenty-one responses provided no information pertinent to this portion of the survey.

SUMMARY 23

Pressure Monitoring

How Is the Storage Condition Parameter of Pressure Monitored?

Polymerizing	Shock-Sensitive	Thermally Decomposing	Isomerizing	Disproportionating	Pyrophoric	Peroxide-Forming	Water-Reactive	
1	1	1	1	1	1	1	1	Pressure is monitored; no details given
1	1	1	1	1	1	1	1	Controlled, measured, recorded, and alarmed when a critical storage parameter
1			1			1	1	Multiple instruments/alarms and interlocks; remote indicators with alarms and interlocks
1					1			Local and control room monitors
1						1		Readings taken several times every shift
1							1	Storage vessels have pressure measurements and alarms in the process DCS
		1				1		Pressure indicating controller (PIC) or pressure control valve (PCV)
1								Pressure gage (Bourdon tube), differential pressure transmitter (with and without chemical seal), vacuum/pressure (weight and area) valves, seal loop with known fluid density (non-freezing, unreactive, and non-evaporating)
1								Pressure indicating controller (PIC)
1								Usually controlled with redundant venting systems; may or may not be indicated
		1						Pressure indicators and low pressure alarm (for loss of nitrogen)
		1						Pressure transmitter transmits pressure to control room; out of range alarm sounds in control room
		1						Standard instrumentation is used
					1			Pressure indicating controller (PIC), low pressure alarm (PAL)
						1		Pressure monitored on storage tanks during filling

Thirteen responses provided no information pertinent to this portion of the survey.

SUMMARY 24

Temperature Monitoring

How Is the Storage Condition Parameter of Temperature Monitored?

Polymerizing	Shock-Sensitive	Thermally Decomposing	Isomerizing	Disproportionating	Pyrophoric	Peroxide-Forming	Water-Reactive	
2	1	1	1	1	1	1		Temperature is monitored; no details given
1	1	1	1	1	1	1	1	Controlled, measured, recorded, and alarmed when a critical storage parameter
1	1	1	1	1	1	1	1	Multiple instruments/alarms and interlocks
1	1	1	1	1	1	1	1	Temperature alarms on storage buildings
1		2				1		Temperature indicator (TI), high temperature alarm (TAH)
1		1			1		1	Temperature monitoring or alarms
		1			1		1	Temperature indicator (TI)
1					1			Local and control room monitors
1						1		Temperature indicating controller (TIC), high temperature alarm (TAH)
1						1		Readings taken several times every shift
1							1	Storage vessels have temperature measurement and alarms in the process DCS
					1		1	Thermocouple
1								Storage tank temperature recorder/alarm in process building
1								Multiple temperature indicators in tank; high temperature alarm
1								Electrical resistance thermocouples; liquid filled capillaries
1								Where monitored, ideally use an indicator installed in a thermowell
1								Redundant temperature sensors in control room and locally where needed; many tanks storing monomers are not monitored
1								Monitored, need local indication as a minimum; some large tanks have multiple remote indication with alarms
		1						Chilled building(s); redundant control room alarms for temperature excursions
		1						Temperature transmitter sends signal to control room; out-of-range alarm sounds in control room
		1						Standard instrumentation is used
							1	Remote indicators with alarms/interlocks

Six responses provided no information pertinent to this portion of the survey.

SUMMARY 25

Temperature Control

What Kind of Systems Are Used for Controlling the Temperature of the Stored Material?

Polymerizing	Shock-Sensitive	Thermally Decomposing	Isomerizing	Disproportionating	Pyrophoric	Peroxide-Forming	Water-Reactive	
								Tank/cylinder cooling
6	1	2	2	1	1	2	2	Heat transfer system used - Chilled/cooling tower water for non-water sensitive materials - Brine system - Refrigerated storage required by some products - Tempered water system to cool during summer - Diketene units have dedicated refrigeration units backed up by central auxiliary refrigeration - Tanks cooled by refrigerated water using circulation system - Mechanical refrigeration units; circulation to external cooler - Non-reactive intermediate fluid
2	1	3	1	2	2	1	1	Tank/cylinder cooling used; no details given
2						1		Emergency fire monitors if needed; water spray/deluge to cool storage tank in hot summer conditions
1								Tanks painted white or aluminum
								Tank/cylinder heating
6	1	2	2	1	1	2	1	Heat transfer system used - Steam or heated oil - Brine system - Tempered water system to prevent freezing - Tempered water with no direct contact with steam source - Tanks heated by tempered water using circulation system - Hot water with high temperature alarms and/or interlocks, limited heat transfer area - Non-reactive intermediate fluid
1	1	1	1	1	2	1	1	Tank/cylinder heating used; no details given
1								In the case of styrene storage, tops of tankage are painted black to ensure high surface temperature to reduce condensation and hence polymerization sites
2								Heating only to prevent material freezing; special procedures for thawing frozen monomer in drums, trucks, railcars, pipes to ensure re-mixing of inhibitors and minimum exposure to heat
		1						Backup temperature control is used; no details given
								Building cooling or refrigeration
1	1	1	1	1	1	1	1	Standard heating/ventilation/air conditioning systems designed to comply with specific electrical classification
		4						Peroxides stored in refrigerated areas/refrigerated building/cold room; peroxide storage building has mechanical refrigeration unit, high temperature alarm
						4		Peroxide-formers are usually refrigerated; cold storage for some products; stored in small quantities in refrigerated buildings
3								Cold rooms for storing uninhibited monomers; recirculated cooling or building air conditioning used when required
					2			Pyrophorics stored in small quantities in refrigerated buildings
	1						1	Refrigeration used on occasion

SUMMARY 25

Temperature Control

What Kind of Systems Are Used for Controlling the Temperature of the Stored Material?

Polymerizing	Shock-Sensitive	Thermally Decomposing	Isomerizing	Disproportionating	Pyrophoric	Peroxide-Forming	Water-Reactive	
								Building heating
1	1	1	1	1	1	1	1	Standard heating/ventilation/air conditioning systems designed to comply with specific electrical classification
1	1	1			1	1	1	Building heating used; no details given
1								For drum storage of monomers that can freeze, have high temperature shut-off (fail-safe) device
1								Building heating system used; must comply with fire code requirements for flammable/combustible materials
								Building ventilation
1	1	2			2	1	2	Building ventilation used; no details given
1	1	1	1	1	1	1	1	Standard heating/ventilation/air conditioning systems designed to comply with specific electrical classification
1			1			1		Process buildings: 15 air turnovers normal, 30 air turnovers at 50% LFL
1								Building ventilation system used; must comply with fire code requirements for flammable/combustible materials
		1						Facilities are in open structures
					1			Building ventilation used only during handling and transfer
								Other temperature control
1					1		1	Open shed or a roof over some materials
1								Self-limiting electric tracing for high-freezing-point monomers
		1						Drive-in (forklift) refrigerators
		1						Backup temperature control is used
				1				Liquid is continuously mixed, filtered, and cooled

Eight responses provided no information pertinent to this portion of the survey.

SUMMARY 26

Handling and Transfer—Manual Handling

What Practices Are Employed for the Manual Handling of Reactive Materials On-Site?

Polymerizing	Shock-Sensitive	Thermally Decomposing	Isomerizing	Disproportionating	Pyrophoric	Peroxide-Forming	Water-Reactive	
								Pallets with hand trucks
3	4	7	1	1	4	4	4	Pallets with hand trucks used; no details given
1								Not by hand trucks but possibly by worksaver type (self-propelled with operator walking)
						1		Material in cartons
								Pallets with forklift trucks
4	3	6	1	1	5	4	5	Pallets with forklift trucks used; no details given
1		1			1	1	1	From warehouse to reactor building depending on site; inside building, may be forklifts or may be hand trucks
2		1			1		1	Drummed materials/almost no containers under 55-gal size used
1		1					1	In rare cases to handle drums
1								Only fully closed and sealed drums are moved
		1						Handled by fork truck to deliver to cold room and transfer to process area when ready to use
						1		Material in cartons
							1	Tote bins handled with fork truck to specially designated storage area
								Roller conveyors
2	1	1			1	1	2	Roller conveyors used; no details given
2		1			1		1	Drummed materials/in drumming facility
1								Only to move drum container from under filling position and 4-drum pallets away from auto drum filling system
						1		Material in cartons
								Other
2	1						2	Reactive materials are <u>not</u> handled manually on-site - Pump/piping transfer; made on-site to prevent handling - Perbenzoates and peroctoates are pressure-transferred; might dry out and become shock/temperature-sensitive
1	1	1	1	1	1	1	1	No special handling practices, specific storage containers and specific storage areas are designated
					1		1	Forklifts for cylinders
					1		1	Container directly on the forks of a fork lift truck
					1		1	Skid tanks/cylinders designed for forklift truck handling
1								Company policy regarding stacking and grouping of drums
		1						Delivered to the facility in chilled trucks; unloading door-to-door or drive-in
				1				Only sampling bombs to the lab, hand carried
					1			Use of DOT-required design and procedures
					1			Large portable tanks with lifting lugs for crane lifting
							1	Super sacks

Ten responses provided no information pertinent to this portion of the survey.

SUMMARY 27

Handling and Transfer—Piping

What Design Specifications Are Used for On-Site Transfer (Piping) Systems?

Polymerizing	Shock-Sensitive	Thermally Decomposing	Isomerizing	Disproportionating	Pyrophoric	Peroxide-Forming	Water-Reactive	
								Design specifications, general
1	1	1	1	1	1	1	1	Industry standards and codes used
1					1			Plant pipe/equipment specification used per material; applicable codes incorporated (ASME, API, ANSI, NFPA, OSHA)
								Fittings and connections, including flexible unloading connections
4		2	1		2	2	2	Flange connections: - Steel braided hoses with flanged connections, blinded when not in use - Flanged connectors used for highly flammable or highly toxic materials - Use as few flange connections as possible - 300# flanges on reactors/process equipment - Because of potential leaks/toxicity problems, pipe is mostly welded with as few flanges as possible
1	1	2	1	1	1	1	1	Dedicated unloading hoses for sensitive materials; dedicated and unique connections and hoses are used to load tankage because some of the thermally sensitive peroxides are also sensitive to acids and unsaturated organics
1		1			1	1	1	ANSI B31.3 piping code usually 316 stainless steel, schedule 40; stainless banded hoses with stainless bellows; toxic materials are hard-piped
1		1			1		1	Quick connectors used for less hazardous materials
1							2	Flexible connections - Flexible connections and sight glasses kept to a minimum - Flexible connections are not permitted on diketene, are allowed on anhydride
1							1	Material of construction specified
		1		1				Stainless steel piping and tubing, Teflon gaskets
					1		1	Carbon steel or stainless steel with Teflon gaskets
1								Carbon steel, schedule 40, welded
1								Dripless coupling type
		1						Hard piping with joints or flexible hoses used with breakaway seal closed, couplings
							1	Water-reactive corrosives may require Teflon-lined pipes
								Piping specification; layout details (e.g., pertaining to dead legs)
4		2			2	1	3	Dead legs avoided in all piping systems/eliminated from design; usually self-draining lines; short legs positioned to be self-draining if possible
3		1			1	1	1	Slope to drain from double block and bleed (no dead legs); sloped headers (no dead legs); slope piping to drain and eliminate need to blow lines (avoid use of nitrogen)
1		1			1	1	1	Lines insulated
1			1			1		Piping layout high enough to prevent any rupture by truck/rail
1								Carbon steel, schedule 40, welded
1								Company standard okay for pipe; because of corrosion issues use stainless steel flanges and bolts
1								Welded pipe connections preferred for dissipating static electricity
1								Sample valves are ram-type to prevent collection of material at sample point

SUMMARY 27

Handling and Transfer—Piping

What <u>Design Specifications</u> Are Used for On-Site Transfer (Piping) Systems?

Polymerizing	Shock-Sensitive	Thermally Decomposing	Isomerizing	Disproportionating	Pyrophoric	Peroxide-Forming	Water-Reactive	
		1						Dedicated and unique connections and hoses are used to load tankage because some of the thermally sensitive peroxides are also sensitive to acids and unsaturated organics
		1						Piping permanently installed; laid out per design and review
				1				Stainless steel piping
								Pumps and pump seals
4	1	4	1	2	3	2	3	Centrifugal pumps - tandem seals/double mechanical seals with barrier fluid - mechanical seals with flush
3		1						Magnetic drive pumps - where leaks are not tolerable - most-preferred option
2	1	1	1	1	1	1	1	Canned pumps - double mechanical seals or canned pumps on most
1			1			1		Simple pusher seals on centrifugals; recommend piston or something different than centrifugal
		1			1		1	Double diaphragm pumps with inert fluid
1								Canned pumps not used
1								Have used sealless pumps where odors are a concern
1					1		1	Special seals; no details given
1							1	Packing is usually not allowed
1								Carbon steel or stainless steel with Teflon packings
1	1	1	1	1	1	1	1	Compatible seal material; no details given
1								For emulsions light oil #10, or ethylene glycol OK for seal fluid
							1	Methanol used on pump seals
1		3			1			Instrumentation for deadhead protection - Temperature and flow indicators on pumps; timers to shut off pump if no flow detected in 30 seconds; high temperature shutoff - Pumps protected against deadheading by temperature, pressure, or flow interlock, depending on situation - Flowmeters (rarely flow switches) are used to monitor flow from pump and interlocked to shutdown pump on low flow
1		1					1	Minimum flow recirculation lines to prevent deadheading
								Valves (control; isolation; other)
1	1	1	1	1	1	1	1	Plugged or blinded terminal valves with compatible materials of construction
1		1			1	1	1	Dead spaces in valves are avoided
1				1	1		1	Carbon steel or stainless steel with Teflon packings
1		1				1		Double block and bleed
1							1	Redundant high-level shutoffs at destination tanks are used to prevent overfilling
					1		1	Automatic shutoff, fire valves
1								Isolation - plug/ball with air to fail close actuator
		1						Valves interlocked with pumps (pumps cannot run if certain valves are closed)
							1	Water-reactive corrosives may require Teflon-lined valves
							1	Diaphragm valves used when possible

SUMMARY 27

Handling and Transfer—Piping

What Design Specifications Are Used for On-Site Transfer (Piping) Systems?

Polymerizing	Shock-Sensitive	Thermally Decomposing	Isomerizing	Disproportionating	Pyrophoric	Peroxide-Forming	Water-Reactive	
								Drain systems
					1		1	Draining not practiced for pyrophoric and water-reactive materials; small mist carryover directed to closed systems (seal drum) with inert liquid seal and a flame arrester on the vent
					1		1	Away from tank to ditch and pond
1								Divinyl benzene tank and lines are refrigerated and water-traced
1								Biological system required prior to dumping drain water into the ecosystem
				1				Typical rainwater drains
						1		To sump
								Cleaning equipment
1								Long lines are designed to allow pigging to clear line
							1	Specially designed cleaning equipment; no details given
								Other
1								Because of toxicity, pump valves/flanges/valve stems should have splash collars to avoid spraying personnel
1								Suppliers' information regarding systems that require the exclusion of copper, asbestos, and other materials
1								Use approved ground strap clamps at flanged joints
		1						Special heat tracing system with shutdown on shorted thermocouples and on separate high temperature switch

Two responses provided no information pertinent to this portion of the survey.

SUMMARY 28

Handling and Transfer—Piping

What __Operating Practices__ Are Used for On-Site Transfer (Piping) Systems?

Polymerizing	Shock-Sensitive	Thermally Decomposing	Isomerizing	Disproportionating	Pyrophoric	Peroxide-Forming	Water-Reactive	
								Operating practices, general
4	2	2	1	1	2	2	2	Procedures; written operating procedures - Piping, valves, pump type and seals are specified by individual procedures as water-reactive corrosives may require Teflon-lined pipes and Teflon valves - Follow manufacturers' recommended practices; where appropriate, request supplier audit of storage facilities
	1	1			1	1	1	Procedures reviewed via HAZOP
	1	1			1	1	1	Compatibility charts and internal standard to ensure safe handling and transfer of material
								Fittings and connections, including flexible unloading connections
1								Procedures to ensure proper materials of construction, especially for gaskets; hose inspection/replacement programs
								Piping specification; layout details (e.g., pertaining to dead legs)
1		1				1		Blow lines after transfer
								Pumps and pump seals
1								Weekly visual inspections; quarterly monitoring of seal leakage
								Drain systems
1		1			1	1	1	Lines usually are flushed, drained, or blown out after transfer
1								Systems drained if going to stand more than 2 weeks (no oxygen to activate inhibitor except dissolved oxygen)
								Cleaning equipment
1							1	Line are flushed with acid and/or acid-water mix to remove reactive materials before cleanout
1				1				Water used; regular hoses for water
1								Decontamination critical; 5% solution sodium bisulfite, 1:1 ratio
								Maintenance procedures related to the above transfer systems
1	1	1			1	1	1	Operations responsible for cleaning by special procedures before maintenance works on system
1	1	1			1	1	1	Manufacturers' recommendations followed
1		1			1	1	1	Thorough or long flush, purge, and test
1			1			1		Mild steel must be pickled before use
		1			1		1	Specific maintenance and decommissioning procedures: washing, purging, testing for absence of material
1				1				Water used; regular hoses for water
							1	Specific washing procedures, gasket tightening, testing procedures (nitrogen, water, then drying)
								Other
1			1			1		Gases are flared by remote actuation
1								All connectors, piping, and tankage labeled

Two responses provided no information pertinent to this portion of the survey.

SUMMARY 29

Last-Resort Safety Systems

Give Details of Any Last-Resort Safety Systems Related to Your Reactive Materials Storage/Handling

Polymerizing	Shock-Sensitive	Thermally Decomposing	Isomerizing	Disproportionating	Pyrophoric	Peroxide-Forming	Water-Reactive	
								Inhibitor injection
1	1	1	1	1	1	1	1	Inhibitor injection used; no details given
3								Compressed gas motive force - Dedicated short-stop tanks that automatically activate on high temperature or can be manually activated - Motive gases from bottle (manually initiated system) - Pressurized tank of concentrated inhibitor solution releases into a mixing eductor in the storage tank
2								Gravity-charged system - Interlocked to temperature - Gravity-feed liquid inhibitor injection systems
2								Inhibitor solution can be pumped to the tank and mixed
2								Manual addition - Operating procedures for manually inhibiting in some plants - Container of inhibitor kept at tank manway
1								Air line interlocked to temperature for air sparge
1								Based on risk and specific facility evaluation, short-stop inhibitor system used
								Quench system
3	2	3	2	2	4	2	2	Quench system used; no details given
2								Water quench - Water dump interlocked to temperature - Dedicated water quench that automatically activates on high temperature or can be manually activated
1								Based on risk evaluation of specific facility
							1	Large water-filled pond to receive water-reactives releases
								Dump system
1	1	2	1	1	2	1	1	Dump system used; no details given
		1			1			Temperature/manual activated dump systems
1								Only used on nitrogen system where volumes are small and vessels could be quickly emptied
		1						High temperature shuts off feed streams; dumps to quench tank at high-high temperature
		1						Eduction system driven by firewater to empty vessel to drain
		1						Out-of-range alarms provide information to operating personnel
		1						Automatic and manual dump systems
						1		Sump in isolated area
								Other last-resort safety systems
							1	Vent system to scrubber with sufficient capacity for upset
1								Frangible roof and emergency burn pit
		1						Dry ice on hand

Six responses provided no information pertinent to this portion of the survey.

SUMMARY 30

Emergency Relief—Configuration

What Is the Standard Configuration for Your Emergency Relief Systems Related to Reactive Materials Storage or Handling?

Polymerizing	Shock-Sensitive	Thermally Decomposing	Isomerizing	Disproportionating	Pyrophoric	Peroxide-Forming	Water-Reactive	
5	2	4	1	1	4	2	3	Standard configuration used; no details given
8		2			4		3	**Rupture disk and relief valve combination** - Rupture disks always placed under relief valves to keep them solids-free - Rupture disk/relief valve configuration and sizing set for each case - Pressure safety valve with a rupture disk underneath used for larger pressure safety valves
1		1			2	2	3	**Rupture disk** - All relief systems on reacting systems have rupture disks
		1		1	1		1	**Relief valve / pressure safety valve** - PSV installed on transfer hoppers - Atmospheric pressure safety valve with discharge to sewer or a dedicated dumpster - PSV with inert gas purge - Department of Transportation containers with PSVs and fusible plugs, other containers with PSVs only
		1					1	Relief valve or Rupture disk
1								Numerous combinations
1								Specific tank designs based on contents, and the kinetics of its polymerization; requirements range from small rupture disks to weak-seam tank roofs

Four responses provided no information pertinent to this portion of the survey.

SUMMARY 31

Emergency Relief—Sizing

What Is Your Current Practice for How Emergency Relief Devices in Reactive Materials Service are Sized?

Polymerizing	Shock-Sensitive	Thermally Decomposing	Isomerizing	Disproportionating	Pyrophoric	Peroxide-Forming	Water-Reactive	
3	1	2	1	2	3	1	3	American Petroleum Institute (API) - API RP 520 (Recommended Practice) for fire under the vessel - Standard fire single phase; have used AIChE multi-phase sizing in some critical cases - API methodology is used with anhydrides, with measures to ensure the predicable nature of the material
1	1	1	1	1	1	1	1	Based on two-phase flow and surrounded by fire heat load unless the runaway reaction case is worst
2	1				2	1	1	Sized based on external fire scenario
1		1	1	1	1	1	1	Runaway reaction
1	1	1			1	1	1	Sized based on ASME Section 8; ASME code for fire sizing
1		1			1	1	1	All sizing is credible worst case run in VSP
1		1			1	1	1	In the past, mostly on fire exposure; being reviewed to include DIERS technology
1							2	Sized using Design Institute for Emergency Relief Systems (DIERS) technology; DIERS methodology is used on diketene polymerization
2								Company venting standards; extensive corporate guideline defines protocol for normal and emergency vent sizing
1		1						Fire case for relief valves; weak-seam roofs for monomer storage tanks
1					1			Generally for fire case, sized by design consultant
1								Worst credible case is identified by design team, then evaluated by Accelerating Rate Calorimeter (ARC), Vent Sizing Package (VSP), or Reactive System Screening Tool (RSST) tests; vent system is designed taking into consideration the predicted level of entrainment
1								Use data from DIERS and from tests of actual runaway polymerizations
1								Frangible roof and tie-down chains
1								Must consider both protecting the tank and consequences of an emission from the relief device; can result in doing a quantitative risk analysis
1								Disk/valve configuration and sizing set for each case
		1						Sizing based on DIERS and fire exposure

Five responses provided no information pertinent to this portion of the survey.

SUMMARY 32

Emergency Relief—Header Systems

What Is Your Current Practice for Use of Emergency Relief Header Systems in Reactive Materials Service?

Polymerizing	Shock-Sensitive	Thermally Decomposing	Isomerizing	Disproportionating	Pyrophoric	Peroxide-Forming	Water-Reactive	
2	1	3	1	1	2	1	1	Relief header systems used; no details given
1	1	2		1	2	2	2	Indicated that relief header systems not used
1	1	1	1	1	1	1	1	Headers designed to handle multiple relieving systems
1		1			1	1	1	Discharge to atmosphere as short as possible
1					1		1	Flame arresters upstream of atmospheric discharge
2								Relief to flare header; relief valves are tied into common header with process to flare
1					1			Some relieve to atmosphere
1								Avoided if possible
1								Normal vents are often connected to a header for vapor emission control; emergency vents from storage tanks rarely connected to headers
1								Not used for emergency vents from monomer storage tanks
							1	Relieves to common catch tank; backwards second rupture disk protects non-relieving vessels from material and pressure released into common system

Thirteen responses provided no information pertinent to this portion of the survey.

SUMMARY 33

Emergency Relief—Treatment Systems

What Systems are Employed for Treating Emergency Relief Discharges in Reactive Materials Service?

Polymerizing	Shock-Sensitive	Thermally Decomposing	Isomerizing	Disproportionating	Pyrophoric	Peroxide-Forming	Water-Reactive	
1		1			1		2	Treatment systems are used to handled reactive material released through relief devices; no details given
3	1	2			2	2	2	Treatment systems not used or are under consideration
1	1	1	1	1	1	1	2	Released material scrubbed to neutralize; caustic scrubbers
1	1	1	1	1	1	1	1	Relief systems go through catch tanks to water/caustic scrubbers
1	1	1	1	1	1	1	1	Where toxicity is a consideration, scrubbers, flares, or incinerators are used; for flammable vapor cloud potential materials, flares are used
1		1			1	1	1	Preliminary testing for catch tanks in progress
				1	2		1	Relieve to safe area via flame arrester; flame arresters upstream of atmospheric discharge
1								Relief valves for major vessels directed to vent stacks, some with water spray systems to direct-contact condense hot vapors and others are used simply to direct effluents to a safe location
1								Vapor flared; liquid treated biologically
1								Collected and reacted by controlled polymerization
1								All vinyl chloride vents to blowdown tank during vessel overpressure
1								Generally not feasible due to large size
		1						Pressure safety valve (PSV) releases to a dedicated dumpster, where decomposition occurs

Twelve responses provided no information pertinent to this portion of the survey.

SUMMARY 34

Last-Resort Safety Features—Explosion Suppression

Are Explosion Suppression Systems Used Inside Any Reactive Chemicals Storage or Handling Equipment?

Polymerizing	Shock-Sensitive	Thermally Decomposing	Isomerizing	Disproportionating	Pyrophoric	Peroxide-Forming	Water-Reactive	
11	3	7	2	1	6	5	9	Explosion suppression systems are not used inside any storage or handling equipment
1	1	2			1	1	1	Indicated not applicable
1	1	1	1	1	1	1	1	Deflagration suppression based on minimum quenching distance
1								Flame arresters used in venting equipment

Fourteen responses provided no information pertinent to this portion of the survey.

SUMMARY 35

Passive Mitigation

What Passive Mitigation Systems Are Used to Reduce the Severity of Consequences of an Incident Involving Stored/Handled Reactive Materials?

Polymerizing	Shock-Sensitive	Thermally Decomposing	Isomerizing	Disproportionating	Pyrophoric	Peroxide-Forming	Water-Reactive	
								Flow-limiting orifices
3	1	2	1	1	3	1	2	Flow-limiting orifices used; no details given
1	1	1	1	1	1	1	1	Minimum C_V of control valves (to limit maximum flow)
1		1						Usually a pipe spool piece is used instead of an orifice because of wear to orifice and resultant loss of protection
					1		1	Flow-control orifices (meters)
1								Try not to use this type due to plugging with polymer
							1	Orifice on catch tank to scrubber to limit flow rate to scrubber
								Fire-resistant construction
6	4	6	2	2	5	4	5	Fire-resistant construction used; no details given
1	1	1	1	1	1	1	1	Fire-safe valves (will not leak if exposed to external fire)
1					1		1	Fireproofing of structural support steel
2								Some tanks insulated for fire exposure protection
1								Rack structural steel always fireproofed, at least past first floor where liquid containment is normally; piping supports over containment areas fireproofed to prevent piping failures
								Explosion-resistant design
4	3	4	2	2	3	2	2	Explosion-resistant design used; no details given
								Weak seams/explosion venting
5	1	2	1	1	3	1	2	Weak seams/explosion venting used; no details given
2								Some tanks and installations have weak-seam roofs
1								Weak seam used on large storage tanks; small tanks use dished heads and relief devices
1								Frangible roof design
1								Weak-seam roofs needed to provide adequate vent area for some monomers

SUMMARY 35

Passive Mitigation

What Passive Mitigation Systems Are Used to Reduce the Severity of Consequences of an Incident Involving Stored/Handled Reactive Materials?

Polymerizing	Shock-Sensitive	Thermally Decomposing	Isomerizing	Disproportionating	Pyrophoric	Peroxide-Forming	Water-Reactive	
								Bunkers/blast walls/barricades
5	3	5	2	2	4	1	1	Bunkers/blast walls/barricades used; no details given
1		1						Used in certain reactors, but not storage
					1		1	Stored in open space (outdoors) or buildings with three walls only
1								Used as last resort in process except in control room
		1						Short barricades
								Secondary containment (double-walled piping, dikes, drainage, retention areas, etc.)
7	3	6	2	2	6	4	6	Secondary containment used; no details given
1	1	1	1	1	1	1	1	Dikes provided for all storage areas
1		1			1	1	1	Secondary containment for all storage (tanks, drums, etc.)
1		1					1	Individual dikes for tankage, usually 110% operating capacity
1		1						Dikes and retention areas
1				1				Diked storage, deluge system
1						1		Diked areas
					1		1	Diked transfer areas
1								Diking/retention at unloading site
1								Dikes for tanks
1								Always secondary storage containment and runoff water containment for sampling before release or disposal
1								New monomer tanks are placed in individual dikes to minimize potential fire exposure
							1	Water-reactive substances have dikes supplemented by covered sumps
							1	Underground receiver for dike contents and ducting around pipes to direct leakage to underground receiver
							1	Dike area with leak detection system
							1	Covered dike to prevent/mitigate release of vapor cloud
								Storage/handling separated from personnel and equipment
3	3	6	1	1	3	3	3	Separation of storage/handling from personnel and equipment used; no details given
2					1			Tank farm layout; tank farm storage sited per hazard
		1			1		1	Storage located at a distance from transfer
1			1					Remote location; most inventory is far from personnel
					1		1	Some separation of storage/handling from personnel/equipment
1								Sphere/weigh tanks as remote as possible
1								Determined in Safety, Health and Environment (SHE) review
1								Isolated locations preferred; potential for fire exposure from other facilities is also a concern
			1					Limited inventory; dedicated storage
							1	Storage separated from handling area
							1	"Regulated areas" with no unauthorized personnel, strict personal protective equipment requirements

Two responses provided no information pertinent to this portion of the survey.

SUMMARY 36

Detection and Warning

What Systems Are In Place to Detect Loss of Containment of Reactive Materials?

Polymerizing	Shock-Sensitive	Thermally Decomposing	Isomerizing	Disproportionating	Pyrophoric	Peroxide-Forming	Water-Reactive	
								Routine operator patrols
17	7	12	3	3	11	7	12	Routine operator patrols used; no details given
1		1			1	1	2	Patrols every 2 hours
1		1			1	1	1	Routine patrols may or may not detect, depending on release location relative to process area
1				1				Twice per shift or more
			1					Checked several times daily
								Area security
10	5	6	3	3	6	5	6	Area security used; no details given
1						1		Checked once per shift
1								Permission from unit manager/foreman to be on site
								Fixed concentration monitors
6	2	4	1	1	3	2	2	Fixed concentration monitors used; no details given
1	1	1	1	1	1	1	1	Some products, where toxicity or vapor cloud potential exists; due to toxicity, detects ppb (10 minute cycle), also instantaneous alarm for higher concentration
1								Wide use of vapor cloud monitors (flammable and toxic) usually set for specific toxic level or alarm at 25% LFL and shutdown and alarm at 50% LFL
1								Monitors located within dikes on selected tankage
1								Diked areas, process building ventilation systems
1								IST (GC-type International sensor technology) used at rail car unloading stations
1								Automatic GC sampling at storage tanks, transfer pump station, and reactor area
				1				Explosion alarms, LFL monitors, flame sensors (UV) around final storage tank
								Remote sensors (televisions, laser, etc.)
3	1	2	1	1	2	1	1	Remote sensors used; no details given
1		1						Television; limited TV use
1							1	TV cameras in one unloading area; TV monitors on unloading HCl cars
1								Thermal fire sensors for deluge system
1								Used in a few cases, but not common

Three responses provided no information pertinent to this portion of the survey.

SUMMARY 37

Detection and Warning

What Systems Are In Place to Warn of a Release of Reactive Materials?

Polymerizing	Shock-Sensitive	Thermally Decomposing	Isomerizing	Disproportionating	Pyrophoric	Peroxide-Forming	Water-Reactive	
								Control room alarms
12	4	8	3	4	6	3	4	Control room alarms used; no details given
3		1			1		2	Process alarms (level, pressure, temperature)
2						1		Triggered by monitors/leak detectors
1								Panel alarms
							1	Due to toxicity, detects ppb (10 minute cycle); also instantaneous alarm for higher concentrations
								Plant-wide warning siren
10	5	8	3	3	7	6	7	Plant-wide warning siren used; no details given
1	1	1	1	1	1	1	1	Available for all reactive chemicals
1		1			1		1	Manually operated from central location/then verbal transmission of alarm signal from site to central location (telephone or radio) [see discussion in Section 5.10 of text]
1		1				1		Emergency alarms and area-specific alarms
1							1	Plant-wide for specific chemicals that have "over-the-fence" potential
1					1			Hydrocarbon detectors alarm at Plant Protection
1								Plant safety system
1								Primarily for fire and toxicity, not reactivity
1								Most plants
								Other
1	1	1	1	1	1	1	1	Excess-change-in-level alarms
1	1	1	1	1	1	1	1	Excess-pressure-change alarms
1								Haze meters under evaluation to provide early detection of acrylic/methacrylic acid polymerization (polymer is insoluble in the monomer)
					1			Solid material, not evaporating, inventory control and log of transferred material
							1	If spill occurs during manual handling, attendant observes

Three responses provided no information pertinent to this portion of the survey.

SUMMARY 38

Release Isolation

What Is In Place to Reduce the Duration of a Release By Isolating the Source?

Polymerizing	Shock-Sensitive	Thermally Decomposing	Isomerizing	Disproportionating	Pyrophoric	Peroxide-Forming	Water-Reactive	
								Excess flow valves
4	1	2	1	1	1	1	2	Excess flow valves used; no details given
1		1						On materials with high vapor pressures; pressurized liquids
		1				1		On sightglass bridle connections; all gauge glasses have excess flow checks in case of failure
					1		1	On large equipment
1								As dictated by the need
								Remotely actuated block valves
10	3	5	2	2	6	5	5	Remotely actuated block valves used; no details given
1		1			2		3	Flow control valves operated from control rooms
2		1			1	1	1	Depends on configuration, accessibility, location; as dictated by the need
1		1						Triggered manually from control rooms
1								Auto-actuated (air) with spring to close, air supply by plastic tubing that melts in fire and automatically closes valve--fail safe
1								On large equipment
								Manual isolation only
2		3	1	1	2	1	3	Manual isolation only; no details given
2		1			1	1	1	Depends on configuration, accessibility, location; as dictated by the need
1	1	1	1	1	1	1	1	Not normally used for critical/reactive process fluids
1								For very small lines
								Other
1								Fire-rated valve with fusible link on bottom valve interlocked to shut off pump
				1				Heat-sensing trip switch

Seven responses provided no information pertinent to this portion of the survey.

SUMMARY 39

Fire Prevention/Protection—Ignition Source Control

Which of the Following Are Used for Ignition Source Control in Reactive Chemical Storage/Handling Areas?

Polymerizing	Shock-Sensitive	Thermally Decomposing	Isomerizing	Disproportionating	Pyrophoric	Peroxide-Forming	Water-Reactive	
								Intrinsically safe/electrically classified equipment
15	5	9	2	3	10	7	10	Intrinsically safe/electrically classified equipment used; no details given
1	1	2	1	1	1	1		Any material with an NFPA electrical classification uses the approved electrical equipment or better; National Electric Code adhered to; intrinsically safe/electrically classified provided where required by the electrical classification
1		1			1	1	1	If Class I Division 1 equipment is not available, then intrinsically safe is used, if available
1			1			1		Explosion-proof classification in storage/process areas
1								Normal motor controls or electrical power supply
1								Acrylonitrile is especially dangerous due to leaks/spills because of broad fire/explosion range (3-17%); heavier than air
1								As dictated by the system if vapor hazard exists
								Purged equipment
8	5	6	2	2	5	5	5	Purged equipment used; no details given
1	1	2	1	1	1	1		Any material with a NFPA electrical classification uses the approved electrical equipment or better; National Electric Code adhered to; purged equipment provided where required by the electrical classification
1		1			1	1	1	Used as last resort if intrinsically safe or Class I Division 1 equipment not available
					1		1	Due to highly flammable diluents (e.g., metal alkyls)
1								Special instruments and computer low-voltage equipment, panel boards, control rooms
1								As dictated by the system if vapor hazard exists
							1	Nitrogen-purged

Three responses provided no information pertinent to this portion of the survey.

SUMMARY 40

Fire Prevention/Protection—Mitigation

Which of the Following Are Used for Mitigating Fires in Reactive Chemicals Storage and Handling Areas?

Polymerizing	Shock-Sensitive	Thermally Decomposing	Isomerizing	Disproportionating	Pyrophoric	Peroxide-Forming	Water-Reactive	
								Fireproofing
6	3	4	1	1	3	3	5	Fireproofing used; no details given
1		1	1	1				Fireproofing for support skirts, legs
				1	1		1	Vessel support columns and structures
1						1		Fireproofing used, but not plant standard
1								Structural steel and buildings
								Insulation for thermal barrier
5	3	4	1	1	3	3	3	Insulation for thermal barrier used; no details given
1								Storage tanks
1								Steel, structural insulated
1								Some tanks insulated, protected in conjunction with sprinklers, separations are fairly large
1								On some tanks where overheating can result in extremely rapid polymerization
								Sprinkler systems
16	4	10	1	1	5	5	2	Sprinkler systems used; no details given
1		1			1	1	1	Determined individually; depends on location, storage density, occupation, compatibility
1		1	1	1				Fixed water spray systems
				1	1		1	On all tanks
					1		1	Deluge tanks to cool; use fine spray on fire to remove heat or as an aid in fighting fire
1								Process equipment in buildings (warehouses) and racks
1								On exterior of tanks for cooling in case of external fire
1								Use remote fire monitor nozzles
		1						Consider explosion hazards (sheds for benzoyl peroxides are sprinklered)
								Carbon dioxide systems
1		1			1	1	1	Determined individually; depends on location, storage density, occupation, compatibility
					1		1	Carbon dioxide systems used; no details given
1								Computer electrical termination cabinets, electrical switch gears
					1			Portable CO_2 units sited in this area
								Foam systems
6	2	2	2	2	3	2	2	Foam systems used; no details given
2		1				1	1	No fixed systems - foam truck; portable foam systems
1		1			1	1	1	Determined individually; depends on location, storage density, occupation, compatibility
	1	1			1	1	1	Fire truck available for area response
1	1	1				1		Manual application only
2								In dikes and tankage; in storage tank dike areas, plus railroad tank car, tank truck, and drum loading/unloading areas
1								Storage tank designed for subsurface foam injection
1								Manufacturing Chemists Association Chemical Safety Data Sheet (SD-31) recommends mechanical foam, alcohol type (burning acrylonitrile may release cyanide gases, especially where O_2 supply is limited)
				1				On main storage tank

Two responses provided no information pertinent to this portion of the survey.

SUMMARY 41

Post-Release Mitigation Systems

What Is In Place To Be Used to Counter a Loss-of-Containment Accident Involving Reactive Materials?

Polymerizing	Shock-Sensitive	Thermally Decomposing	Isomerizing	Disproportionating	Pyrophoric	Peroxide-Forming	Water-Reactive	
								Water curtains
4	3	4	1	2	2	3	1	Water curtains used; no details given
1		1						Water spray systems first choice; backup system of choice is monitors and fire hoses
1								To cool small equipment and pieces of large equipment
1								To put out fires
1								To knock down vapor clouds
							1	Clouds produced by release of water-reactive materials are contained with water curtains
							1	For HCl tank car unloading
							1	Massive amounts of water have been used with water-reactive materials
								Firewater monitors
11	4	8	2	3	8	6	4	Firewater monitors used; no details given
2		1			2	1	2	Mostly to cool adjacent vessels and (large) equipment close to fire
1		1						Water spray systems first choice; backup system of choice is monitors and fire hoses
1								Water cannons activate automatically during release at rail car unloading stations
1								Extensive use of firewater monitors
							1	Massive amounts of water have been used with water-reactive materials
								Fire hoses with fog nozzles
10	4	7	2	3	7	5	5	Fire hoses with fog nozzles used; no details given
2		1			1	1	1	Mainly to disperse vapor clouds; to knock down vapor clouds of flammable and toxic material
	1	1			1	1	1	Fire department equipped with fog nozzles
1		1						Water spray systems first choice, backup system of choice is monitors and fire hoses
					1		1	For cooling adjacent vessels and equipment
							1	Clouds produced by release of water-reactive materials are knocked down with fog nozzles
							1	Water reactives that can be diluted are flushed to sewers using fire hoses
								Spill absorbent
8	5	8	2	2	6	5	6	Spill absorbent used; no details given
1		1			1	1	1	For small spills, there are compatibility problems with some adsorbents
	1	1			1	1	1	Fire department equipped with spill absorbent
1		1						Adsorbents used by emergency response crew
					1		1	Sand or soil for containing flow
1								Spill absorbent used to pick up small releases
					1			Diked transfer and storage area
							1	Shovel off spilled solid after diluting with diatomite

SUMMARY 41

Post-Release Mitigation Systems

What Is In Place To Be Used to Counter a Loss-of-Containment Accident Involving Reactive Materials?

Polymerizing	Shock-Sensitive	Thermally Decomposing	Isomerizing	Disproportionating	Pyrophoric	Peroxide-Forming	Water-Reactive	
								Other
1		1				1	1	All plant wastewater contained onsite
1		1						All tanks in bunded (diked/curbed) areas
1		1						Large amounts of water to flush spills if not contained
1								Foam to reduce pool evaporation
							1	PCl_3 storage tanks in containment building
							1	Secondary containment (three layers of packaging—double plastic bags plus steel drum)
							1	Special soap/water mix to react and neutralize spill stored on site, due to high toxicity

Five responses provided no information pertinent to this portion of the survey.

SUMMARY 42

Post-Release Mitigation—Personal Protective Equipment

What Personal Protective Equipment Is Used to Protect Against a Reactive Chemical Spill or Release?

Polymerizing	Shock-Sensitive	Thermally Decomposing	Isomerizing	Disproportionating	Pyrophoric	Peroxide-Forming	Water-Reactive	
								Respiratory Protection
9	2	7	1	2	6	5	7	SCBA/SCBE (self-contained breathing apparatus/self-contained breathing equipment) - always used when going into a spill
5	2	5	2	2	5	4	4	Escape respirators; 5-min escape packs
2	2	2	1	1	2	2	2	Air-line respirators; fresh air masks
2	1	1	1	1	2	1	1	Respirators, unspecified type
1		1				1	1	Organic half-mask respirators issued to all personnel
1								Chemical or air-purifying respirators
1								Air packs with full face shields
1								Respiratory protection, unspecified type
							1	Escape bottles on rack structures
							1	Respiratory protection for materials releasing toxic vapors
								Suits
2	1	2	1	1	1	2	3	Full gear / full encapsulation suit (for mitigation of an event)
1	1	1	1	1	1	1	1	Slicker suit
1		1			1	1	1	Protective suit, unspecified type
					2		1	Aluminized suit
				1	1		1	Bunker gear/bunker suit
1		1						Acid suit
1								Vinyl suit
					1			Fire-retardant suit (used to unload and to respond to leaks or potential leaks)
							1	Tyvek™ suit
								Other personal protective equipment
3	1	2	1	1	2	2	2	Footwear; boots; safety shoes
3	1	2	1	1	1	1	2	Gloves; chemical gloves
3	1	2	1	1	1	1	2	Face shield; goggles; safety glasses
								Other comments
	1	1			1	1	1	According to Material Safety Data Sheets (MSDSs), which includes Level A protection
1	1	1				1		Loss of containment generally involves "normal" fire and health hazards but not "reactivity" hazard; the suitable NFPA level of protective equipment for emergency response teams is used
1							1	Personal protective equipment is available and used per site Standard Operating Procedure for the release in question
					1		1	Fires are allowed to burn out
1								Specific requirements vary with monomer; MSDSs specify need

Five responses provided no information pertinent to this portion of the survey.

SUMMARY 43

Post-Release Mitigation—Emergency Response

What Features Of Your Site's Emergency Response Plans Are Specifically Associated With the Storage and Handling of Reactive Chemicals On-Site?

Polymerizing	Shock-Sensitive	Thermally Decomposing	Isomerizing	Disproportionating	Pyrophoric	Peroxide-Forming	Water-Reactive	
5		3		1	2	2	2	No features of site emergency response plan are specifically associated with reactive materials storage and handling
1	1	1	1	1	1	1	1	Emergency plan deals with credible scenarios pertaining to reactive chemicals
1	1	1	1	1	1	1	1	Safety Emergency Security Manual followed
1	1	1	1	1	1	1	1	Annual drills/critique/discussion of "what-if" scenarios
2		1			1	1	1	Dispersion models developed for a range of releases
2		1			1	1	1	Community involvement: - When community Emergency Response Planning Committees exist, response plans have been developed together and emergency drills have been held - Fire protection by on- and off-site departments - Community notification and warning - Evacuation plans - Media interaction
1	1	1			1	1	1	"Highly Hazardous Chemicals" are listed with locations and hazards along with emergency response information
	1	1			1	1	1	Detailed Action Plans developed for emergency response to extremely hazardous chemicals
1		1			1		1	Site emergency response restricted to trained fire brigades who use personal protective equipment (available and used per site procedure for the release in question)
		1			1		1	Emergency response team trained in handling fires and releases of materials
1								Remedial action plans (for both on-site and off-site effects)
1								Evacuation plans/on-site, possibly off-site
1								Extensive emergency plans; particular concern is the "safe distance" for emergency response personnel
							1	Specific evacuation plans
							1	Specially trained emergency response team
							1	Weekly drills

Twelve responses provided no information pertinent to this portion of the survey.

SUMMARY 44

Hazard Reviews—Severity Categories

If Different Categories of Hazard Severity Are Used, How Are They Defined?

Number of Responses	
	Five Severity Levels ***(respondents reporting a five-category system)***
1	Level 5 - multiple fatality/serious community impact Level 4 - single fatality/serious community impact Level 3 - serious injury potential Level 2 - minor injury potential Level 1 - zero to slight injury
	Four Severity Levels ***(respondents reporting a four-category system)***
2	Classes as described in corporate guidelines: Class D - Catastrophic event Class C - Serious event Class B - Major event Class A - Moderate event
1	High, Medium, Moderate, Low (NFPA Criteria)
1	High, Medium, Low, Very Low
1	High, High-Moderate, Moderate-Low, Low
	Three Severity Levels ***(respondents reporting a three-category system)***
4	High, Medium, Low
1	High - high toxic, reactive, exothermic, flammable Medium - medium toxic, reactive exothermic, flammable Low - combustible
1	High (off-site or site-wide), Local, Nuisance
1	High, Medium, Low (site evaluation)
1	High, Medium, Low (combination of history, perceived risk, and use of Dow FEI, CEI)
	Two Severity Levels ***(respondents reporting a two-category system)***
1	High, Medium
1	HHM (Highly Hazardous Material—a material present on a site in sufficient quantity that if it were released at a worst-credible rate, a threat to personnel at the plant boundary would exist) All others
1	Major - Hazards that have potential for multiple injuries and/or fatalities and/or significant effects off-site Other - Not much distinction made between other hazard classes
	One Severity Level Used or Levels Not Differentiated
2	High
1	High (large spills/release; tank storage located close to battery limits of plant)
1	All hazard evaluation are handled the same; the only change is timing of re-evaluation
1	Off-site and on-site risk potential; on-site risks posed by toxics and flammables stored or used in volumes of 10,000 pounds or more; sub-categorization is made by evaluating age of process equipment, approximate numbers of workers, complexity of process, incident history, etc.
1	Based on consequences of identified incidents (categories as outlined on page 194 of July 1990 *Plant/Operations Progress*)
1	New PHA Program and Management-of-Change program to provide
1	All from tank farm area (DOW F&E Index used to categorize areas of plant)
1	All materials, designs, and procedures are reviewed by a What If PHA; if problems are found, then more complex techniques such as HAZOP are used and a pre-startup safety review may be required

Two responses provided no information pertinent to this portion of the survey.

Number of Responses (reviews for all severity categories unless noted otherwise)	**SUMMARY 45** **Hazard Reviews—Types of Reviews** ***What Reviews are Conducted for Each Hazard Severity Category?***
	Design reviews
16	Design reviews conducted; no details given
2	HAZOP review for all capital projects / for tank farm area
2	What-If review at design stage
1	Screening (pre-authorization) review conducted for all areas
1	Reviews conducted at several stages, beginning early in design
1	*High-hazard processes:* design reviews conducted *Medium and Moderate:* design reviews at initial stage of design and when major modifications made *Low-hazard processes:* no design reviews
	Reviews for adequate safety instrumentation redundancy
8	Instrumentation redundancy reviews conducted; no details given
1	Corporate guidelines on alarm/interlock design specifies level of redundancy required and acceptable instrument hardware based on incident consequences
1	Screening (pre-authorization) review conducted for all areas
1	One design process hazard analysis conducted for all areas; methodology depends on perceived hazard(s) and previous operating experience (generally What-If or HAZOP method)
1	HAZOP review for all capital projects
1	Failure modes and effects analysis
1	As part of quantitative risk analysis, when quantitative risk analysis required
1	*High:* reviewed at time of design and construction *Medium, Moderate, and Low:* no instrumentation redundancy review
1	*Catastrophic and Serious:* reviews conducted *Major and Moderate:* no instrumentation redundancy review
1	*High and Medium:* reviews conducted *Low:* no instrumentation redundancy review
1	*High:* reviews conducted *Medium:* reviews probably conducted *Low:* no instrumentation redundancy review
	Process hazard analyses; methodologies used
9	Hazard and Operability (HAZOP) studies - large spills/releases - HAZOP review for all capital projects - *High:* HAZOP every 2 years *Medium:* HAZOP every 3 years *Low:* HAZOP every 5 years
3	Major Accident Prevention Program (described in *Plant Operations Progress 9*(3), July 1990)
2	Plant consequences using consequence analysis software
2	Quantitative Risk Analysis
2	Fault-Tree Analysis
2	Consultant / outside technical assessment
1	HAZOP is primary methodology; also Checklist using pre-approved checklist
1	All materials, designs, and procedures reviewed by What-If; if problems are found, then more complex techniques such as HAZOP studies are used
1	HAZOP/What-If/Checklist if on-site risk posed by toxics and flammables stored or used in volumes of 10,000 pounds or more
1	What-If/Checklist is primary tool; HAZOP, FMEA used in some specific instances
1	What-If/Checklist

SUMMARY 45

Hazard Reviews—Types of Reviews

What Reviews are Conducted for Each Hazard Severity Category?

Number of Responses (reviews for <u>all</u> severity categories unless noted otherwise)	
1	What-If
1	Checklist
1	Failure Modes and Effects Analysis
1	Event-Tree Analysis
1	All hazard evaluations handled the same; frequency of re-evaluation set on 2- to 4-year cycle after hazard evaluation completed
1	Transportation risk
1	Safety reviews
1	*High:* Major Accident Prevention Program *Medium:* no hazard reviews
1	*High, Medium, and Low:* HAZOP *Low:* What-If
1	*High:* HAZOP, Quantitative Risk Analysis *Medium and Low:* no hazard reviews
1	*High:* HAZOP *Medium and Low:* HAZOP, What-If, Checklist
1	*High and Medium:* HAZOP *High, Medium, and Low:* Preliminary Hazard Analysis, What-If, Checklist
1	*Catastrophic to Serious:* HAZOP *Major to Moderate:* What-If
1	*High and Medium:* HAZOP *Moderate and Low:* no hazard reviews
1	*High:* HAZOP, FMEA/FTA *Medium:* HAZOP, What-If/Checklist *Low and Very Low:* What-If/Checklist
1	*Highly hazardous material:* What-If followed by a HAZOP *All others:* What-If
	Operability reviews
12	Operability reviews conducted; no details given
4	Addressed during HAZOP reviews
2	*High and Medium:* operability reviews conducted *Low:* no operability reviews
1	*Catastrophic to Serious:* operability reviews conducted *Major to Moderate:* no operability reviews
1	As part of overall reviews
	Personnel safety reviews
14	Personnel safety reviews conducted; no details given
2	As part of Job Safety Analyses
2	Addressed during HAZOP review
1	Safety (OSHA)
1	As part of overall reviews
1	As part of safety review of all minor plant facility/equipment modifications
1	*High:* personnel safety reviews conducted *Medium, Low, and Very Low:* no personnel safety reviews
	Maintenance safety reviews
11	Maintenance safety reviews conducted; no details given
2	Addressed during HAZOP review
1	As required by OSHA
1	As part of overall reviews
1	As part of safety review of all minor plant facility/equipment modifications

Number of Responses (reviews for <u>all</u> severity categories unless noted otherwise)	**SUMMARY 45** **Hazard Reviews—Types of Reviews** ***What Reviews are Conducted for Each Hazard Severity Category?***
	Human factors reviews
5	Human factors reviews conducted; no details given
3	Addressed during HAZOP/process hazard analysis
1	As part of overall reviews
1	As part of safety review of all minor plant facility/equipment modifications
1	*High:* human factors reviews conducted *Medium and Low:* no human factors reviews
	Facility siting reviews
7	Facility siting reviews conducted; no details given
2	Facility siting considerations addressed as part of process hazard analyses
2	Major Accident Prevention Program review considers siting issues
2	*High:* facility siting reviews done / might be done *Medium and Low:* no facility siting reviews
1	As part of overall reviews
1	Some
1	*High and Medium:* facility siting reviews done *Low:* no facility siting reviews
1	*High:* facility siting reviews conducted at initial design stage and for major modifications *Medium, Low, and Very Low:* no facility siting reviews
	Pre-startup safety reviews
17	Pre-startup safety reviews conducted; no details given
2	Pre-startup check on all new equipment / for capital projects
1	Detailed checklist used
1	Review before start-up; usually a walkthrough
1	As part of overall reviews
1	All materials, designs, and procedures reviewed by a What-If process hazard analysis; if problems are found, a pre-startup safety review may be required
1	*High and Medium:* pre-startup safety reviews conducted *Low and Very Low:* no pre-startup safety reviews
	Other
5	Management of change reviews
3	Safety, Health and Environmental reviews
1	Reactive chemical reviews are conducted during process design, prior to start-up, or whenever process changes or change in critical personnel occur, or every 2 years
1	Process safety assessments
1	Responsible Care® (initiative of the Chemical Manufacturers Association, Washington, D.C.)
1	Periodic reviews
1	Product integrity review might be conducted

Two responses provided no information pertinent to this portion of the survey.

SUMMARY 46

Codes and Standards

What Publicly Available Consensus Standards Do You Utilize in the Design of Reactive Chemicals Storage and Handling Facilities?

Polymerizing	Shock-Sensitive	Thermally Decomposing	Isomerizing	Disproportionating	Pyrophoric	Peroxide-Forming	Water-Reactive	
17	6	12	5	5	12	9	10	NFPA (National Fire Protection Association), including National Electric Code
14	6	11	3	4	13	9	11	ASME (American Society of Mechanical Engineers)
15	7	10	4	4	11	8	11	Suppliers' recommendations
15	5	8	3	4	10	7	10	API (American Petroleum Institute)
10	5	7	4	5	8	6	9	State ordinances
8	5	8	3	4	7	5	6	FM (Factory Mutual)—limited availability; contact FM for availability
8	4	6	3	4	7	5	7	Local ordinances
7	4	7	2	3	8	6	7	Uniform Building Code
1	1	2	1	2	2	2	2	American Tables of Distances (explosives, separation distances)
1	2	2	1	1	2	2	2	OSHA (U.S. Occupational Safety and Health Administration)
4	1	1	1	1	1	1	1	BS (British Standards)
3	1	1	1	1	1	1	1	DIN Standards (Deutsches Institut für Normung, from Germany) *Additional response, not associated with particular categories of reactive chemicals:* Only on foreign procured equipment; otherwise, ASME Section VIII
3	1	1	1	1	1	1	1	VDI (Verein Deutscher Ingenieure, German engineering standards)
1	1	1	1	1	1	1	1	EPA (U.S. Environmental Protection Agency)
1		1			1			J.I.S. (Japanese standards)
1		1			1			Japanese local requirements (In Japan, local requirements added to the normal requirements of the first safety laws are negotiated as required)
1					1			Available literature

Five responses provided no information pertinent to this portion of the survey.

CCPS INDUSTRY PRACTICE SURVEY
REACTIVE CHEMICALS STORAGE AND HANDLING GUIDELINES

INSTRUCTIONS

1. **Purpose**: To get information on current good practices in the process industries related to the storage and handling of reactive materials. This information will be compiled into a chapter on General Design Considerations in the CCPS Project *REACTIVE CHEMICALS: Guidelines for Safe Storage and Handling.*

2. **Scope**: The questions are divided into the following categories:

Your answers should only pertain to the storage and handling (not processing) of reactive materials in greater than laboratory quantities. As a general guide, "reactive materials" for the purposes of this survey can be considered as those having an NFPA reactivity hazard rating of 2, 3, or 4, plus the special categories of *isomerizing, disproportionating,* and *peroxide-forming* materials that are not part of the NFPA rating definitions.

3. **Approach**: The example on the next page shows how the survey should be completed. Since different design/operating practices may be employed for different types of reactive materials, the columns allow you to mark your answers by reactive chemical category. Use more paper where needed; attach to survey.

4. **Priorities**: If you are unable to provide your company's or facility's current practices in detail for any page, please at least *check the boxes pertaining to all practices that you do employ.*

5. **Anonymity:** The information submitted to CCPS will be made anonymous, so that the final guidelines will not use company names or indicate the source of the information. However, you may request that credit be given to your company or site for any particular information.

Thank you for helping CCPS make the storage and handling of reactive chemicals safer.

EXAMPLE

Say a facility receives methyl methacrylate (polymerizing) in tank truck quantities and dichlorosilane (pyrophoric) in drums.

COMPATIBILITY	Reactivity Hazard Type							
	Polymerizing	Shock-Sensitive	Thermally Decomposing	Isomerizing	Disproportionating	Pyrophoric	Peroxide-Forming	Water-Reactive
1 How are potential incompatibility problems identified? *(give only a brief summary)*	**A**					**A**		
2 What design and/or operating practices are used for preventing incompatibility incidents related to the following?								
• process materials / reagents	**B**					**C**		
• impurities	**D**					**D**		
• heat transfer fluids	**n/a**					**n/a**		
• materials of construction	:					:		
• corrosion products	:					:		
• insulation								
• any other materials *(please specify)*								
3 Any other practices related to compatibility issues?								

Details/Rationale:

A: We use a chemical interaction matrix to identify potential incompatibilities as part of all hazard reviews.

B: Dedicated unloading spot; no cross-connecting piping; storage tank inspected before each filling operation.

C: Stored in drums; depend on labeling and operator training for safe transfer to batch reactor under inert blanket.

D: All incoming reactive materials sampled for purity before transferring to storage.

(**n/a**: not applicable)

COMPATIBILITY	Reactivity Hazard Type							
	Polymerizing	Shock-Sensitive	Thermally Decomposing	Isomerizing	Disproportionating	Pyrophoric	Peroxide-Forming	Water-Reactive
4 How are potential incompatibility problems identified? *(give only a brief summary)*								
5 What design and/or operating practices are used for preventing incompatibility incidents related to the following?								
• process materials / reagents								
• impurities								
• heat transfer fluids								
• materials of construction								
• corrosion products								
• insulation								
• any other materials *(please specify)*								
6 Any other practices related to compatibility issues?								

Details/Rationale:

STORAGE TIME / SHELF LIFE	Reactivity Hazard Type							
	Polymerizing	Shock-Sensitive	Thermally Decomposing	Isomerizing	Disproportionating	Pyrophoric	Peroxide-Forming	Water-Reactive
7 What limitations are there on the shelf life of reactive materials stored at your facilities?								
8 What design and/or operating practices are employed for increasing shelf life?								
• inhibitors								
• sampling systems								
• inventory records								
• any other practices *(please specify)*								
9 How is too-old material handled and disposed of?								

Details/Rationale:

STORAGE QUANTITY AND CONFIGURATION	Reactivity Hazard Type							
	Polymerizing	Shock-Sensitive	Thermally Decomposing	Isomerizing	Disproportionating	Pyrophoric	Peroxide-Forming	Water-Reactive
10 What determines maximum inventories for reactive materials stored at your facilities?								
11 In what configuration are the reactive materials stored (tanks, cylinders, cartons, etc.)?								
12 If tank storage is used, are outlet nozzle(s) on the top or bottom of the tank(s)?								
13 What are your siting practices for reactive materials storage, such as tank spacing or distance to plant boundaries?								
14 Are there any restrictions on container shape or configuration, such as the following?								
• free space								
• length-to-diameter ratio								
• other *(please specify)*								
15 Are the storage container contents circulated or mixed? If so, how and why?								

Details/Rationale:

AIR AND MOISTURE EXCLUSION

	Reactivity Hazard Type							
	Polymerizing	Shock-Sensitive	Thermally Decomposing	Isomerizing	Disproportionating	Pyrophoric	Peroxide-Forming	Water-Reactive
16 What design and/or operating practices are used to exclude air from the reactive materials containment?								
• container filling procedures								
• inert gas padding								
• other *(please specify)*								
17 What design and/or operating practices are used to keep moisture out of the reactive materials containment?								
• start-up procedures								
• air drying systems								
• package waterproofing								
• building waterproofing								
• building drainage								
• flood control								
• other *(please specify)*								

Details/Rationale:

MONITORING AND CONTROL	Reactivity Hazard Type							
	Polymerizing	Shock-Sensitive	Thermally Decomposing	Isomerizing	Dispropor-tionating	Pyrophoric	Peroxide-Forming	Water-Reactive
18 How are the following storage condition parameters monitored?								
• oxygen concentration								
• humidity/moisture content								
• pressure								
• temperature								
• other *(please specify)*								
19 What kind of systems, in the following categories, are used for controlling the temperature of the stored material?								
• tank/cylinder cooling								
• tank/cylinder heating								
• building cooling or refrigeration								
• building heating								
• building ventilation								
• other temperature control *(please specify)*								

Details/Rationale:

HANDLING AND TRANSFER	Reactivity Hazard Type							
	Polymerizing	Shock-Sensitive	Thermally Decomposing	Isomerizing	Disproportionating	Pyrophoric	Peroxide-Forming	Water-Reactive
20 What practices are employed for the manual handling of reactive materials on-site?								
• pallets with hand trucks								
• pallets with forklift trucks								
• roller conveyors								
• other *(please specify)*								
21 What design specifications and operating practices are used for on-site transfer (piping) systems?								
• fittings and connections, including flexible unloading connections								
• piping specifications; layout details (e.g., pertaining to dead legs)								
• pumps and pump seals								
• valves (control; isolation, other)								
• drain systems								
• cleaning equipment								
• maintenance procedures related to the above transfer systems								
• other *(please specify)*								

Details/Rationale:

LAST-RESORT SAFETY FEATURES	Reactivity Hazard Type							
	Polymerizing	Shock-Sensitive	Thermally Decomposing	Isomerizing	Dispropor-tionating	Pyrophoric	Peroxide-Forming	Water-Reactive
22 Please give details of any last-resort safety systems related to your reactive materials storage or handling, such as:								
• inhibitor injection								
• quench system								
• dump system								
• other *(please specify)*								
23 What is your current practice for your **emergency relief systems** related to your reactive materials storage or handling, in the following categories?								
• standard configuration (relief valve; rupture disk; combination; fusible plug; etc.)								
• how emergency relief devices are sized; sizing basis								
• use of header systems								
• treatment systems used to handle reactive material released through relief devices								
24 Are **explosion suppression** systems used inside any storage or handling equipment?								

Details/Rationale:

PASSIVE MITIGATION	Reactivity Hazard Type							
	Polymerizing	Shock-Sensitive	Thermally Decomposing	Isomerizing	Dispropor-tionating	Pyrophoric	Peroxide-Forming	Water-Reactive
25 What passive mitigation systems are used to reduce the severity of consequences of an incident involving stored/handled reactive chemicals? (*Passive* mitigation systems require no human or mechanical action to work.)								
• flow-limiting orifices								
• fire-resistant construction								
• explosion-resistant design								
• weak seams; explosion venting								
• bunkers; blast walls; barricades								
• secondary containment (double-walled piping, dikes, drainage, retention areas, etc.)								
• storage/handling separated from personnel and equipment								
• other *(please specify)*								

Details/Rationale:

DETECTION, WARNING, AND ISOLATION	Reactivity Hazard Type							
	Polymerizing	Shock-Sensitive	Thermally Decomposing	Isomerizing	Dispropor-tionating	Pyrophoric	Peroxide-Forming	Water-Reactive
26 What systems are in place to detect loss of containment of reactive materials and warn of their release?								
• routine operator patrols								
• area security								
• fixed concentration monitors								
• remote sensors (television, laser, etc.)								
• control room alarms								
• plant-wide warning siren								
• other *(please specify)*								
27 What is in place to reduce the duration of a release by isolating the source?								
• excess flow valves								
• remote block valves								
• manual isolation only								
• other *(please specify)*								

Details/Rationale:

FIRE PREVENTION / FIRE PROTECTION	Reactivity Hazard Type							
	Polymerizing	Shock-Sensitive	Thermally Decomposing	Isomerizing	Dispropor-tionating	Pyrophoric	Peroxide-Forming	Water-Reactive
28 Which of the following are used for ignition source control in reactive chemicals storage and handling areas?								
• intrinsically safe electrical equipment								
• purged equipment								
29 Which of the following are used for mitigating fires in reactive chemicals storage and handling areas?								
• fireproofing								
• insulation for thermal barrier								
• sprinkler systems								
• carbon dioxide systems								
• foam systems								
30 What potential hazards (if any) have been identified with respect to compatibility of extinguishing agents with on-site reactive materials?								

Details/Rationale:

POST-RELEASE MITIGATION	Reactivity Hazard Type							
	Polymerizing	Shock-Sensitive	Thermally Decomposing	Isomerizing	Dispropor-tionating	Pyrophoric	Peroxide-Forming	Water-Reactive
31 What is in place to be used to counter a loss-of-containment accident involving reactive materials?								
• water curtains								
• firewater monitors								
• fire hoses with fog nozzles								
• spill absorbent								
• other *(please specify)*								
32 What personal protective equipment (including e.g. escape respirators) is used to protect against a reactive chemical spill or release?								
33 What features of your site's emergency response plans are specifically associated with the storage and handling of reactive chemicals on-site?								

Details/Rationale:

HAZARD REVIEWS	Hazard Category (fill in category names; e.g., HIGH, MEDIUM, LOW or 1 - 4)			
34 If different categories of hazard severity are used, how are they defined?				
35 What reviews are conducted for each hazard severity category?				
• design reviews				
• reviews for adequate safety instrumentation redundancy				
• process hazard analyses *(what methodologies are used?)*				
• operability reviews				
• personnel safety reviews				
• maintenance safety reviews				
• human factors reviews				
• facility siting reviews				
• pre-startup safety reviews				
• other *(please specify)*				

Details/Rationale:

CODES AND STANDARDS	Reactivity Hazard Type							
	Polymerizing	Shock-Sensitive	Thermally Decomposing	Isomerizing	Dispropor-tionating	Pyrophoric	Peroxide-Forming	Water-Reactive
36 What publicly available consensus standards do you utilize in the design of reactive chemicals storage and handling facilities? *(check off all used, or rank in order of importance with 1 = highest)*								
• **API**: American Petroleum Institute								
• American Table of Distances								
• **ASME**: American Society of Mechanical Engineers								
• **BS**: British Standards								
• **DIN** Standards (from Germany)								
• **FM**: Factory Mutual								
• local ordinances								
• **NFPA**: National Fire Protection Association								
• state ordinances								
• suppliers' recommendations								
• **TÜV**								
• Uniform Building Code								
• **VDI** (Europe)								
• other *(please specify)*								

Details/Rationale:

Glossary

Acute effect: An adverse effect on a human or animal body, with severe symptoms developing rapidly and coming quickly to a crisis. If symptoms develop slowly over a long period of time, and are prolonged and lingering, it is termed a *chronic effect.*

Autodecomposition: The sustained decomposition of a substance without introduction of any other apparent ignition source besides thermal energy and without air or other oxidant present. Autodecomposition is the result of a thermal self-decomposition reaction for given initial conditions (temperature, pressure, volume) at which the rate of heat evolution exceeds the rate of heat loss from the reacting system, thus resulting in an increasing reaction temperature and reaction rate.

Autodecomposition temperature: The minimum temperature for a specified test method, test apparatus (including material of construction and test volume), and initial pressure required to initiate self-sustained decomposition of a solid, liquid, or gaseous substance without any other apparent source of ignition and without air or other oxidant present.

Autoignition: The ignition and sustained combustion of a substance without introduction of any other apparent ignition source besides thermal energy. Autoignition is the result of oxidation for given initial conditions (temperature, pressure, volume) at which the rate of heat gain exceeds the rate of heat loss from the reacting system, thus resulting in an increasing reaction temperature and reaction rate.

Autoignition temperature: The minimum temperature for a specified test method, test apparatus (including material of construction and test volume), and initial pressure required to initiate self-sustained combustion of a solid, liquid, or gaseous substance without any other apparent source of ignition. This is the hot flame ignition temperature. Autoignition temperatures (AITs) have been measured and tabulated for many materials; however, they are only indicative of the range of test conditions at which they were measured.

Increased test volume and pressure will usually lower the AIT, and a given test apparatus may have important surface effects. At atmospheric pressure, only slow oxidation is found up to the AIT in the absence of other ignition sources, although a cool flame region may be traversed as temperature is increased toward the AIT. This is not an intrinsic property of a material.

Autoxidation: A slow, easily initiated, self-catalyzed reaction, generally by a free-radical mechanism, between a substance and atmospheric oxygen. Initiators of autoxidation include heat, light, catalysts such as metals, and free-radical generators. Davies (1961) defines autoxidation as interaction of a substance with molecular oxygen at below 120°C without flame. Possible consequences of autoxidation include pressure buildup by gas evolution, autoignition by heat generation with inadequate heat dissipation, and the formation of peroxides. Also termed autooxidation or auto-oxidation.

Combustible: In the broadest meaning of the term, a combustible substance is any solid, liquid, or gaseous substance that will burn in air after being ignited, with or without preheating of the substance. As usually defined, the term "combustible" refers to solids that are relatively difficult to ignite and that burn relatively slowly, and to liquids having a flash point greater than or equal to 100°F (37.8°C). It should be noted that some solids (e.g., aluminum) are considered noncombustible in bulk, but can burn very rapidly if finely subdivided and dispersed in air.

Cool flame ignition: A relatively slow, self-sustaining, barely luminous gas-phase reaction of the sample or its decomposition products with an oxidant. Cool flames are visible only in a darkened area (NFPA 325M, 1994). Cool flames are self-sustaining, exothermic ignition reactions that, under proper circumstances, may act as the initiator of more energetic hot flame reactions.

Decomposition temperature: See Autodecomposition temperature.

Deflagration: A release of energy caused by the propagation of a chemical reaction in which the reaction front advances into the unreacted substance at less than sonic velocity in the unreacted material. Where a blast wave is produced that has the potential to cause damage, the term *explosive deflagration* may be used.

Detonation: A release of energy caused by the propagation of a chemical reaction in which the reaction front advances into the unreacted substance at greater than sonic velocity in the unreacted material.

Disproportionation: A chemical reaction in which a single compound serves as both oxidizing and reducing agent and is thereby converted into a more oxidized and a more reduced derivative. For example, a hypochlorite upon appropriate heating yields a chlorate and a chloride, and ethyl radicals formed as intermediates are converted into ethane and ethylene (Sax and Lewis, 1987, 434).

Flammable: Broadly, a flammable material is any solid, liquid, or gas that will ignite easily and burn in air rapidly.

Flammable liquid: A liquid with a flash point less than 100°F (37.8°C) as specified by NFPA 321 (1991). Class IA flammable liquids have flash points below 73°F (22.8°C) and boiling points below 100°F (37.8°C). Class IB flammable liquids have flash points below 73°F (22.8°C) and boiling points at or above 100°F (37.8°C). Class IC flammable liquids have flash points at or above 73°F (22.8°C) and below 100°F (37.8°C).

Fireball: A burning fuel–air cloud whose energy is emitted primarily in the form of radiant heat. The inner core of the cloud consists almost completely of fuel, whereas the outer layer (where ignition first occurs) consists of a flammable fuel–air mixture. As the buoyancy forces of hot gases increase, the burning cloud tends to rise, expand, and assume a spherical shape.

Flash fire: The combustion of a flammable gas or vapor and air mixture in which the flame propagates through that mixture in a manner such that negligible or no damaging overpressure is generated.

Hot flame ignition: A rapid, self-sustaining, sometimes audible gas-phase reaction of the sample or its decomposition products with an oxidant. A readily visible yellow or blue flame usually accompanies the reaction (NFPA 325M, 1994).

Hypergolic: Hypergolic behavior is characterized by immediate, spontaneous ignition of an oxidation reaction upon mixing of two or more substances.

Isomerization: The conversion of a chemical with a given molecular formula to another compound with the same molecular formula but a different molecular structure, such as from a straight-chain to a branched-chain hydrocarbon or an alicyclic to an aromatic hydrocarbon. Examples include the isomerization of ethylene oxide to acetaldehyde (both C_2H_4O) and butane to isobutane (both C_4H_{10}).

Jet discharge: A pressurized discharge of liquid, vapor, and/or gas into free space from an orifice, the momentum of which induces the surrounding atmosphere to mix with the discharged material.

Jet fire: Fire type resulting from ignition of a jet discharge.

Oxidation: Depending on the context, oxidation can either refer to (a) a reaction in which oxygen combines chemically with another substance or (b) any reaction in which electrons are transferred. For the latter definition, *oxidation* and *reduction* always occur simultaneously (redox reactions), and the substance that gains electrons is termed the *oxidizing agent*. It should be noted that electrons may also be displaced within a molecule without being completely transferred away from it. Such partial loss of electrons likewise constitutes oxidation in its broadest sense and leads to the application of the term to a large number of processes that at first sight might not be considered to be oxidations (Sax and Lewis, 1987, 863). The term *partial oxidation* refers to the combination of oxygen with a material in an oxygen-deficient atmosphere that results in carbon monoxide being one of the combustion products.

Oxidizer: A solid, liquid, or gas that yields oxygen or another oxidizing gas during the course of a chemical reaction, or that readily reacts to oxidize combustible materials (NFPA 49, 1994). More generally, an oxidizer is any reactant that receives electrons from the substance undergoing oxidation. *See* Oxidation.

Oxidizing agent: See Oxidizer.

Peroxide: A chemical compound that contains the peroxy (–O–O–) group, which may be considered to be a derivative of hydrogen peroxide (HOOH). An organic (or inorganic) peroxide is one in which some organic (or inorganic) substituent has replaced one or both hydrogens. *Ozonides,* having three oxygen atoms, are in the same class as peroxides.

Peroxide former: A material that reacts with oxygen or hydrogen peroxide to produce a peroxide of the reactant. The common structural feature in organic peroxidizable compounds is the presence of a hydrogen atom that is susceptible to autoxidative conversion to the hydroperoxy group –OOH (Bretherick, 1986, 72–73) or to an ozonide.

Polymerization: A chemical reaction in which a large number of molecules combine to form a macromolecule. The combining units are called monomers.

Pool fire: The combustion of material evaporating from a layer of liquid at the base of the fire.

Pyrophoric: A readily oxidizable material that ignites spontaneously upon short exposure to air under ordinary ambient conditions (CCPS/AIChE, 1993, 341–342). Temperature and test methods for pyrophoric liquids are defined in DOT Part 173 Subpart E.

Reactive material: Under the concepts of this book, a reactive material is one that can liberate sufficient energy for the occurrence of a hazardous event by readily self-reacting (polymerizing, decomposing, or rearranging), oxidizing in air without an ignition source (pyrophoric or peroxide forming), and/or reacting with water. Initiation of the reaction can be spontaneous, by input of thermal or mechanical energy, or by catalytic action increasing the reaction rate.

Self-accelerating decomposition temperature: Certain compounds, such as organic peroxides and [some] swimming pool chemicals, when held at moderate ambient temperatures for an extended period of time, may undergo an exothermic reaction that accelerates with increase in temperature. If the heat liberated by this reaction is not lost to the environment, the bulk material increases in temperature, which leads to an increase in the rate of decomposition. Unchecked, the temperature grows exponentially to a point at which the decomposition cannot be stopped or slowed. The minimum temperature at which this exponential growth occurs in a material packed in its largest standard shipping container is defined as the *self-accelerating decomposition temperature* (NFPA 49, 1994). The result applies only to the specific container used.

Shock-sensitive: A relatively unstable material, the energetic decomposition of which can be initiated by merely the input of mechanical energy at normal ambient conditions. Materials are considered as shock-sensitive if they are more easily initiated than dinitrobenzene in a standard drop-weight test.

Tautomerizing: Converting from one isomer into one another in organic compounds that differ from one another in the position of a hydrogen atom and a double bond.

Thermally unstable: A material that will undergo an exothermic, self-sustaining or accelerating self-reaction (decomposition, polymerization, or rearrangement) when heated to a specific temperature for given conditions of pressure, volume, composition, and containment. Thus, the self-reaction can be initiated by thermal energy alone.

Water-reactive: A material that will react upon contact with water under normal ambient conditions. Includes materials that react violently with water so they would have an NFPA reactivity rating of 2 or higher based on water reactivity alone, and other materials that react slower but can generate heat and/or gases that can result in elevated pressure if contained.

References

CCPS/AIChE (1993). *Guidelines for Engineering Design for Process Safety.* American Institute of Chemical Engineers, New York.

Bretherick, L. (1986). *Reactive Chemical Hazards.* Butterworths, London.

Davies, A. G. (1961). *Organic Peroxides.* Butterworths, London.

NFPA 321 (1991). *Basic Classification of Flammable and Combustible Liquids.* National Fire Protection Association, Quincy, Massachusetts.

NFPA 325M (1994). *Fire Hazard Properties of Flammable Liquids, Gases, and Volatile Solids.* National Fire Protection Association, Quincy, Massachusetts.

NFPA 49 (1994). *Hazardous Chemical Data.* National Fire Protection Association, Quincy, Massachusetts.

Sax, N. I. and Lewis, R. J. (1989). *Dangerous Properties of Industrial Materials* (3 volumes), 7th ed. Van Nostrand Reinhold, New York.

Index

Q

R

T

U

V

W

Y